Alink权威指南

基于Flink的机器学习实例入门（Java）

杨旭 著

电子工业出版社
Publishing House of Electronics Industry
北京·BEIJING

内 容 简 介

Alink 是阿里巴巴开源的机器学习算法平台，提供了丰富、高效的算法及简便的使用方式，可帮助用户快速构建业务应用。Alink 非常适合工业级的实际应用，支持在个人计算机上快速进行原型研发，支持分布式计算处理海量的数据，支持流式数据的场景，同时机器学习流程与模型可以方便地嵌入用户的应用系统或预测服务中。

本书是根据机器学习的知识点由浅入深来逐层讲述的，这样可降低阅读的门槛，让读者能对所学的内容有一个清晰的印象，并可熟练地运用到实践中。本书重点介绍算法的使用，每节结合实际的数据和典型的场景，通过 Alink 算法组件形成完整的解决方案，可帮助读者理解各类算法所擅长处理的问题，同时本书的方案还可以被推广、应用到类似的场景中。

本书适合机器学习算法的初学者及中级用户快速入门，也可供数据分析师、算法工程师等专业人员参考阅读。

未经许可，不得以任何方式复制或抄袭本书之部分或全部内容。
版权所有，侵权必究。

图书在版编目（CIP）数据

Alink 权威指南：基于 Flink 的机器学习实例入门：Java／杨旭著. —北京：电子工业出版社，2021.10
ISBN 978-7-121-42058-0

Ⅰ. ①A… Ⅱ. ①杨… Ⅲ. ①机器学习－指南②JAVA 语言－程序设计－指南 Ⅳ. ①TP181-62 ②TP312.8-62

中国版本图书馆 CIP 数据核字（2021）第 188761 号

责任编辑：刘皎
印　　刷：三河市良远印务有限公司
装　　订：三河市良远印务有限公司
出版发行：电子工业出版社
　　　　　北京市海淀区万寿路 173 信箱　　　邮编：100036
开　　本：787×980　1/16　　印张：29.5　　字数：637 千字
版　　次：2021 年 10 月第 1 版
印　　次：2021 年 10 月第 1 次印刷
定　　价：149.00 元

凡所购买电子工业出版社图书有缺损问题，请向购书店调换。若书店售缺，请与本社发行部联系，联系及邮购电话：（010）88254888，88258888。
质量投诉请发邮件至 zlts@phei.com.cn，盗版侵权举报请发邮件至 dbqq@phei.com.cn。
本书咨询联系方式：010-51260888-819，faq@phei.com.cn。

序

算法、算力、数据三个要素对于机器学习的重要性已经深入人心。前几年，深度学习算法突飞猛进，给工业界带来了非常丰富的神经网络算法，尤其是带来了感知类的模型。同时，我们也意识到，在产业互联网中，很多传统的非深度学习算法，比如线性回归、决策树、K 最近邻算法等，依然是智能化改造的必备工具。

如何更容易地将这些算法和实际的生产数据结合起来，如何让机器学习"从算法到模型再到应用"的道路变得更加普惠，是我们开发 Alink 之初就会考虑的问题。基于同样开源的流计算引擎 Flink，Alink 将众多的机器学习算法以标准组件的方式结合在一起，力图让那些对机器学习感兴趣的业务工程师可以迅速将这些算法和生产数据集结合在一起，验证效果，进行参数的调优，并最终将这些算法嵌入业务流程中。

今天我们说"一切业务数据化"，而数据体量的井喷让我们看到了不断增长的算力需求。一方面，机器学习的算法实现需要和大数据体系无缝结合，支持海量数据的分析计算；另一方面，每一位开发者和每一家企业都希望通过云上的 AI 工程体系来更容易地触达更高弹性的算力。从第一天起，Alink 就秉承了大数据与 AI 一体化、云原生的设计模式。作为一家技术企业，同时作为云服务的提供者，阿里巴巴集团希望能够通过开源和云的结合，帮助大家实现更多业务场景的数据化和智能化。

贾扬清
阿里巴巴集团副总裁
阿里巴巴开源技术委员会负责人

前　　言

　　Alink 是阿里巴巴开源的机器学习平台，在阿里巴巴集团内外经历了各种项目的实际验证，非常适合工业级的机器学习应用。该机器学习平台降低了我们使用机器学习技术的门槛——其将各个算法封装成组件，即使读者不完全了解背后的理论知识，也可以仿照书中的实例，将组件连接起来解决一些实际问题，在实践中学习。

　　在 Alink 开源后，陆续有一些网友分享自己学习 Alink 的心得。我自己也写过一系列的文章，每次都聚焦在一个技术点，尽量讲清楚、讲透。在 Alink 的文档中，对每个算法都有专门的介绍，详细说明了参数，并给出了使用示例。但是，各算法是分散讲解的，我们无法窥其全貌。所以，我希望通过本书将分散的知识点串联起来。这样读者读完本书，就可以对 Alink 有一个整体的清晰认识，了解各个算法组件的功能与优缺点，并在面临实际问题的时候，联想到本书的相关章节，能有一个大致的解决问题的思路，知道需要哪方面的组件。

　　本书内容的脉络如下：

- 第 1 章为 Alink 快速上手，通过示例演示多个场景的应用，可使读者对 Alink 有一个大致的印象。随后的第 2 章介绍了 Alink 的系统概况与核心概念。
- 机器学习需要基于数据，但在实际应用中，数据的来源是五花八门的，我们如何读/写各种文件系统中的数据文件、如何读/写各种类型的数据库？第 3、4 章系统地介绍相关内容。
- 完整的机器学习流程中一定包括数据处理流程，第 5、6、7 章分别介绍了三种重要的方式：使用 SQL 语句、使用用户定义函数，以及使用 Alink 定义好的基本数据处理组件（比如各种采样操作、数据标准化、缺失值填充等）。
- 第 8 章以纸钞的真假判断为例，介绍了线性二分类模型的使用及二分类评估方法。第 9

前　言

章以判断蘑菇是否有毒为例，介绍了两种常用的非线性模型：朴素贝叶斯模型与决策树模型。

- 第 10、11 章分别以信用预测及商品购买行为预测为例，介绍了特征的转化和生成方法，介绍了常用的随机森林和 GBDT 算法。
- 第 12、13 章以鸢尾花分类及手写数字识别为例，介绍了常用的多分类算法：二分类器组合算法、Softmax 算法、多层感知器分类器算法、K 最近邻算法。
- 第 14 章针对在线学习场景，以广告点击率预估为例，介绍了流式训练及预测流程。
- 第 15、16 章从回归的由来讲起，以身高预测及葡萄酒品质预测为例，介绍了常用的回归算法。
- 第 17、18 章以鸢尾花数据、经纬度数据、手写数字数据为例，介绍了常用的聚类算法，并介绍了流式 K-Means（K 均值）聚类。
- 第 19、20 章介绍了重要的参数降维方法——主成分分析；介绍了超参数搜索的工具，以大幅减少模型调参所需的时间。
- 第 21、22 章介绍文本分析的常用方法以及单词向量化方法，对一些典型文本数据，包括新闻标题数据集、相声《报菜名》的内容、《三国演义》全文进行了分析。
- 第 23 章以电影评论数据集为例，构建情感分析方案，并通过调整算法模型、优化特征工程，不断改进预测效果。
- 第 24 章介绍了常用的推荐算法，并以影片推荐为例，展示了如何构建推荐系统。

本书提供了完整的源代码，读者在个人计算机中就能直接尝试、验证书中的方法和算法。书中所介绍的是业界正在使用的工具，其支持分布式计算处理海量的数据、支持流式数据的场景，同时机器学习流程及模型可以方便地嵌入用户的应用系统或预测服务中。

感谢一直支持 Alink 发展的各位同事和朋友，衷心希望 Alink 能够帮助更多的用户！感谢家人的理解和支持！

杨　旭
2021 年 9 月

目　　录

第 1 章　Alink 快速上手 1
1.1　Alink 是什么 1
1.2　免费下载、安装 1
1.3　Alink 的功能 2
1.3.1　丰富的算法库 2
1.3.2　多样的使用体验 3
1.3.3　与 SparkML 的对比 3
1.4　关于数据和代码 4
1.5　简单示例 5
1.5.1　数据的读/写与显示 5
1.5.2　批式训练和批式预测 7
1.5.3　流式处理和流式预测 9
1.5.4　定义 Pipeline，简化操作 10
1.5.5　嵌入预测服务系统 12

第 2 章　系统概况与核心概念 14
2.1　基本概念 14
2.2　批式任务与流式任务 15
2.3　Alink=A+link 18
2.3.1　BatchOperator 和 StreamOperator 19

目　录

2.3.2	link 方式是批式算法/流式算法的通用使用方式	20
2.3.3	link 的简化	23
2.3.4	组件的主输出与侧输出	23

2.4 Pipeline 与 PipelineModel 24
  2.4.1 概念和定义 24
  2.4.2 深入介绍 25
2.5 触发 Alink 任务的执行 28
2.6 模型信息显示 29
2.7 文件系统与数据库 34
2.8 Schema String 36

第 3 章　文件系统与数据文件 38

3.1 文件系统简介 38
  3.1.1 本地文件系统 39
  3.1.2 Hadoop 文件系统 41
  3.1.3 阿里云 OSS 文件系统 43
3.2 数据文件的读入与导出 45
  3.2.1 CSV 格式 47
  3.2.2 TSV、LibSVM、Text 格式 53
  3.2.3 AK 格式 56

第 4 章　数据库与数据表 60

4.1 简介 60
  4.1.1 Catalog 的基本操作 60
  4.1.2 Source 和 Sink 组件 61
4.2 Hive 示例 62
4.3 Derby 示例 65
4.4 MySQL 示例 67

第 5 章　支持 Flink SQL 70

5.1 基本操作 70
  5.1.1 注册 70

VII

5.1.2　运行 71
　　5.1.3　内置函数 74
　　5.1.4　用户定义函数 74
5.2　简化操作 75
　　5.2.1　单表操作 76
　　5.2.2　两表的连接（JOIN）操作 80
　　5.2.3　两表的集合操作 82
5.3　深入介绍 Table Environment 86
　　5.3.1　注册数据表名 87
　　5.3.2　撤销数据表名 88
　　5.3.3　扫描已注册的表 89

第 6 章　用户定义函数（UDF/UDTF） 90

6.1　用户定义标量函数（UDF） 90
　　6.1.1　示例数据及问题 91
　　6.1.2　UDF 的定义 91
　　6.1.3　使用 UDF 处理批式数据 92
　　6.1.4　使用 UDF 处理流式数据 93
6.2　用户定义表值函数（UDTF） 95
　　6.2.1　示例数据及问题 95
　　6.2.2　UDTF 的定义 96
　　6.2.3　使用 UDTF 处理批式数据 96
　　6.2.4　使用 UDTF 处理流式数据 99

第 7 章　基本数据处理 101

7.1　采样 101
　　7.1.1　取"前" N 个数据 102
　　7.1.2　随机采样 102
　　7.1.3　加权采样 104
　　7.1.4　分层采样 105
7.2　数据划分 106
7.3　数值尺度变换 108

	7.3.1 标准化	109
	7.3.2 MinMaxScale	111
	7.3.3 MaxAbsScale	112
7.4	向量的尺度变换	113
	7.4.1 StandardScale、MinMaxScale、MaxAbsScale	113
	7.4.2 正则化	115
7.5	缺失值填充	116

第 8 章 线性二分类模型 ... 119

- 8.1 线性模型的基础知识 ... 119
  - 8.1.1 损失函数 ... 119
  - 8.1.2 经验风险与结构风险 ... 121
  - 8.1.3 线性模型与损失函数 ... 122
  - 8.1.4 逻辑回归与线性支持向量机（Linear SVM） ... 123
- 8.2 二分类评估方法 ... 125
  - 8.2.1 基本指标 ... 126
  - 8.2.2 综合指标 ... 128
  - 8.2.3 评估曲线 ... 131
- 8.3 数据探索 ... 136
  - 8.3.1 基本统计 ... 138
  - 8.3.2 相关性 ... 140
- 8.4 训练集和测试集 ... 144
- 8.5 逻辑回归模型 ... 145
- 8.6 线性 SVM 模型 ... 147
- 8.7 模型评估 ... 149
- 8.8 特征的多项式扩展 ... 153
- 8.9 因子分解机 ... 157

第 9 章 朴素贝叶斯模型与决策树模型 ... 160

- 9.1 朴素贝叶斯模型 ... 160
- 9.2 决策树模型 ... 162
  - 9.2.1 决策树的分裂指标定义 ... 165

9.2.2 常用的决策树算法	167
9.2.3 指标计算示例	169
9.2.4 分类树与回归树	172
9.2.5 经典的决策树示例	173
9.3 数据探索	176
9.4 使用朴素贝叶斯方法	179
9.5 蘑菇分类的决策树	185

第 10 章 特征的转化 191

10.1 整体流程	195
10.1.1 特征哑元化	197
10.1.2 特征的重要性	198
10.2 减少模型特征的个数	200
10.3 离散特征转化	202
10.3.1 独热编码	202
10.3.2 特征哈希	204

第 11 章 构造新特征 207

11.1 数据探索	208
11.2 思路	210
11.2.1 用户和品牌的各种特征	211
11.2.2 二分类模型训练	212
11.3 计算训练集	213
11.3.1 原始数据划分	213
11.3.2 计算特征	214
11.3.3 计算标签	222
11.4 正负样本配比	224
11.5 决策树	226
11.6 集成学习	227
11.6.1 Bootstrap aggregating	228
11.6.2 Boosting	229
11.6.3 随机森林与 GBDT	232

11.7　使用随机森林算法 233
11.8　使用 GBDT 算法 234

第 12 章　从二分类到多分类 235

12.1　多分类模型评估方法 235
　　12.1.1　综合指标 237
　　12.1.2　关于每个标签值的二分类指标 238
　　12.1.3　Micro、Macro、Weighted 计算的指标 239
12.2　数据探索 241
12.3　使用朴素贝叶斯进行多分类 244
12.4　二分类器组合 246
12.5　Softmax 算法 249
12.6　多层感知器分类器 253

第 13 章　常用多分类算法 256

13.1　数据准备 256
　　13.1.1　读取 MNIST 数据文件 257
　　13.1.2　稠密向量与稀疏向量 258
　　13.1.3　标签值的统计信息 261
13.2　Softmax 算法 262
13.3　二分类器组合 264
13.4　多层感知器分类器 265
13.5　决策树与随机森林 267
13.6　K 最近邻算法 270

第 14 章　在线学习 273

14.1　整体流程 273
14.2　数据准备 275
14.3　特征工程 277
14.4　特征工程处理数据 279
14.5　在线训练 280
14.6　模型过滤 283

第 15 章 回归的由来 286

15.1 平均数 287
15.2 向平均数方向的回归 288
15.3 线性回归 289

第 16 章 常用回归算法 292

16.1 回归模型的评估指标 292
16.2 数据探索 294
16.3 线性回归 297
16.4 决策树与随机森林 300
16.5 GBDT 回归 301

第 17 章 常用聚类算法 303

17.1 聚类评估指标 304
    17.1.1 基本评估指标 304
    17.1.2 基于标签值的评估指标 306
17.2 K-Means 聚类 308
    17.2.1 算法简介 308
    17.2.2 K-Means 实例 310
17.3 高斯混合模型 314
    17.3.1 算法介绍 314
    17.3.2 GMM 实例 316
17.4 二分 K-Means 聚类 317
17.5 基于经纬度的聚类 320

第 18 章 批式与流式聚类 324

18.1 稠密向量与稀疏向量 324
18.2 使用聚类模型预测流式数据 326
18.3 流式聚类 329

第 19 章 主成分分析 331

19.1 主成分的含义 333

	19.2	两种计算方式	337
	19.3	在聚类方面的应用	339
	19.4	在分类方面的应用	343

第 20 章　超参数搜索 ... 347

  20.1　示例一：尝试正则系数 ... 348
  20.2　示例二：搜索 GBDT 超参数 ... 349
  20.3　示例三：最佳聚类个数 ... 350

第 21 章　文本分析 ... 353

  21.1　数据探索 ... 353
  21.2　分词 ... 355
    21.2.1　中文分词 ... 356
    21.2.2　Tokenizer 和 RegexTokenizer ... 359
  21.3　词频统计 ... 363
  21.4　单词的区分度 ... 365
  21.5　抽取关键词 ... 367
    21.5.1　原理简介 ... 367
    21.5.2　示例 ... 369
  21.6　文本相似度 ... 371
    21.6.1　文本成对比较 ... 372
    21.6.2　最相似的 TopN ... 375
  21.7　主题模型 ... 387
    21.7.1　LDA 模型 ... 388
    21.7.2　新闻的主题模型 ... 390
    21.7.3　主题与原始分类的对比 ... 392
  21.8　组件使用小结 ... 396

第 22 章　单词向量化 ... 398

  22.1　单词向量预训练模型 ... 399
    22.1.1　加载模型 ... 399
    22.1.2　查找相似的单词 ... 400

  22.1.3 单词向量 ... 402
 22.2 单词映射为向量 ... 406

第 23 章 情感分析 ... 412

 23.1 使用提供的特征 ... 413
  23.1.1 使用朴素贝叶斯方法 ... 416
  23.1.2 使用逻辑回归算法 ... 419
 23.2 如何提取特征 ... 423
 23.3 构造更多特征 ... 426
 23.4 模型保存与预测 ... 430
  23.4.1 批式/流式预测任务 ... 430
  23.4.2 嵌入式预测 ... 431

第 24 章 构建推荐系统 ... 433

 24.1 与推荐相关的组件介绍 ... 434
 24.2 常用推荐算法 ... 437
  24.2.1 协同过滤 ... 437
  24.2.2 交替最小二乘法 ... 438
 24.3 数据探索 ... 439
 24.4 评分预测 ... 444
 24.5 根据用户推荐影片 ... 446
 24.6 计算相似影片 ... 452
 24.7 根据影片推荐用户 ... 454
 24.8 计算相似用户 ... 457

Alink 快速上手

随着大数据时代的到来和人工智能的崛起,机器学习所能处理的场景更加广泛和多样。算法工程师们不单要处理好批式数据的模型训练与预测,也要能处理好流式数据,并需要具备将模型嵌入企业应用和微服务上的能力。为了取得更好的业务效果,算法工程师们需要尝试更多、更复杂的模型,并需要处理更大的数据集,因此他们使用分布式集群已经成为常态。为了及时应对市场的变化,越来越多的业务选用在线学习方式来直接处理流式数据、实时更新模型。

Alink 就是为了更好地满足这些实际应用场景而研发的机器学习算法平台,以帮助数据分析和应用开发人员轻松地搭建端到端的业务流程。

1.1 Alink是什么

Alink 是阿里巴巴计算平台事业部 PAI(Platform of Artificial Intelligence)团队基于 Flink 计算引擎研发的批流一体的机器学习算法平台,该平台提供了丰富的算法组件库和便捷的操作框架。借此,开发者可以一键搭建覆盖数据处理、特征工程、模型训练、模型预测的算法模型开发全流程。Alink 的名称取自相关英文名称,即 Alibaba、Algorithm、AI、Flink 和 Blink 中的公共部分。Alink 提供了 Java 接口和 Python 接口(PyAlink),开发者不需要 Flink 的技术背景也可以轻松构建算法模型。

Alink 在 2019 年 11 月的 Flink Forword Asia 2019 大会上宣布开源。Alink 所在的 GitHub 地址如链接 1-1 所示,欢迎大家下载使用、反馈意见、提出建议,以及贡献新的算法。

1.2 免费下载、安装

可以在 Alink 开源网站获取其最新版本。为了方便用户查看 Alink 文档,解决 Alink 的本地安装、使用问题，以及解决 Alink 在集群上部署、运行等方面的问题，我们提供了如下专门的资料供读者参考。相关的网址如下：
- 【主页，参见链接 1-1】完整的开源内容：代码、函数说明、注意事项、安装包、历史版本。
- 【文档，参见链接 1-2】优化文档显示，便于查询、阅读。
- 【指南，参见链接 1-3】侧重介绍 Alink 的安装、运行、部署等方面的内容。
- 【技巧，参见链接 1-4】作者的知乎主页，内容丰富，涉及的话题比较发散。

另外，在开源网站的首页中有 Alink 开源用户钉钉群的二维码，欢迎大家加入该钉钉群。若有问题，可以随时在该钉钉群里与大家沟通、交流。注意：本书提供的额外参考资料，如文中的"链接 1-1""链接 1-2"等，可从封底的"读者服务"处获取。

1.3 Alink的功能

1.3.1 丰富的算法库

Alink 拥有丰富的批式算法和流式算法，可帮助数据分析和应用开发人员端到端地完成从数据处理、特征工程、模型训练到预测的整个流程工作。Alink 开源算法如图 1-1 所示，在 Alink 提供的开源算法模块中，每一个模块都包含流式算法和批式算法。比如线性回归，包含批式线性回归训练、流式线性回归预测和批式线性回归预测。

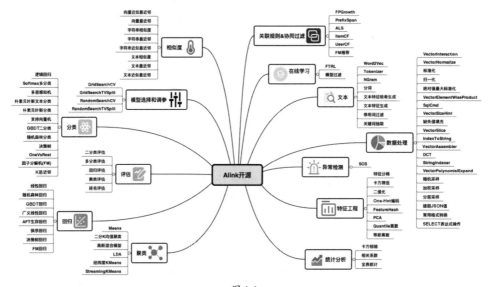

图 1-1

1.3.2 多样的使用体验

Alink 使用 Java 研发，原生提供了整套 Java 调用接口，可以在单机中编辑、调试、运行 Alink 任务，也可以将 Alink 任务编译、发布到 Flink 集群中运行。此外，Alink 也提供了 Python 版本：PyAlink。可以通过 Jupyter 等 Notebook 的方式使用 Alink，提升其交互式和可视化体验；PyAlink 既支持单机运行，又支持集群提交。PyAlink 打通了 Operator（Alink 组件）和 DataFrame 的接口，因此，Alink 的整个算法流程可无缝融入 Python。PyAlink 也提供了使用 Python 函数来调用 UDF 或者 UDTF 的方法。PyAlink 在 Notebook 中的使用如图 1-2 所示。图 1-2 中展示了一个模型训练预测，并打印出了预测结果的过程。

图 1-2

1.3.3 与SparkML的对比

在离线学习算法方面，Alink 与 SparkML 的性能基本相当。图 1-3 给出的是一些经典算法的性能对比：对于同一算法，采用相同的数据集、同样的迭代次数等参数。其中的加速比指的是，SparkML 所用的时间除以 Alink 所用的时间之比。若加速比的值为 1x（1 倍），则说明 Alink 与 SparkML 的性能相当。加速比的值越大，说明 Alink 的性能越好。

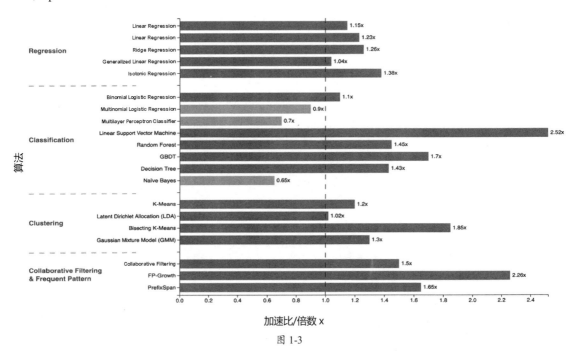

图 1-3

通过图 1-3 可以看出，Alink 在大部分算法方面的性能优于 SparkML，在个别算法方面的性能比 SparkML 弱，二者在整体上是一个相当的水平。但是，在功能的完备性方面，Alink 更有优势，Alink 除了覆盖了 SparkML 的算法，还包含流式算法、流批混跑、在线学习、中文分词等功能。

1.4　关于数据和代码

本书的全部实验都提供了 Java 源代码，所用数据集均可以通过 Web 下载。

Java 源代码的地址：参见链接 1-5。

该代码在 MacBook Pro（Intel Core i7-4770HQ CPU @ 2.20 GHz 四核八线程, 16GB 1600MHz DDR3）计算机上运行通过，本书记录的实验运行时间也是在该计算机上的运行时间。

本书所使用的数据集都是可以通过 Web 获取的免费数据集。这些免费的数据集在书中都有相应的介绍，并提供了数据来源。其大部分可以直接点击下载；不过，若要下载来自 Kaggle 的数据集，则需要注册 Kaggle 账户，但下载是免费的。

每个章节的示例代码都在使用不同的数据，建议用户将数据下载到本地使用，并在本地建立一个数据存放的总文件夹。作者在本地的数据总文件夹为"/Users/yangxu/alink/data/"。随后，在数据总文件夹下建立不同的子文件夹，用于存放不同的数据集。子文件夹的命名可以参考示例代码中的命名。这样示例的代码下载到本地后，只需修改一处数据总文件夹路径变量的设置，便可直接运行了。

数据总文件夹的路径变量设置在 Utils 中，为静态变量 ROOT_DIR：

```
public static final String ROOT_DIR = "/Users/yangxu/alink/data/";
```

在各个章节的代码中，所用的数据路径如下：

```
static final String DATA_DIR = Chap01.ROOT_DIR + "iris" + File.separator;
```

该路径由 ROOT_DIR 和当前章节所用的子文件夹名称拼接而成。注意：代码最后使用的 File.separator 是为跨平台而设的。

1.5 简单示例

这里将通过几个简单的 Alink 示例，让读者留下一个初步的 Alink 印象。

1.5.1 数据的读/写与显示

我们以常用的鸢尾花（iris）数据集为例，演示一下如何读取数据。该数据可以通过链接 1-6 直接下载。Alink 的 CSV 格式数据源读取组件，不但可以读取本地文件，还可以直接读取网络文件；在该组件读取文件的时候需要指定各列数据的名称和类型。下面是具体的代码。在此，读取出数据后，选择前 5 条数据进行打印输出：

```
CsvSourceBatchOp source =
  new CsvSourceBatchOp()
    .setFilePath("http://archive.ics.uci.edu/ml/machine-learning-databases"
      + "/iris/iris.data")
    .setSchemaStr("sepal_length double, sepal_width double, petal_length double, "
      + "petal_width double, category string");
source.firstN(5).print();
```

运行结果如下：

```
sepal_length|sepal_width|petal_length|petal_width|category
------------|-----------|------------|-----------|--------
5.1000|3.5000|1.4000|0.2000|Iris-setosa
4.9000|3.0000|1.4000|0.2000|Iris-setosa
4.7000|3.2000|1.3000|0.2000|Iris-setosa
4.6000|3.1000|1.5000|0.2000|Iris-setosa
5.0000|3.6000|1.4000|0.2000|Iris-setosa
```

这些数据输出的字符形式，满足 Markdown 表格的格式要求。在此，可以通过 Markdown 编辑器将这些数据的格式转换为表格形式。这样可以更好地展现数据，如表 1-1 所示。

表 1-1 格式转换后的数据

sepal_length	sepal_width	petal_length	petal_width	category
5.1000	3.5000	1.4000	0.2000	Iris-setosa
4.9000	3.0000	1.4000	0.2000	Iris-setosa
4.7000	3.2000	1.3000	0.2000	Iris-setosa
4.6000	3.1000	1.5000	0.2000	Iris-setosa
5.0000	3.6000	1.4000	0.2000	Iris-setosa

下一步，我们对数据进行采样，采样数为 10 条。然后连接到 CSV 格式输出组件 CsvSinkBatchOp，并设置相应的输出路径。可以设置覆盖写参数为 true。

```
source
  .sampleWithSize(10)
  .link(
    new CsvSinkBatchOp()
      .setFilePath(DATA_DIR + "iris_10.data")
      .setOverwriteSink(true)
  );

BatchOperator.execute();
```

采样出来的数据被保存到文件 iris_10.data。使用文本编辑器显示这些数据，如图 1-4 所示，正好 10 行，数据间使用逗号进行分隔，这是标准的 CSV 格式。

```
iris_10.data
4.7,3.2,1.3,0.2,Iris-setosa
4.9,3.1,1.5,0.1,Iris-setosa
6.7,3.1,4.7,1.5,Iris-versicolor
6.4,3.2,4.5,1.5,Iris-versicolor
5.8,2.7,5.1,1.9,Iris-virginica
5.6,2.7,4.2,1.3,Iris-versicolor
5.2,3.5,1.5,0.2,Iris-setosa
7.3,2.9,6.3,1.8,Iris-virginica
5.1,3.8,1.9,0.4,Iris-setosa
6.1,2.6,5.6,1.4,Iris-virginica
```

图 1-4

1.5.2 批式训练和批式预测

从本节开始,我们以预测某电商平台"双 11"的成交总额(GMV, Gross Merchandise Volume)为例,演示各种操作。

我们很容易查到某电商平台历年"双 11"的成交总额,下面是其 2009—2017 年的记录,成交总额的单位为万亿元,左边年份列记为 x,右边成交总额列记为 gmv。

```
BatchOperator <?> train_set = new MemSourceBatchOp(
    new Row[] {
        Row.of(2009, 0.5),
        Row.of(2010, 9.36),
        Row.of(2011, 52.0),
        Row.of(2012, 191.0),
        Row.of(2013, 350.0),
        Row.of(2014, 571.0),
        Row.of(2015, 912.0),
        Row.of(2016, 1207.0),
        Row.of(2017, 1682.0),
    },
    new String[] {"x", "gmv"}
);

BatchOperator <?> pred_set
    = new MemSourceBatchOp(new Integer[] {2018, 2019}, "x");
```

我们需要对这些数据进行建模,从而预测该电商平台 2018 年和 2019 年的值。由于 2018 年和 2019 年已经过去,我们知道实际的数值:2018 年该电商平台的成交总额为 2135 万亿元,2019 年该电商平台的成交总额为 2684 万亿元。

从年份和成交总额的数据上看,二者显然不成线性关系,这就需要构造新的特征,更好地拟合历史数据,进行预测。我们可以定义 x 的平方为新的特征,这样就相当于计算 gmv 与关于 x 的二次多项式之间的关系。

```
train_set = train_set.select("x, x*x AS x2, gmv");

LinearRegTrainBatchOp trainer
    = new LinearRegTrainBatchOp()
        .setFeatureCols("x", "x2")
        .setLabelCol("gmv");

train_set.link(trainer);

trainer.link(
    new AlinkFileSinkBatchOp()
        .setFilePath(DATA_DIR + "gmv_reg.model")
        .setOverwriteSink(true)
);
```

```
BatchOperator.execute();
```

需要说明的是：

- train_set.select("x, x*x AS x2, gmv")实际上是对数据表 train_set 执行 SQL 语句 "SELECT x, x*x AS x2, gmv FROM train_set"。更详细的内容，读者可以参阅本书第 5 章有关 SQL 语句的具体操作。
- 定义 LinearRegTrainBatchOp 类型的组件 trainer，即训练器，并设置其特征参数和标签参数。
- train_set.link(trainer)可将训练数据与训练器连接（link）起来。
- 随后，训练器 trainer 的输出为模型。将训练器 trainer 连接到数据导出组件 AlinkFileSinkBatchOp，就会将模型数据导出到文件 gmv_reg.model。并且由于设置了 setOverwriteSink(true)，因此如果目标文件存在，则会将其覆盖。
- 最后，调用 BatchOperator.execute()，执行批式任务。

运行结束后，可以看到生成了 gmv_reg.model 文件。下面，我们导入此模型，并进行批式预测：

```
BatchOperator <?> lr_model
    = new AlinkFileSourceBatchOp().setFilePath(DATA_DIR + "gmv_reg.model");

pred_set = pred_set.select("x, x*x AS x2");

LinearRegPredictBatchOp predictor
    = new LinearRegPredictBatchOp().setPredictionCol("pred");

predictor
    .linkFrom(lr_model, pred_set)
    .print();
```

预测流程分为如下几步：

（1）通过 AlinkFileSourceBatchOp 读取文件 gmv_reg.model，得到模型数据，并记为 lr_model。

（2）预测的原始数据只有一列 x，需要执行 SQL SELECT 方案，生成新的特征 x2。

（3）定义线性回归的批式预测组件（LinearRegPredictBatchOp）为 predictor，指定输出结果列的名称。

（4）将这些组件组合为预测流程。因为 predictor 需要模型数据及预测特征数据，所以使用 linkFrom 方法，同时连接两个上游组件。之后链式调用 print 方法，输出预测结果。注意：在批式场景中的 print 方法本身就会触发批式任务执行，这里就不用再调用 BatchOperator.execute()了。

表 1-2 为模型预测结果，其最后一列为预测的成交额（单位：万亿元）。我们对比前面提到的实际数值（2018 年该电商平台的成交总额为 2135 万亿元，2019 年该电商平台的成交总额为 2684 万亿元），可以发现该预测结果与实际数值非常接近。

表 1-2　模型预测结果

x	x2	pred
2018	4072324	2142.4048
2019	4076361	2682.2263

1.5.3　流式处理和流式预测

前面介绍了批式模型训练，以及针对批式数据的预测。在实际的场景中，需要预测的数据常常以流的方式陆续到来，因此需要随时进行预测，即进行流式预测。

我们还是以某电商平台"双 11"的成交总额预测为例，演示一个流式预测的简单场景：预测模型已经通过批式训练的方式生成，使用这个固定的预测模型，构建流式任务，预测流式数据。

Alink 批式任务的代码与流式任务的代码很相似，将批式组件的名称后缀"BatchOp"改为"StreamOp"即可，参数基本不用调整；在一些使用方式上需要进行微调。流式预测的代码如下：

```
MemSourceStreamOp pred_set
  = new MemSourceStreamOp(new Integer[] {2018, 2019}, "x");

BatchOperator <?> lr_model
  = new AlinkFileSourceBatchOp().setFilePath(DATA_DIR + "gmv_reg.model");

LinearRegPredictStreamOp predictor
  = new LinearRegPredictStreamOp(lr_model).setPredictionCol("pred");

pred_set
  .select("x, x*x AS x2")
  .link(predictor)
  .print();

StreamOperator.execute();
```

可以对比批式预测的代码，理解流式预测的代码：

（1）在数据源获取方面，通过内存数据构造出一个流式数据源。这里的流式组件 MemSourceStreamOp 与前面的批式组件 MemSourceBatchOp 只有名称有区别，组件参数的设置类似。这就为我们进行流式操作和批式操作的互相转换提供了方便。

（2）因为使用批式的模型，这里还是通过 AlinkFileSourceBatchOp 读取文件 gmv_reg.model，得到模型数据，并记为 lr_model。

（3）定义线性回归的流式预测组件（LinearRegPredictStreamOp）为 predictor，指定输出结果列的名称。注意：因为流式任务的流程中无法接入批式的数据，所以将批式模型数据 lr_model 作为 LinearRegPredictStreamOp 构造函数的参数传入。在流式任务执行前，会先完成批式模型数据的导入。

（4）组装预测流程。由于流式预测组件 predictor 已经在构造函数中导入了模型数据，因此只需连入一个流式预测数据即可，不要使用 linkFrom 方法。整个流程可以写得更简单，可直接把预测数据作为源头，接入流式 SQL SELECT 操作，生成 x2（注意这里代码的写法，与批式调用时完全一致），然后连接流式预测组件 predictor，并对预测结果进行打印。

（5）调用 StreamOperator.execute()，执行流式任务。

流式预测结果，如表 1-3 所示，其与批式预测的结果完全相同。

表 1-3　流式预测的结果

x	x2	pred
2018	4072324	2142.4048
2019	4076361	2682.2263

1.5.4　定义 Pipeline，简化操作

前面介绍了 Alink 在批式模型训练、批式预测和流式预测中的例子，可以看到这三个流程中都涉及两个步骤（使用 SQL SELECT 构造新的特征项，以及线性回归的训练或者预测），因此，可以对该功能进行抽象，形成 Pipeline（管道）。其具体概念和用法在本书后面会介绍，这里就不展开说明了。这里只是通过示例，让读者有一个初步的印象：可以通过 Pipeline 简化操作。

示例代码如下：

```
MemSourceBatchOp train_set = ...

Pipeline pipeline = new Pipeline()
    .add(
        new Select()
            .setClause("*, x*x AS x2")
    )
    .add(
        new LinearRegression()
            .setFeatureCols("x", "x2")
```

```
        .setLabelCol("gmv")
        .setPredictionCol("pred")
);

File file = new File(DATA_DIR + "gmv_pipeline.model");
if (file.exists()) {
    file.delete();
}

pipeline.fit(train_set).save(DATA_DIR + "gmv_pipeline.model");

BatchOperator.execute();
```

这段代码由四部分构成：
- 训练数据 train_set 生成的代码在前面出现过，这里不再重复。
- 核心为 Pipeline 的构成，其需要以下两个操作：
  - Select
    设置了 SQL 子语句 "*, x*x AS x2"。注意第一个字符为 "*"，因此会匹配输入数据表中出现的所有列。因为在批式训练场景中，输入的列为 "x, gmv"，而在预测场景中输入的只有一列 "x"，所以使用 "*" 会同时适用这两个场景。
  - LinearRegression
    设置了训练时需要的参数 FeatureCols 和 LabelCol，也设置了预测时需要的 PredictionCol。
- 有了 Pipeline 的定义后，对训练数据执行 fit 方法，就会得到 PipelineModel(管道模型)。可以直接使用 PipelineModel；也可选择保存 PipelineModel，后用在不同场景中。Alink 提供了简单的保存 PipelineModel 的方法，提供文件路径，运行 save 方法即可。注意，save 方法将 PipelineModel 对应的模型连接到了 sink 组件，还需要等到执行 BatchOperator.execute() 时，才会真正写出模型。
- 最后使用 BatchOperator.execute()，执行批式任务。

下面我们将通过读取模型文件 gmv_pipeline.model，得到 PipelineModel，并使用其 transform 方法，对批式数据和流式数据进行预测。

读取模型文件，得到 PipelineModel 的代码很简单：

```
PipelineModel pipelineModel = PipelineModel.load(DATA_DIR + "gmv_pipeline.model");
```

对于批式数据的预测代码如下：

```
BatchOperator <?> pred_batch
    = new MemSourceBatchOp(new Integer[] {2018, 2019}, "x");

pipelineModel
```

```
  .transform(pred_batch)
  .print();
```

结果如表 1-4 所示。

表 1-4 Pipeline 批式预测的结果

x	x2	pred
2018	4072324	2142.4048
2019	4076361	2682.2263

对于流式数据的预测代码如下，我们看到 pipelineModel 对于批式数据和流式数据处理所用的方法名称都是一样的，使用方式也是一样的；但 transform 方法的输出结果究竟是批式的还是流式的，取决于输入数据是批式的还是流式的：

```
MemSourceStreamOp pred_stream
  = new MemSourceStreamOp(new Integer[] {2018, 2019}, "x");

pipelineModel
  .transform(pred_stream)
  .print();

StreamOperator.execute();
```

使用 Pipeline 方式，流式预测的结果如表 1-5 所示，与批式预测的结果（见表 1-4）相同。

表 1-5 Pipeline 流式预测的结果

x	x2	pred
2019	4076361	2682.2263
2018	4072324	2142.4048

1.5.5　嵌入预测服务系统

除了使用 Alink 算法组件直接对批式的数据或者流式的数据进行预测，用户也希望我们能提供 SDK 的方式，即，由参数或模型数据直接构建一个本地的 Java 实例，可以对单条数据进行预测，我们称之为 LocalPredictor。如此一来，预测不再必须由 Flink 任务完成，而可以嵌入提供 RestAPI 的预测服务系统中，或者嵌入用户的业务系统里。

有了存储好的 PipelineModel 模型，可以直接使用 LocalPredictor 的构造函数来构建实例。构造函数需要两个参数，一个是 PipelineModel 模型文件的路径，另一个是所要预测数据的各数据列名称和类型，即输入一个 Alink Schema String 格式（详见 2.8 节的内容）的参数。这里的参数为"x int"，表示输入的预测数据只有 1 列，名称为 x，类型为整型。具体代码如下：

```
LocalPredictor predictor
  = new LocalPredictor(DATA_DIR + "gmv_pipeline.model", "x int");
```

LocalPredictor 输入的预测数据是 Row 格式的，输出的预测结果也是 Row 格式的。Row 格式本身并没有列名和类型的定义，需要通过在构造函数中输入预测数据的 Schema 来获取预测数据各数据列的名称和类型；LocalPredictor 可以根据 PipelineModel 模型的计算过程，推导出预测结果的 Schema（结果数据列名称及类型）。对于我们刚构建的 localPredictor，使用 getOutputSchema 方法获取 Schema 信息并打印显示 Schema 信息，具体代码如下：

```
System.out.println(predictor.getOutputSchema());
```

运行结果如下：

```
root
 |-- x: INT
 |-- x2: INT
 |-- pred: DOUBLE
```

可以看出，LocalPredictor 的预测结果与批式和流式预测结果一样，其预测输出共 3 列，最关键的分类预测结果列"pred"在最后。

LocalPredictor 使用 map 方法进行预测，具体代码如下：

```
for (int x : new int[] {2018, 2019}) {
   System.out.println(predictor.map(Row.of(x)));
}
```

计算结果如下：

```
2018,4072324,2142.404761955142
2019,4076361,2682.2262857556343
```

最右面的数值是预测结果，对其保留小数点后 4 位有效数字，得到的结果与前面批式/流式任务计算的结果相同。

2 系统概况与核心概念

本章将从 Alink 涉及的基本概念开始,逐步深入,帮助读者系统地了解 Alink。在介绍 Alink 基本概念的同时,也介绍了一些小技巧,以帮助读者写出更简练、高效的代码。

2.1 基本概念

数据集(DataSet)与数据流(DataStream)的区别在于用户是否可以假定数据有界。数据集(DataSet)的数据有界就意味着,数据是静止的、确定的,内容和个数都不会变化。

在逐条读取数据时,数据集(DataSet)的数据是有界的、个数确定,我们可以执行某些操作。比如,统计该数据集的记录总数,计算平均值、方差等统计量;在逻辑上,可以看作把该数据集按照某个操作处理完,得到一个新的数据集,然后对该新数据集进行另外一个操作,得到另一个新的数据集。

批式(Batch)数据对应着数据集(DataSet),流式(Stream)数据对应着数据流(DataStream)。

数据处理的基本流程就是三部分:数据源(Source)、算法组件(Operator)和数据导出(Sink),如图 2-1 所示。

考虑到数据集和数据流之间的区别,处理操作被分为批式处理(Batch Processing)和流式处理(Stream

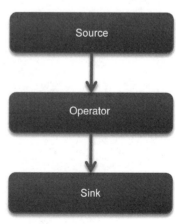

图 2-1

Processing），如图 2-2 所示。在批式处理中，批式数据源组件（Batch Source）读入的数据为数据集；批式导出组件（Batch Sink）将数据集导出到文件系统或数据库；批式算法组件（BatchOperator）的输入和计算结果都为数据集。对于流式处理，流式数据源组件（Stream Source）用来接入数据流；流式导出组件（Stream Sink）负责将数据流导出，即将数据流导出到文件系统或数据库；流式算法组件（StreamOperator）的处理粒度是单条数据，即从输入的数据流中逐条获取数据进行计算，该计算结果为若干条数据，这些数据会进入输出数据流。连接流式数据源组件、各个流式算法组件和流式导出组件，构成数据流的管道。

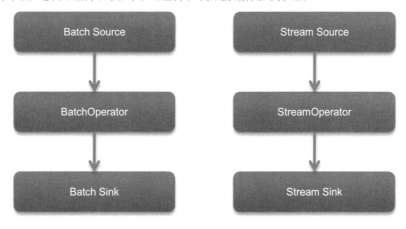

图 2-2

Alink 将 Flink Table 作为数据集和数据流的统一表示,其中数据行的类型为 org.apache.flink.types.Row。

Alink 为了统一各算法模块间交换数据的格式，确定将 Row 格式作为各算法处理的单条数据的标准类型；相应的批数据与流数据的类型为 Flink Table。如此一来，SQL 操作与算法模块间可以通过 Flink Table 来传递数据，这样 SQL 操作与算法操作就可以出现在同一个工作流中，方便用户使用。

2.2　批式任务与流式任务

批式任务与流式任务看待数据的方式不同，批式任务将数据当作一个整体（数据集）来看待，流式任务将每条数据作为一个基本单位。

我们先看一个典型的批式任务场景，包括读取数据、数据预处理、特征生成、模型训练、预测、评估环节。各个组件依次对上游的结果数据集进行处理。图 2-3 为实际某算法平台的一个批式任务运行状态截图。

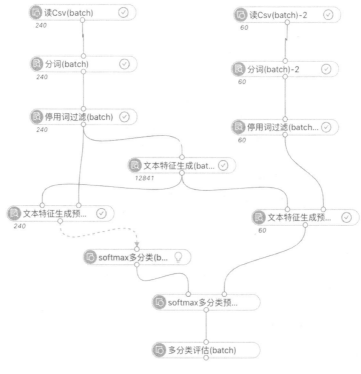

图 2-3

整个流程可用一个有向无环图（Directed Acyclic Graph，缩写为 DAG）来表示。在典型的工作流场景中，每个组件是在其所依赖的前序组件都完成后，才能开始执行的。图 2-4 中正在运行的是"softmax 多分类"组件，关注此处细节。

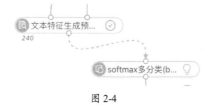

图 2-4

接入的数据连线是动态的虚线，表明从前序组件获取处理数据进行训练；而计算完成的组件的结果数据集是固定的，在图标左下方显示了该结果数据集的记录条数。"文本特征生成预处理"组件生成的带特征的训练数据为 240 条。

我们再来看流式任务，和"流水线"的操作类似，每个组件就相当于一名工人，每经过一名工人，产品上就会多一个组件；各个计算组件是同时启动的，每条数据在前一个组件处理完成后，会自动推到下一个组件。对于流式任务，经常用单位时间处理数据的条数来衡量效率。评估需要并发配置的计算资源，以便对线上请求及时响应。

如图 2-5 所示，我们看到流式任务的所有组件都在运行状态，每个组件左下角显示的是其每秒处理的数据条数。

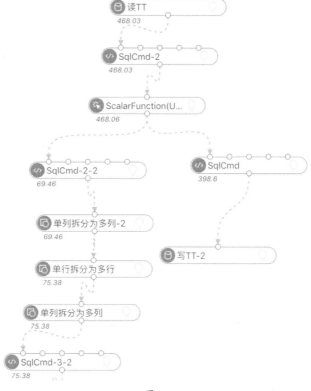

图 2-5

图 2-6 还展示了"流批混跑"的情形，右上方和右侧中部的三个组件为批式任务组件，左侧及下方的组件为流式任务组件。先运行批式任务，其最终输出结果会作为流式任务的初始化数据；批式任务都运行结束后，启动流式任务。

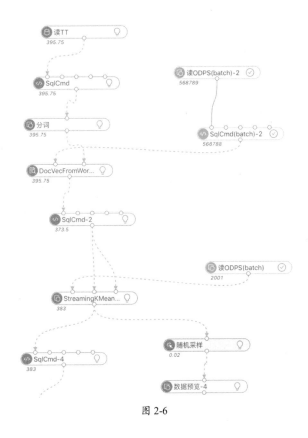

图 2-6

2.3 Alink=A+link

由前面关于批式任务与流式任务的介绍可知，批式任务与流式任务的数据计算及处理操作都发生在组件中，各组件间的连线就是数据的通路。批式任务和流式任务中有关组件与连接的描述是通用的。

Alink 定义了组件的抽象基类 AlgoOperator，规范了组件的基本行为。由 AlgoOperator 派生出了两个基类：用于批式计算及处理场景的批式算法组件（BatchOperator）和用于流式计算及处理场景的流式算法组件（StreamOperator）。

组件间的连接是通过定义 link 方法实现的，比如，组件 algoA 的输出是组件 algoB 的输入，而组件 algoB 的输出又是组件 algoC 的输入，则可以通过组件的 link 方法表示：

```
algoA.link(algoB).link(algoC)
```

Alink 的名称可以从这个角度进行解读:"Alink=A+link"。这里的 A 代表 Alink 的全部算法组件,其都是由抽象基类 AlgoOperator 派生出来的;link 是 AlgoOperator 各派生组件间的连接方法。

2.3.1 BatchOperator和StreamOperator

由算法组件的抽象基类(AlgoOperator)派生出两个基类:批式算法组件(BatchOperator,或称为批式处理组件、批式组件)和流式算法组件(StreamOperator,或称为流式处理组件、流式组件),它们的 UML 类图如图 2-7 所示。

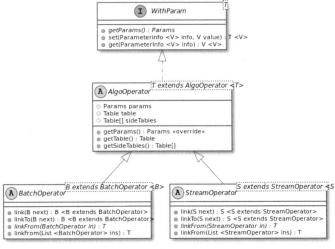

图 2-7

由 UML 类图,可以看到如下信息:
- WithParam 定义了参数设定和获取的接口。
- 抽象基类 AlgoOperator 中有 Params 类型的成员变量 params,并实现了参数设定和获取的接口;还定义了 Table 类型的变量 table,以及 Table 类型数组的变量 sideTables,用来存放算法的结果,并提供方法供后续组件读取。
- AlgoOperator 下面有 2 个派生泛型基类:BatchOperator(批式算法组件)和 StreamOperator(流式算法组件)。可以看到,这两种算法组件都支持 link 的操作;但批式算法组件只能连接一个或多个批式算法组件,流式算法组件只能连接一个或多个流式算法组件。对于需要批式数据和流式数据混合处理的算法,我们会将其作为流式算法组件,一般会将批式数据通过流式算法组件的构造函数传入。

Alink 将每个批式操作定义为一个批式组件（BatchOperator），每个批式组件在命名上都以"BatchOp"为后缀；同样，将每个流式操作定义为一个流式组件（StreamOperator），每个流式组件在命名上都以"StreamOp"为后缀。

通过这样的定义，批式任务和流式任务都可以用相同的方式进行描述，这样就可以大大降低批式任务和流式任务转换的代价。若需要将一个批式任务改写为流式任务，只需要将批式组件后面的"BatchOp"后缀变为"StreamOp"，相应的 link 操作便可转换为针对流式数据的操作。也正是因为 Alink 的批式组件和流式组件有如此密切的联系，所以才能将机器学习的管道（Pipeline）操作推广到流式场景。

算法的输入数据，在大多数情况下可以用一个表（Table）表示，但也有不少情况下需要用多个表（Table）才能表示。比如 Graph 数据，一般包括 Edge Table 和 Node Table，这两个表在一起才是完整的表示。算法的输出也是这样的情况。在大多数情况下可以用一个表（Table）表示，但也有不少情况下需要用多个表（Table）才能表示。比如，Graph 操作的结果还是 Graph，仍然需要用 2 个 Table 分别表示结果图中各条边和各个顶点的信息。自然语言方面的常用算法 LDA（Latent Dirichlet Allocation）的计算结果为 6 个 Table。其算法组件包含了一个 Table 类型的成员变量 table，用来放置该组件的主输出结果（大多数情况下，算法计算的结果只有一个 Table，输出到该变量即可）。该算法组件也定义了一个 Table 类型数组的变量 sideTables，该变量用来存储在多表（Table）输出的情况下，除主表外的所有其他表。

2.3.2　link方式是批式算法/流式算法的通用使用方式

简单地说，link 方式指的是在工作流中通过连线的方式，串接起不同的组件。link 方式给我们带来的一个简化是，前序组件的产出结果可能比较复杂，比如描述的是一个机器学习模型，我们不必了解其细节、不必详细描述它，只要通过"link"的方式将两个组件建立连接，后面的组件即可通过该连接（link）来获取前序组件的处理结果数据、数据的列数，以及各列的名称和类型。

1. link、linkTo和linkFrom

连接（link）是有方向的，组件 A 连接组件 B，即先执行组件 A，然后将计算结果传给组件 B 继续执行，则组件间的关系可以通过以下三种方式表示：

- A.linkTo(B)
- B.linkFrom(A)
- A.link(B)

这里，可将 link 看作 linkTo 的简写。

关于两个组件 A、B 之间的连接（link）关系，很容易理解。在实际应用中，我们还会遇到更复杂的情况，但使用 link 方法仍可以轻松处理。

（1）一对多的情况

组件 B1、B2、B3 均需要组件 A 的计算结果，组件间的关系可以通过以下多种方式表示：
- A.linkTo(B1)、A.linkTo(B2)、A.linkTo(B3)
- A.link(B1)、A.link(B2)、A.link(B3)
- B1.linkFrom(A)、B2.linkFrom(A)、B3.linkFrom(A)
- A.linkTo(B1)、A.link(B2)、B3.linkFrom(A)

从上述表示方式上可以看出，表示方式可以很灵活。因为组件 B1、B2、B3 与组件 A 的关系是独立的，所以可以分别选用表示方式。

（2）多对一的情况

组件 B 同时需要组件 A1、A2、A3 的计算结果，表示方式只有一种：

```
B.linkFrom(A1, A2, A3)
```

即，linkFrom 可以同时接入多个组件。

2. 深入理解

批式处理组件 BatchOperator 的相关代码如下：

```
public abstract class BatchOperator {
    ... ...
public BatchOperator link(BatchOperator f) {
    return linkTo(f);
}

public BatchOperator linkTo(BatchOperator f) {
    f.linkFrom(this);
    return f;
}

abstract public BatchOperator linkFrom(BatchOperator in);

public BatchOperator linkFrom(List<BatchOperator> ins) {
    if (null != ins && ins.size() == 1) {
        return linkFrom(ins.get(0));
    } else {
        throw new RuntimeException("Not support more than 1 inputs!");
    }
}
    ... ...
}
```

流式处理组件 StreamOperator 的相关代码如下：

```
public abstract class StreamOperator {
    ... ...
    public StreamOperator link(StreamOperator f) {
        return linkTo(f);
    }

    public StreamOperator linkTo(StreamOperator f) {
        f.linkFrom(this);
        return f;
    }

    abstract public StreamOperator linkFrom(StreamOperator in);

    public StreamOperator linkFrom(List <StreamOperator> ins) {
        if (null != ins && ins.size() == 1) {
            return linkFrom(ins.get(0));
        } else {
            throw new RuntimeException("Not support more than 1 inputs!");
        }
    }
    ... ...
}
```

从上述代码中，我们可以看出 link、linkTo 与 linkFrom 的关系。首先看看 link 与 linkTo 的关系：

```
public BatchOperator link(BatchOperator f) {
    return linkTo(f);
}

public StreamOperator link(StreamOperator f) {
    return linkTo(f);
}
```

显然，link 等同于 linkTo，可以将 link 看作 linkTo 的简写。

然后，我们再将注意力转向 linkTo 与 linkFrom：

```
public BatchOperator linkTo(BatchOperator f) {
    f.linkFrom(this);
    return f;
}

public StreamOperator linkTo(StreamOperator f) {
    f.linkFrom(this);
    return f;
}
```

显然，A.linkTo(B)等效于 B.linkFrom(A)，在 linkTo 组件具体实现时，只要实现 linkFrom 方法即可。

基类 BatchOperator 和 StreamOperator 均定义了输入参数为一个组件的抽象方法 linkFrom，该方法需要继承类进行实现；抽象类 BatchOperator 和 StreamOperator 同时也实现了一个输入是组件列表的方法 linkFrom，在组件列表中只含有一个组件的时候，该方法会调用前面的抽象方法 linkFrom；在其他情况下，则会抛出异常。

在我们实现的组件中，大部分组件只支持输入一个组件，即只要实现输入参数为一个组件的抽象方法 linkFrom 就可以了；有的组件支持输入多个组件，则需要重载输入为组件列表的方法 linkFrom，并将输入为一个组件的情况看作输入为算法列表时，列表中的组件个数为一个的情况。

2.3.3 link的简化

link 组件是 Alink 的基本使用方式；但对于一些常用的功能，比如取前 N 条数据、随机采样、SQL SELECT、数据过滤等，Alink 定义了相关的方法（方法内部的实现过程也是 link 相应的组件），这样代码写起来会更简练。

对比下面两段代码，执行的是同样的功能；但是很明显，右边的代码更简练，也更易懂：

```source.link(new SelectBatchOp() .setClause("petal_width, category")).link(new FilterBatchOp() .setClause("category='Iris-setosa'")).link(new SampleBatchOp() .setRatio(0.3)).link(new FirstNBatchOp() .setSize(5)).print();```	```source.select("petal_width, category").filter("category='Iris-setosa'").sample(0.3).firstN(5).print();```

2.3.4 组件的主输出与侧输出

组件可能需要一个或多个输入，通过 linkFrom 方法便可将多个上游组件的输出连接到该组件。组件也会产生一个或多个输出。对大部分算法组件来说，结果只有一个数据表，输出是唯一的，但是有些算法组件会产生多个数据表。这时，就需要确认一个数据表作为主输出，其余数据表作为侧输出（Side Output）。

侧输出（Side Output）有两个重要方法：
- getSideOutputCount()获得该组件侧输出的个数。
- getSideOutput(int index)方法通过索引号获取具体的侧输出，每个侧输出是 BatchOperator 或者 StreamOperator。

比如，我们在做机器学习实验的时候，经常要把原始数据分为训练集和测试集，数据划分组件就会对应两个输出：主输出为训练集；侧输出（Side Output）只有一个，即输出测试集。详细的例子可以参考 7.2 节。

2.4　Pipeline与PipelineModel

Alink 提供了 Pipeline/PipelineModel，其在功能和使用方式上与 Scikit-learn 和 SparkML 的 Pipeline/PipelineModel 类似。保持训练和预测过程中数据处理的一致性，这样调用过程清晰、简练，也降低了用户的学习成本。Alink Pipeline/PipelineModel 与批式/流式组件（BatchOperator/StreamOperator）一样，将 Flink Table 作为计算输入和输出结果的数据类型。

Scikit-learn 和 SparkML 的 Pipeline 是针对批式数据的训练和预测设计的。Alink 不仅支持批式数据的训练和预测，也支持将批式训练出的模型用于预测流式数据。

2.4.1　概念和定义

管道（Pipeline）的概念源于 Scikit-learn。可以将数据处理的过程看成数据在 "管道" 中流动。管道分为若干个阶段（PipelineStage），数据每通过一个阶段就发生一次变换，数据通过整个管道，也就依次经历了所有变换。

如果要在管道中加入分类器，对数据进行类别预测，就涉及模型的训练和预测，这需要分两个步骤完成。所以，管道也会被细分为管道定义与管道模型（PipelineModel）。在管道定义中，每个 PipelineStage 会按其是否需要进行模型训练，分为估计器（Estimator）和转换器（Transformer）。随后，可以对其涉及模型的部分，即对估计器（Estimator）进行估计，从而得到含有模型的转换器。该转换器被称为 Model，并用 Model 替换相应的估计器，从而每个阶段都可以直接对数据进行处理。我们将该 Model 称为 PipelineModel。

上面介绍了几个概念，图 2-8 会用图形表示它们之间的关系，并进行深入介绍。

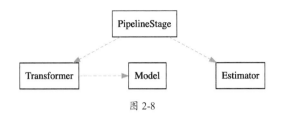

图 2-8

- 转换器（Transformer）：Pipeline 中的处理模块，用于处理 Table。输入的是待处理的数据，输出的是处理结果数据。比如，向量归一化是一个常用的数据预处理操作。它就是一个转换器，输入向量数据，输出的数据仍为向量，但是其范数为 1。
- Model：派生于转换器（Transformer）。其可以存放计算出来的模型，用来进行模型预测。其输入的是预测所需的特征数据，输出的是预测结果数据。
- 估计器（Estimator）：估计器是对输入数据进行拟合或训练的模型计算模块，输出适合当前数据的转换器，即模型。输入的是训练数据，输出的是 Model。

Pipeline 与 PipelineModel 构成了完整的机器学习处理过程，可以分为三个子过程：定义过程，模型训练过程，数据处理过程。

- 定义过程：按顺序罗列 Pipeline 所需的各个阶段。Pipeline 由若干个 Transformer 和 Estimator 构成，按用户指定的顺序排列，并在逻辑上依次执行。
- 模型训练过程：使用 Pipeline 的 fit 方法，对 Pipeline 中的 Estimator 进行训练，得到相应的 Model。Pipeline 执行 fit 方法后得到的结果是 PipelineModel。
- 数据处理过程：该过程指的是通过 PipelineModel 直接处理数据。

比如，LR 分类算法作为一个 Estimator，可以在构建 Pipeline 的时候进行定义，之后在 fit 的过程中，会使用 LR 的训练算法，得到 LR model，并将其作为整个 PipelineModel 的一部分；在使用 PipelineModel 处理数据时，会相应地调用 LR 算法的预测部分。

2.4.2 深入介绍

以使用朴素贝叶斯算法进行多分类为例，演示三个经典场景：批式训练、批式预测与流式预测。为了更好地体现它们之间的关联，我们用一张图表示了出来，如图 2-9 所示。

图 2-9 有如下特点：

（1）左边这列组件展示了批式训练的流程，中间那列组件展示了批式预测的流程，右边那列组件展示了流式预测的流程。

（2）训练和预测都经历了四个阶段：缺失值填充、MultiStringIndexer（将字符串数据用索引值替换）、VectorAssembler（将各数据字段组装为向量）、使用朴素贝叶斯算法进行训练/预测。

（3）在 VectorAssembler 阶段，运行时不需要额外的信息。在批式训练、批式预测和流式预测的过程中，组件均可直接使用。

（4）除 VectorAssembler 外的三个阶段，都需要对整体数据进行扫描或者迭代训练，得到模型后，才能处理（模型预测）数据。训练过程需要拿到所有的模型，缺失值填充和 MultiStringIndexer 阶段需要在得到模型后，对当前数据进行预测，将预测结果数据传给后面的阶段。

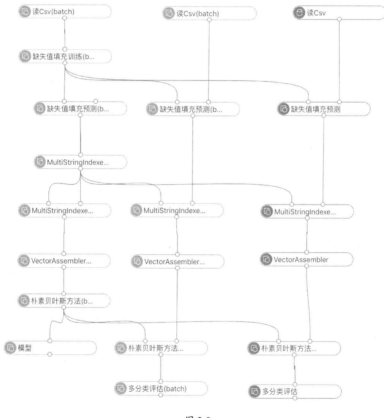

图 2-9

上面的四个阶段分别对应四个 PipelineStage，这些 PipelineStage 构成了 Pipeline。对应的 Java 代码如下：

```
Pipeline pipeline = new Pipeline()
    .add(
```

```
    new Imputer()
        .setSelectedCols("...")
        .setStrategy(Strategy.VALUE)
        .setFillValue("null")
)
.add(
    new MultiStringIndexer()
        .setSelectedCols("...")
)
.add(
    new VectorAssembler()
        .setSelectedCols("...")
        .setOutputCol("vec")
)
.add(
    new NaiveBayesTextClassifier()
        .setVectorCol("vec")
        .setLabelCol("label")
        .setPredictionCol("pred")
);
```

Pipeline 可多次复用，不同的训练数据通过同样的流程可得到不同的模型：

```
CsvSourceBatchOp trainData1 = new CsvSourceBatchOp();
CsvSourceBatchOp trainData2= new CsvSourceBatchOp();

PipelineModel model = pipeline.fit(trainData1);
PipelineModel model2 = pipeline.fit(trainData2);
```

得到的 PipelineModel 可重复应用于不同的预测数据：

```
CsvSourceBatchOp predictData1= new CsvSourceBatchOp();
CsvSourceBatchOp predictData2= new CsvSourceBatchOp();

BatchOperator<?> predict1 = model.transform(predictData1);
BatchOperator<?> predict2 = model.transform(predictData2);
```

PipelineModel 可以无区别地应用于批式预测和流式预测：

```
CsvSourceBatchOp batchData = new CsvSourceBatchOp();
CsvSourceStreamOp streamData= new CsvSourceStreamOp();

BatchOperator<?> predictBatch = model.transform(batchData);
StreamOperator<?> predictStream = model.transform(streamData);
```

从 PipelineModel 可以得到用于本地预测的 LocalPredictor，直接嵌入 Java 应用服务：

```
String inputSchemaStr = "...";
LocalPredictor localPredictor = model.collectLocalPredictor(inputSchemaStr);

Row inputRow = Row.of("...");
Row pred = localPredictor.map(inputRow);
```

2.5 触发Alink任务的执行

Alink 的批式任务/流式任务本质上是通过触发 Flink 的 ExecutionEnvironment/StreamExecutionEnvironment 的 execute 方法来执行的。

为了调用起来更简练、方便，Alink 包装了如下两个方法：

- BatchOperator.execute()触发批式任务的执行。
- StreamOperator.execute()触发流式任务的执行。

在实际使用中，经常需要触发一些小的批式任务，获取执行的结果。Flink 的 DataSet 提供了 print 方法、collect 方法和 count 方法，Alink 的批式组件也提供了相应的方法。这些方法为用户了解中间过程的数据提供了便利，但是其会被多次触发执行，各个任务之间会有许多重复执行部分。尤其当数据量较大时，每个任务的执行时间较长，而重复计算会消耗大量的计算资源。针对这样的问题，Alink 引入了 Lazy 方式，并相应地定义了 lazyPrint()、lazyCollect()等方法。

Lazy方式

"Lazy"的概念与"Eager"的概念相对，在许多开发语言中都会应用。Lazy 方式被称作懒操作方式，Eager 方式被称作急操作或者实时操作方式。Eager 方式与 Lazy 方式的区别在于用户发出数据请求（显示、获取、加载等）时，是立即行动，还是可以慢一些。

对于 Alink 的批式处理场景，我们在执行任务时，需要看到原始数据、中间数据，以及统计、评估指标等。若每一个任务均采用 Eager 方式，则每次数据请求都会触发一个 Flink 任务，而且由于 Flink 机制的原因，后续的 Flink 任务还会重复执行前序任务的部分计算。如果采用 Lazy 方式，则每个任务并不急于显示，可以通过一个 Flink 任务完成整个流程，在任务结束后，再返回各项数据。从数据的获取速度上看，采用 Lazy 方式比较慢，且有延迟；但是从整体的计算时间和资源的消耗方面看，采用 Lazy 方式更优。

Flink 提供了三种方法执行批式任务：print、collect 和 execute。它们都是 Eager 方式的。比如，运行到 print 方法，立即会启动 Flink 任务进行执行，并将结果打印出来。Alink 的批式组件（形如***BatchOp）提供了打印的方法；统计和评估类组件提供了 collect 方法，用户可以立即获取统计和评估结果，这些方法的命名规则为 collect+结果名称，比如 collectSummary。执行批式组件，可以使用 BatchOperator.execute()。

Alink 的 Lazy 方式需要在操作前加上"lazy"前缀，主要有如下两类操作：

- 采用 Lazy 方式打印数据、统计结果或模型信息的操作，方法定义为 lazyPrint***()。

- 采用 Lazy 方式获取结果的操作，方法定义为 lazyCollect***()。

图 2-10 详细列举了在各个使用场景下对应的方法。

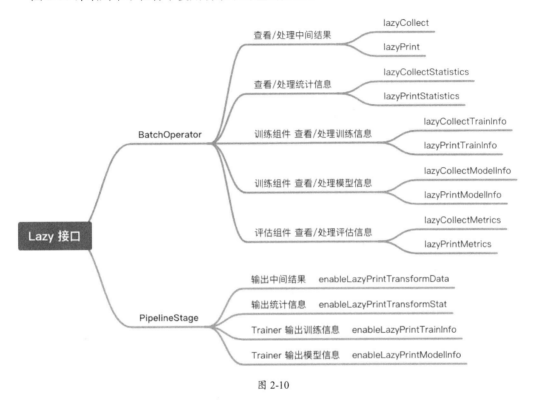

图 2-10

2.6　模型信息显示

我们在训练过程中需要了解训练所得模型的信息；对于已经存储的模型数据或者 PipelineModel 数据，需要采用某种方法来了解其具体内容。Alink 对各场景提供了相应的方案，如表 2-1 所示。

表 2-1 模型显示的场景

	训练过程	模型数据
单个批式训练组件	【场景】在训练出来模型的同时，显示模型信息 【用法】 调用组件相应的方法： • lazyPrintModelInfo：打印模型信息概要 • lazyCollectModelInfo：获取完整模型信息，并调用回调方法做处理	【场景】从单个模型数据中提取模型信息 【用法】 选择相应的模型信息组件（形如***ModelInfoBatchOp），随后调用模型信息组件相应的方法： • lazyPrintModelInfo：打印模型信息概要 • lazyCollectModelInfo：获取完整模型信息，并调用回调方法做处理
Pipeline（多个批式训练组件）	【场景】在 Pipeline 进行 fit 处理的时候，允许某些阶段显示相应的模型信息 【用法】 设置 PipelineStage 的行为。若需要输出模型信息，则必须调用如下方法，并设置布尔型参数值为 True：enableLazyPrintModelInfo(True)。调用 Trainer 的 fit 方法时，输出模型信息	【场景】获取显示 PipelineModel 中各个阶段的信息、各个模型的信息 【用法】 （1）显示各个 PipelineStage 的信息 （2）获取 PipelineStage 的模型数据 （3）根据 PipelineStage 的信息，确定对应的模型信息组件（形如***ModelInfoBatchOp），随后调用模型信息组件相应的方法： • lazyPrintModelInfo：打印模型信息概要 • lazyCollectModelInfo：获取完整的模型信息，并调用回调方法做处理

在训练过程中如何显示模型信息呢？可参阅本书的 8.5 节、8.6 节、9.2.5 节、9.4 节等处的示例。本节着重介绍如何针对存储的模型文件，提取模型信息。

比如，已知某个模型文件为单棵决策树模型文件，我们可以使用文件数据源读取模型数据，再连接决策树模型信息组件（DecisionTreeModelInfoBatchOp），并使用 lazyPrintModelInfo 方法打印模型信息。使用 lazyCollectModelInfo 方法，自定义抽取模型信息，并将决策树可视化导出到图片。具体代码如下：

```
new AkSourceBatchOp()
    .setFilePath(DATA_DIR + TREE_MODEL_FILE)
    .link(
        new DecisionTreeModelInfoBatchOp()
            .lazyPrintModelInfo()
            .lazyCollectModelInfo(new Consumer <DecisionTreeModelInfo>() {
                @Override
                public void accept(DecisionTreeModelInfo decisionTreeModelInfo) {
                    try {
                        decisionTreeModelInfo.saveTreeAsImage(
                            DATA_DIR + "tree_model.png", true);
```

```
            } catch (IOException e) {
                e.printStackTrace();
            }
        }
    })
);
BatchOperator.execute();
```

运行结果如下：

```
Classification trees modelInfo:
Number of trees: 1
Number of features: 4
Number of categorical features: 2
Labels: [no, yes]

Categorical feature info:
|feature|number of categorical value|
|-------|---------------------------|
|outlook|                          3|
|  Windy|                          2|

Table of feature importance Top 4:
|    feature|importance|
|-----------|----------|
|   Humidity|    0.4637|
|      Windy|    0.4637|
|    outlook|    0.0725|
|Temperature|         0|
```

导出的决策树可视化结果如图 2-11 所示。

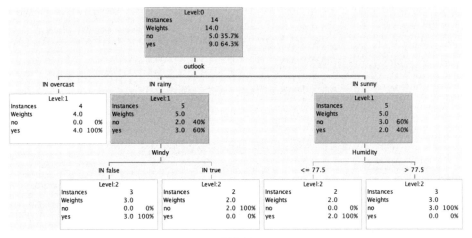

图 2-11

我们再看一下如何从 PipelineModel 中提取信息，显示模型。首先载入 PipelineModel，然后获取 Pipeline 中各阶段（PipelineStage）的信息，相关代码如下：

```
PipelineModel pipelineModel = PipelineModel.load(DATA_DIR + PIPELINE_MODEL_FILE);

TransformerBase <?>[] stages = pipelineModel.getTransformers();

for (int i = 0; i < stages.length; i++) {
    System.out.println(String.valueOf(i) + "\t" + stages[i]);
}
```

运行结果如下：

```
0    com.alibaba.alink.pipeline.sql.Select@19c1f6f4
1    com.alibaba.alink.pipeline.regression.LinearRegressionModel@46fa2a7e
```

这里共有两个阶段，第一个阶段执行 Select 操作，第二个阶段为使用 LinearRegressionModel 进行预测操作。这样，我们就确定了可以使用线性回归（Linear Regression）对应的模型信息组件来查看索引号为 1 的 PipelineStage 的信息，具体代码如下：

```
((LinearRegressionModel) stages[1]).getModelData()
   .link(
      new LinearRegModelInfoBatchOp()
         .lazyPrintModelInfo()
   );
BatchOperator.execute();
```

运行结果如下。这里显示了模型的 meta 信息，在此还可以看到各个权重参数的取值：

```
-------------------------- model meta info --------------------------
{hasInterception: true, model name: Linear Regression, num feature: 2}
-------------------------- model weight info --------------------------
|     intercept|              x|            x2|
|--------------|---------------|--------------|
|122194787.2123|-121612.28370990|30.25813853|
```

最后，我们将模型信息（ModelInfo）组件、批式训练组件和 PipelineStage 的对照关系整理为表格，详见表 2-2。

表 2-2 模型信息组件、批式训练组件和 PipelineStage 对照表

	批式训练组件	PipelineStage	模型信息组件
分类算法	LinearSvmTrainBatchOp	LinearSvm	LinearSvmModelInfoBatchOp
	LogisticRegressionTrainBatchOp	LogisticRegression	LogisticRegressionModelInfoBatchOp
	SoftmaxTrainBatchOp	Softmax	SoftmaxModelInfoBatchOp
	NaiveBayesTrainBatchOp	NaiveBayes	NaiveBayesModelInfoBatchOp

续表

	批式训练组件	PipelineStage	模型信息组件
分类算法	NaiveBayesTextTrainBatchOp	NaiveBayesTextClassifier	NaiveBayesTextModelInfoBatchOp
	FmClassifierTrainBatchOp	FmClassifier	FmClassifierModelInfoBatchOp
	DecisionTreeTrainBatchOp	DecisionTreeClassifier	DecisionTreeModelInfoBatchOp
	GbdtTrainBatchOp	GbdtClassifier	GbdtModelInfoBatchOp
	RandomForestTrainBatchOp	RandomForestClassifier	RandomForestModelInfoBatchOp
	C45TrainBatchOp	C45	C45ModelInfoBatchOp
	CartTrainBatchOp	Cart	CartModelInfoBatchOp
	Id3TrainBatchOp	Id3	Id3ModelInfoBatchOp
回归算法	CartRegTrainBatchOp	CartReg	CartRegModelInfoBatchOp
	DecisionTreeRegTrainBatchOp	DecisionTreeRegressor	DecisionTreeRegModelInfoBatchOp
	RandomForestRegTrainBatchOp	RandomForestRegressor	RandomForestRegModelInfoBatchOp
	GbdtRegTrainBatchOp	GbdtRegressor	GbdtRegModelInfoBatchOp
	LassoRegTrainBatchOp	LassoRegression	LassoRegModelInfoBatchOp
	RidgeRegTrainBatchOp	RidgeRegression	RidgeRegModelInfoBatchOp
	LinearRegTrainBatchOp	LinearRegression	LinearRegModelInfoBatchOp
	AftSurvivalRegTrainBatchOp	AftSurvivalRegression	AftSurvivalRegModelInfoBatchOp
	FmRegressorTrainBatchOp	FmRegressor	FmRegressorModelInfoBatchOp
聚类算法	KMeansTrainBatchOp	KMeans	KMeansModelInfoBatchOp
	GmmTrainBatchOp	GaussianMixture	GmmModelInfoBatchOp
	BisectingKMeansTrainBatchOp	BisectingKMeans	BisectingKMeansModelInfoBatchOp
	LdaTrainBatchOp	Lda	LdaModelInfoBatchOp
数据处理	ImputerTrainBatchOp	Imputer	ImputerModelInfoBatchOp
	MaxAbsScalerTrainBatchOp	MaxAbsScaler	MaxAbsScalerModelInfoBatchOp
	MinMaxScalerTrainBatchOp	MinMaxScaler	MinMaxScalerModelInfoBatchOp
	StandardScalerTrainBatchOp	StandardScaler	StandardScalerModelInfoBatchOp
	VectorImputerTrainBatchOp	VectorImputer	VectorImputerModelInfoBatchOp
	VectorMaxAbsScalerTrainBatchOp	VectorMaxAbsScaler	VectorMaxAbsScalerModelInfoBatchOp
	VectorMinMaxScalerTrainBatchOp	VectorMinMaxScaler	VectorMinMaxScalerModelInfoBatchOp
	VectorStandardScalerTrainBatchOp	VectorStandardScaler	VectorStandardScalerModelInfoBatchOp
	EqualWidthDiscretizerTrainBatchOp	EqualWidthDiscretizer	EqualWidthDiscretizerModelInfoBatchOp
	OneHotTrainBatchOp	OneHotEncoder	OneHotModelInfoBatchOp
	PcaTrainBatchOp	PCA	PcaModelInfoBatchOp
	QuantileDiscretizerTrainBatchOp	QuantileDiscretizer	QuantileDiscretizerModelInfoBatchOp

2.7 文件系统与数据库

我们处理的数据主要存储在文件系统和数据库中。本书的第 3 章将专门介绍文件系统，第 4 章会详细介绍与数据库、数据表相关的操作。Alink 采用插件的方式来管理各种文件系统、数据库以及各个版本所需的函数库。本节还会专门介绍如何利用 Alink 的插件（Plugin）工具自动下载相关的函数库。

图 2-12 为 Alink 文件系统（File System）与数据库（Catalog）的架构图，具体说明如下：

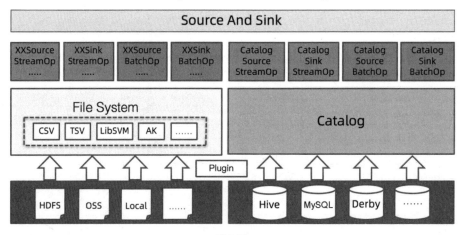

图 2-12

- Alink 通过定义统一的 File System，规范常用的文件操作；Alink 定义统一的 Catalog，抽象常用的数据库（Database）和表（Table）操作。
- 我们在实际应用中，不会同时用到所有文件系统和数据库。另外，我们还要考虑版本的问题。使用插件机制，方便大家选择适合自己的方式与版本。
- 在数据源和导出方面，数据文件和数据库均利用统一定义的 File System 和 Catalog 来统一操作流程，避免逐个定义带来大量的类似接口。
- 数据库方面的操作比较直接。确定好 Catalog 以及数据表的路径信息（数据库→表）。此外，还要统一定义数据表的批式/流式数据源组件（CatalogSourceBatchOp/Catalog-SourceStreamOp），以及定义数据表的批式/流式导出组件（CatalogSinkBatchOp/Catalog-SinkStreamOp）。
- 对于数据文件，需要考虑两个方面：文件的格式和所在文件系统的路径。由于每种文件格式所需的参数不同，因此 Alink 按文件格式定义相应的数据源和导出组件。对于 *XX*

文件格式，定义批式/流式数据源组件（*XX*SourceBatchOp/ *XX*SourceStreamOp），同样定义数据表的批式/流式导出组件（*XX*SinkBatchOp/ *XX*SinkStreamOp）。
- 各组件的相同之处是，文件路径参数的设置方式相同；基本形式是使用 FilePath 类进行设置，包括两类信息：所在文件系统的信息和文件路径的信息。另外，对于本地路径，可以直接设置路径字符串。

插件下载

Alink 能够支持不同第三方库（例如 OSS、Hive、Derby、MySQL 等）的不同版本（例如 Hive 的 2.3.4 版本、2.3.6 版本等）。为了更好地管理插件（外部的第三方库），我们提供了插件下载器（PluginDownloader）来管理不同插件的多个版本。

插件下载器封装了插件的常见功能，如下所示：
- 枚举仓库中的所有插件。
- 枚举某个插件的所有版本。
- 下载某个插件的特定版本/默认版本。
- 下载所有插件的默认版本。
- 升级所有的插件。

在 Java 代码中可以这样使用插件下载器：

```
// 设置插件下载的位置。当路径不存在时，会自行创建路径
AlinkGlobalConfiguration.setPluginDir("/Users/xxx/alink_plugins/");

// 获得 Alink 插件下载器
PluginDownloader pluginDownloader = AlinkGlobalConfiguration.getPluginDownloader();

// 从远程加载插件的配置项
pluginDownloader.loadConfig();

// 展示所有可用的插件名称
List<String> plugins = pluginDownloader.listAvailablePlugins();
// 输出结果：[oss, hive, derby, mysql, hadoop, sqlite]

// 显示第 0 个插件的所有版本
String pluginName = plugins.get(0); // oss
List<String> availableVersions = pluginDownloader.listAvailablePluginVersions(pluginName);
// 输出结果：[3.4.1]

// 下载某个插件的特定版本
String pluginVersion = availableVersions.get(0);
pluginDownloader.downloadPlugin(pluginName, pluginVersion);
// 运行结束后，插件会被下载到 "/Users/xxx/alink_plugins/" 目录中
```

```
// 下载某个插件的默认版本
pluginDownloader.downloadPlugin(pluginName);
// 运行结束后,插件会被下载到 "/Users/xxx/alink_plugins/" 目录中

// 下载配置文件中所有插件的默认版本
pluginDownloader.downloadAll();

// 插件升级
// 在升级的过程中,会先对旧的插件进行备份,备份文件名称的后缀为.old。等到插件更新完毕后,
// 会统一删除旧的插件包
// 若插件更新中断,则用户可以从.old 文件中恢复旧版插件
pluginDownloader.upgrade();
```

2.8　Schema String

Alink 在进行表数据的读取和转换时,有时需要显式声明数据表的列名和列类型信息,即 Schema 信息。Schema String 就是将此信息使用字符串的方式进行描述的,这样便于将该信息作为 Java 函数或者 Python 函数的参数输入。

其定义格式如下:

```
colname coltype[, colname2 coltype2[, ...]]
```

例如,"f0 string, f1 bigint, f2 double",表明数据表共有 3 列,名称分别为 f0、f1 和 f2;其对应的类型分别为字符串类型、长整型和双精度浮点型。熟悉 SQL 的读者会发现,该格式与 CREATE TABLE 的数据列名称和类型的定义格式相同。

关于各种列类型的写法,可以参照 Flink Type 与 Type String 的对应表(见表 2-3)。注意:为了适应不同用户的习惯,同一个 Flink Type 可能对应着多种 Type String 的写法。

表 2-3　Flink Type 与 Type String 的对应表

Flink Type	Type String
Types.STRING	string 或者 varchar
Types.BOOLEAN	boolean
Types.BYTE	byte 或者 tinyint
Types.SHORT	short 或者 smallint
Types.INT	int
Types.LONG	long 或者 bigint

续表

Flink Type	Type String
Types.FLOAT	float
Types.DOUBLE	double
Types.SQL_DATE	sql_date 或者 date
Types.SQL_TIMESTAMP	sql_timestamp 或者 timestamp
Types.SQL_TIME	sql_time 或者 time

3 文件系统与数据文件

基于 Flink 的 FileSystem 类，Alink 进一步定义和实现了常用的文件系统，比如，本地文件系统、Hadoop 分布式文件系统、阿里云 OSS（Object Storage Service）文件系统。本章将先介绍各文件系统的通用操作，随后通过具体的示例，详细介绍常用的 LocalFileSystem、HadoopFileSystem、OssFileSystem 等。

我们使用的数据常常以数据文件的形式存在，而借助统一的文件系统接口，我们就可以方便地读/写各文件系统的数据文件。但要知道具体的文件内容，还需要了解数据文件的格式，比如，CSV、TSV、LibSVM 等都是机器学习领域常用的数据文件格式。本章的后半部分将介绍 Alink 针对常用数据文件格式提供的组件，以帮助用户轻松地在批式/流式场景中读入和写出数据。

3.1 文件系统简介

Alink 的 FileSystem 类是文件系统的抽象父类，定义了一些文件系统共有的功能。文件操作如表 3-1 所示。

表 3-1 文件操作

函数名	返回类型	说明
getFileStatus(Path f)	FileStatus	返回指定路径文件的元数据
listStatus(Path f)	FileStatus[]	列出给定路径文件夹下的所有子文件夹和子文件的元数据

续表

函数名	返回类型	说明
exists(Path f)	boolean	检查文件或目录是否存在
rename(Path src, Path dst)	boolean	重命名文件或文件夹
delete(Path f, boolean recursive)	boolean	删除文件或文件夹。如果 Path 对应的是文件，则无论 recursive 为 true 或者 false，该文件都会被直接删除；如果 Path 对应的是文件夹，则只有在 recursive 为 true 的情况下该文件夹才会被删除，否则会抛出异常
mkdirs(Path f)	boolean	建立文件夹（如果父文件夹不存在，则建立父文件夹）
create(Path f, boolean overwrite)	FSDataOutputStream	创建指定 Path 的一个文件，返回用于写入数据的输出流，参数 overwrite 控制是否强制覆盖已有的文件
open(Path f)	FSDataInputStream	返回读取指定文件的数据输入流

其中，FileStatus 类型封装了文件或目录的元数据，包括文件长度、块大小、文件访问、文件修改时间、是否为文件夹等信息。通过 create 方法获取输出流，以及通过 open 方法获取输入流，可以实现在同一文件系统内部进行文件的 I/O 操作，也可以做到跨文件系统进行文件的 I/O 操作。

下面将通过示例详细介绍本地文件系统、Hadoop 分布式文件系统和阿里云 OSS 文件系统。

3.1.1 本地文件系统

Alink 定义了 LocalFileSystem，用来操作本地文件系统。想必大家对本地文件系统的操作比较熟悉。这里直接通过 Java 示例来演示文件接口的用法。

先定义一个操作文件夹，后面的操作都会在该文件夹下进行：

```
static final String LOCAL_DIR = Utils.ROOT_DIR + "filesys" + File.separator;
```

下面的代码演示了如何查看本地文件系统的一些信息，其中 getKind 方法返回的是文件系统的类型：

```
LocalFileSystem local = new LocalFileSystem();
System.out.println(local.getHomeDirectory());
```

```
System.out.println(local.getKind());
System.out.println(local.getWorkingDirectory());
```

运行结果如下：

file:/Users/yangxu/
FILE_SYSTEM
file:/Users/yangxu/Code/alink/

实现一个常用操作：判断目标文件夹是否存在。若该文件夹不存在，则新建一个文件夹，并显示目标文件夹下子文件夹和子文件的信息。具体代码如下：

```
if (!local.exists(LOCAL_DIR)) {
    local.mkdirs(LOCAL_DIR);
}
for (FileStatus status : local.listStatus(LOCAL_DIR)) {
    System.out.println(status.getPath().toUri()
        + " \t" + status.getLen()
        + " \t" + new Date(status.getModificationTime())
    );
}
```

接下来演示与数据读/写相关的操作，设定目标文件为"hello.txt"，可以通过创建文件输出流的方式，将字符串"Hello Alink!"写入其中。随后显示该文件的状态。具体代码如下：

```
String path = LOCAL_DIR + "hello.txt";

OutputStream outputStream = local.create(path, WriteMode.OVERWRITE);
outputStream.write("Hello Alink!".getBytes());
outputStream.close();

FileStatus status = local.getFileStatus(path);
System.out.println(status);
System.out.println(status.getLen());
System.out.println(new Date(status.getModificationTime()));
```

注意：在创建文件输出流的时候选择了"WriteMode.OVERWRITE"，即覆盖写。如果目标文件存在，则清除其原有数据，之后写入该次执行的结果数据。

获得的输出结果如下：

LocalFileStatus{file=/Users/yangxu/alink/data/temp/hello.txt, path=file:/Users/yangxu/alink/data/temp/hello.txt}
12
Thu Jul 23 11:33:20 CST 2020

最后，我们使用以下文件输入流，将文件内容读取到一个字符串：

```
InputStream inputStream = local.open(path);
String readString = IOUtils.toString(inputStream);
```

```
System.out.println(readString);
```

其输出结果为"Hello Alink!"。

3.1.2 Hadoop文件系统

Alink 定义了 HadoopFileSystem，用来操作 Hadoop 分布式文件系统。

首先看如何构建 HadoopFileSystem 实例。构建该实例需要 HDFS 服务的 IP 地址和 Port（端口），且使用 URI 方式给出。

```
static final String HDFS_URI = "hdfs://10.*.*.*:9000/";
```

> 注意：这里用"*"略去了具体的数字，读者在尝试此代码时，可以根据自己 HDFS 服务的 IP 地址和 Port（端口）进行替换。

使用 HadoopFileSystem 的构建实例，具体代码如下：

```
HadoopFileSystem hdfs = new HadoopFileSystem(HDFS_URI);
```

随后，调用 getKind 方法；并判断目标文件夹是否存在。若该文件夹不存在，则新建一个文件夹。

```
System.out.println(hdfs.getKind());

if (!hdfs.exists(hdfsDir)) {
   hdfs.mkdirs(hdfsDir);
}
```

设定目标文件为"hello.txt"，并判断该文件是否存在。若该文件存在，则删除该文件；然后通过创建文件输出流的方式，将字符串"Hello Alink!"写入其中。具体代码如下：

```
String path = hdfsDir + "hello.txt";

if (hdfs.exists(path)) {
   hdfs.delete(path, true);
}
OutputStream outputStream = hdfs.create(path, WriteMode.NO_OVERWRITE);
outputStream.write("Hello Alink!".getBytes());
outputStream.close();
```

注意：因为判断和删除操作，保证了目标文件不存在，所以在创建文件输出流的时候，可以使用"WriteMode.NO_OVERWRITE"。

最后，我们使用以下文件输入流，将文件内容读取到一个字符串：

```
InputStream inputStream = hdfs.open(path);
String readString = IOUtils.toString(inputStream);
System.out.println(readString);
```

其输出结果为"Hello Alink!"，说明整套写入、读取流程运行正常。

有了文件系统的基本功能，很容易包装出新的方法。比如下面的函数，通过输入/输出流，完成了文件的复制操作：

```
static void copy(InputStream in, OutputStream out) throws IOException {
   byte[] buffer = new byte[1024 * 1024];
   int len = in.read(buffer);
   while (len != -1) {
      out.write(buffer, 0, len);
      len = in.read(buffer);
   }
   in.close();
   out.close();
}
```

我们使用这个函数，就可完成 Hadoop 文件系统和本地文件系统的文件复制操作，并显示文件夹中的内容。具体代码如下：

```
LocalFileSystem local = new LocalFileSystem();

HadoopFileSystem hdfs = new HadoopFileSystem(HDFS_URI);

copy(
   hdfs.open(HDFS_URI + "user/yangxu/alink/data/temp/hello.txt"),
   local.create(LOCAL_DIR + "hello_1.txt", WriteMode.OVERWRITE)
);

copy(
   local.open(LOCAL_DIR + "hello_1.txt"),
   hdfs.create(HDFS_URI + "user/yangxu/alink/data/temp/hello_2.txt", WriteMode.OVERWRITE)
);

for (FileStatus status : hdfs.listStatus(HDFS_URI + "user/yangxu/alink/data/temp/")) {
   System.out.println(status.getPath().toUri()
      + " \t" + status.getLen()
      + " \t" + new Date(status.getModificationTime())
   );
}
```

运行结果如下：

```
hdfs://10.*.*.*:9000/user/yangxu/alink/data/temp/hello.txt       12     Thu Jul 23 13:24:45 CST 2020
hdfs://10.*.*.*:9000/user/yangxu/alink/data/temp/hello_2.txt     12     Thu Jul 23 13:24:46 CST 2020
```

现在我们就可以使用 Hadoop Web UI，查看目标文件夹下的数据情况了。如图 3-1 所示，可以看到新增了两个文件：hello.txt 和 hello_2.txt。

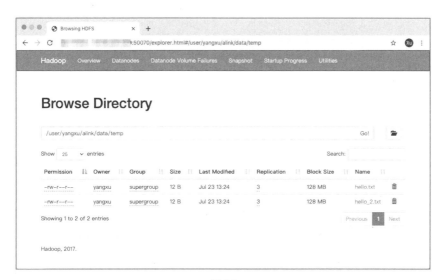

图 3-1

3.1.3　阿里云OSS文件系统

Alink 定义了 OssFileSystem，用来操作阿里云 OSS 文件系统。

首先看如何构建 OssFileSystem 实例，代码如下。构建该实例需要 4 个参数：OSS_END_POINT、OSS_BUCKET_NAME、OSS_ACCESS_ID、OSS_ACCESS_KEY。OSS_ACCESS_ID、OSS_ACCESS_KEY 相当于用户名和密码；OSS_END_POINT 为所要使用的 OSS 服务地址。用户可以设置若干个 Bucket，每个 Bucket 都可以被看作一个根目录，在使用OSS时需要选定一个OSS_BUCKET_NAME 进行。在本示例中，选定的OSS_BUCKET_NAME 名称为"yangxu-bucket"。

```
OssFileSystem oss =
    new OssFileSystem(
        OSS_END_POINT,
        OSS_BUCKET_NAME,
        OSS_ACCESS_ID,
        OSS_ACCESS_KEY
    );
```

调用 getKind 方法，代码如下，得到的结果为"OBJECT_STORE"：

```
System.out.println(oss.getKind());
```

OSS URI 都是以"oss://"+ Bucket 开头的，设置 OSS URI 的前缀如下：

```
static final String OSS_PREFIX_URI = "oss://" + OSS_BUCKET_NAME + "/";
```

定义目标文件夹，判断目标文件夹是否存在。若该文件夹不存在，则新建一个文件夹。具体代码如下：

```
final String ossDir = OSS_PREFIX_URI + "alink/data/temp/";

if (!oss.exists(new Path(ossDir))) {
   oss.mkdirs(new Path(ossDir));
}
```

设定目标文件为"hello.txt"，并判断该文件是否存在。若该文件存在，则删除该文件；然后通过创建文件输出流的方式，将字符串"Hello Alink!"写入其中。相关代码如下：

```
String path = ossDir + "hello.txt";

OutputStream outputStream = oss.create(path, WriteMode.OVERWRITE);
outputStream.write("Hello Alink!".getBytes());
outputStream.close();
```

最后，我们使用以下文件输入流，将文件内容读取到一个字符串：

```
InputStream inputStream = oss.open(path);
String readString = IOUtils.toString(inputStream);
System.out.println(readString);
```

其输出结果为"Hello Alink!"，说明整套写入、读取流程运行正常。

下面演示文件复制操作的例子。使用前面定义的文件复制函数 copy(InputStream in, OutputStream out)来进行 OSS 文件系统和本地文件系统之间的文件复制操作，并显示文件夹内的内容。代码如下：

```
LocalFileSystem local = new LocalFileSystem();

OssFileSystem oss =
   new OssFileSystem(
      OSS_END_POINT,
      OSS_BUCKET_NAME,
      OSS_ACCESS_ID,
      OSS_ACCESS_KEY
   );
copy(
   oss.open(OSS_PREFIX_URI + "alink/data/temp/hello.txt"),
   local.create(LOCAL_DIR + "hello_1.txt", WriteMode.OVERWRITE)
);
copy(
   local.open(LOCAL_DIR + "hello_1.txt"),
   oss.create(OSS_PREFIX_URI + "alink/data/temp/hello_2.txt", WriteMode.OVERWRITE)
```

```
);
for (FileStatus status : oss.listStatus(new Path(OSS_PREFIX_URI + "alink/data/temp/"))) {
   System.out.println(status.getPath().toUri()
      + " \t" + status.getLen()
      + " \t" + new Date(status.getModificationTime())
   );
}
```

运行结果如下：

```
oss://yangxu-bucket/alink/data/temp/hello.txt    12    Thu Jul 23 14:30:00 CST 2020
oss://yangxu-bucket/alink/data/temp/hello_2.txt  12    Thu Jul 23 14:30:01 CST 2020
```

现在，我们就可以使用 OSS 的浏览器软件来查看目标文件夹下的数据情况了。如图 3-2 所示，可以看到新增了两个文件：hello.txt 和 hello_2.txt。

图 3-2

打开文件 hello_2.txt，如图 3-3 所示，这里显示的内容与预期结果一致。

图 3-3

3.2 数据文件的读入与导出

基于统一的文件系统，数据文件的访问更加便捷。在本节中，我们将深入介绍如何从各种文件系统中获取和导出机器学习计算所需的结构化数据（数据表）。

使用数据文件会涉及以下四点：

- 使用场景：批式处理和流式处理。
- 打开方式：从文件中读取数据（用 Source 标记），将数据导出到文件中（用 Sink 标记）。
- 文件格式：常见的 CSV、TSV、LibSVM 等格式；以及 Alink 专门定义的 AK 格式。
- 文件存储路径及所在文件系统的读/写权限。

上面提到的四点，彼此间是相互独立的，组合起来就对应一个具体的功能。Alink 将前面三点体现在了组件名称上，最后一点是通过参数输入的。

先介绍与 Alink 数据文件相关的组件名称，由三部分构成：

【文件格式】+【打开方式】+【使用场景】

比如，CsvSourceBatchOp 就是以批式处理的形式，从数据文件中读取 CSV 格式数据的。文件数据读入与导出组件的构成如表 3-2 所示。

表 3-2　数据文件读入与导出组件的构成

	可选值
文件格式	CSV、TSV、LibSVM、Text、AK 等
打开方式	Source（从文件中读取数据）、Sink（将数据导出到文件中）
使用场景	BatchOp（批式处理）、StreamOp（流式处理）

类 FilePath，包含文件存储路径及所在文件系统的读/写权限信息。FilePath 的基本构造函数如下：

```
public FilePath(String path, BaseFileSystem fileSystem)
```

文件存储路径可以用字符串表示，本地文件可以直接写为 "/Users/yangxu/iris.csv"；也可以使用 URI 方式，比如 "hdfs://namenode:50070/data/iris.csv" 就对应着分布式文件系统 HDFS 中的文件；而阿里云 OSS 中的文件路径可以写为 "oss://yangxu-bucket/data/iris.csv"。执行阿里云 OSS 的文件访问，除了需要文件路径，还需要 OSS_ACCESS_ID、OSS_ACCESS_KEY 等信息，这些内容是在 OssFileSystem 的实例中，从第二个参数传给 FilePath 的构造函数的。

对于本地文件系统和 Hadoop 文件系统，由于无须指定权限等信息，并且可以直接从字符串判断出其属于哪种文件系统，因此可以使用更简单的 FilePath 构造函数：

```
public FilePath(String path)
```

与 Alink 数据文件相关的组件都有如下参数设置接口，以便用来设置 FilePath；也就是设置文件存储路径及所在文件系统的读/写权限信息：

```
setFilePath(String value)
```

```
setFilePath(FilePath filePath)
```

综上,组件名称和参数设置有明显的规律性,便于用户记忆和使用。

注意,有的读者可能会有疑问:为什么不把文件格式也作为参数输入呢?这样组件的个数会减少很多。这是因为每种格式都有自己独特的参数,整合为一个组件时,该组件中要包含所有的参数,用户在使用该组件时就需要区分具体要填入哪些参数,这样容易有困扰。

3.2.1 CSV格式

CSV(Comma-Separated Values)格式的最基本含义是,每条记录以逗号分隔各字段值,记录之间使用换行符分隔,即每条记录为一行。在实际使用中,CSV 格式被赋予了更广泛的含义:字段间的分隔符可以使用其他分隔符替换,各记录间的分隔符也可以设置,第一行可能记录的是各字段的名称。CSV 格式在机器学习领域中被广泛使用,比如,UCI 数据集、Kaggle 中的数据集,很多都是 CSV 格式的。

1. 批式/流式数据源

CSV 格式的批式和流式数据源(Source)组件的参数是一样的,如表 3-3 所示。

表 3-3 CSV 格式数据源组件的参数

参数名称	参数说明
filePath	【必填】CSV 格式文件的路径信息
schemaStr	【必填】数据各字段的名称和类型。格式为 "name1 type1, name2 type2, name3 type3",名称和类型间以空格分隔,各组名称类型对之间使用逗号分隔(详见 2.8 节的内容)
fieldDelimiter	【可选】每行数据各字段的分隔符,默认为逗号,即","
rowDelimiter	【可选】行数据分隔符,默认为换行符,即"\n"
quoteChar	【可选】引号字符,默认为双引号,即"
skipBlankLine	【可选】是否忽略空行,默认值为 true
ignoreFirstLine	【可选】是否忽略第一行数据,默认值为 false

我们先下载一个 CSV 文件用作后面的测试数据。将数据文件(参见链接 3-1)下载到本地,使用文本编辑器打开该文件。如图 3-4 所示,每行为一条数据,每条数据包括四个数值字段和一个字符串字段,各字段间使用逗号分隔。

图 3-4

首先使用 CsvSourceBatchOp 读取本地数据，并取前 5 条数据打印出来。具体代码如下：

```
CsvSourceBatchOp source_local = new CsvSourceBatchOp()
    .setFilePath(LOCAL_DIR + "iris.data")
    .setSchemaStr("sepal_length double, sepal_width double, "
        + "petal_length double, petal_width double, category string");

source_local.firstN(5).print();
```

注意：CsvSourceBatchOp 组件参数 filePath 和 schemaStr 为必填的。filePath 为本地文件的存储路径 LOCAL_DIR + "iris.data"。schemaStr 为 iris 数据集的列名和类型信息，共有 5 个字段：sepal_length、sepal_width、petal_length、petal_width、category，其数据类型分别为 double、double、double、double、string。

与上面介绍的读取本地 CSV 数据相比，我们只需将数据的存储路径参数 filePath 赋值为 http 路径地址，即可直接读取网络数据：

```
CsvSourceBatchOp source_url = new CsvSourceBatchOp()
    .setFilePath("http://archive.ics.uci.edu/ml/machine-learning-databases"
        + "/iris/iris.data")
    .setSchemaStr("sepal_length double, sepal_width double, "
        + "petal_length double, petal_width double, category string");

source_url.firstN(5).print();
```

我们看到两次的结果相同，如下所示：

```
sepal_length|sepal_width|petal_length|petal_width|category
------------|-----------|------------|-----------|--------
```

```
5.1000|3.5000|1.4000|0.2000|Iris-setosa
4.9000|3.0000|1.4000|0.2000|Iris-setosa
4.7000|3.2000|1.3000|0.2000|Iris-setosa
4.6000|3.1000|1.5000|0.2000|Iris-setosa
5.0000|3.6000|1.4000|0.2000|Iris-setosa
```

我们再以流的方式读取数据，只需将组件名称换为 CsvSourceStreamOp，参数设置不动，就得到了流式数据源。流式数据无法指定前几条数据，但为了控制打印显示的数据量，我们可以对数据进行过滤，选取满足条件"sepal_length < 4.5"的数据，并打印输出，具体代码如下：

```
CsvSourceStreamOp source_stream = new CsvSourceStreamOp()
   .setFilePath("http://archive.ics.uci.edu/ml/machine-learning-databases"
      + "/iris/iris.data")
   .setSchemaStr("sepal_length double, sepal_width double, "
      + "petal_length double, petal_width double, category string");

source_stream.filter("sepal_length < 4.5").print();
StreamOperator.execute();
```

得到的结果如下：

```
sepal_length|sepal_width|petal_length|petal_width|category
------------|-----------|------------|-----------|--------
4.4000|3.2000|1.3000|0.2000|Iris-setosa
4.4000|2.9000|1.4000|0.2000|Iris-setosa
4.4000|3.0000|1.3000|0.2000|Iris-setosa
4.3000|3.0000|1.1000|0.1000|Iris-setosa
```

接下来，我们尝试更复杂的例子。对于葡萄酒品质数据集（参见链接 3-2），我们将其下载到本地，可以看到其文件内容，如图 3-5 所示。注意：第一行的数据太长，右边的显示被截断了。

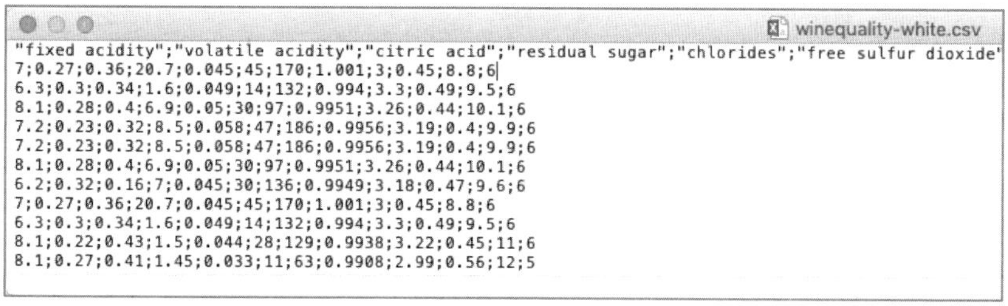
winequality-white.csv

```
"fixed acidity";"volatile acidity";"citric acid";"residual sugar";"chlorides";"free sulfur dioxide"
7;0.27;0.36;20.7;0.045;45;170;1.001;3;0.45;8.8;6
6.3;0.3;0.34;1.6;0.049;14;132;0.994;3.3;0.49;9.5;6
8.1;0.28;0.4;6.9;0.05;30;97;0.9951;3.26;0.44;10.1;6
7.2;0.23;0.32;8.5;0.058;47;186;0.9956;3.19;0.4;9.9;6
7.2;0.23;0.32;8.5;0.058;47;186;0.9956;3.19;0.4;9.9;6
8.1;0.28;0.4;6.9;0.05;30;97;0.9951;3.26;0.44;10.1;6
6.2;0.32;0.16;7;0.045;30;136;0.9949;3.18;0.47;9.6;6
7;0.27;0.36;20.7;0.045;45;170;1.001;3;0.45;8.8;6
6.3;0.3;0.34;1.6;0.049;14;132;0.994;3.3;0.49;9.5;6
8.1;0.22;0.43;1.5;0.044;28;129;0.9938;3.22;0.45;11;6
8.1;0.27;0.41;1.45;0.033;11;63;0.9908;2.99;0.56;12;5
```

图 3-5

第一行为数据列名的说明，从第二行开始是数据，可以看到这些数据都是数值类型的，各个数值之间用分号“;”进行分隔。

我们可以通过将参数 ignoreFirstLine 设置为 true，略过第一行；并且可以将字段分隔符参数 fieldDelimiter 设置为分号“;”。另外，因为列名不能包含空格，所以由文件第一行转化而来的列名需要进行相应的处理。在此，我们将列名写成驼峰形式；在每个列名后加上数据类型，这里的数据都是 double 类型的，所有的列名和类型构成了数据集的 schemaStr。具体的脚本如下：

```
CsvSourceBatchOp wine_url = new CsvSourceBatchOp()
    .setFilePath(LOCAL_DIR + "winequality-white.csv")
    .setSchemaStr("fixedAcidity double,volatileAcidity double,citricAcid double,"
        + "residualSugar double, chlorides double,freeSulfurDioxide double,"
        + "totalSulfurDioxide double,density double, pH double,"
        + "sulphates double,alcohol double,quality double")
    .setFieldDelimiter(";")
    .setIgnoreFirstLine(true);

wine_url.firstN(5).print();
```

数据打印的结果如表 3-4 所示。

表 3–4 数据打印的结果

fixedAcidity	volatileAcidity	citricAcid	residualSugar	chlorides	freeSulfurDioxide	totalSulfurDioxide	density	pH	sulphates	alcohol	quality
7.0000	0.2700	0.3600	20.7000	0.0450	45.0000	170.0000	1.0010	3.0000	0.4500	8.8000	6.0000
6.3000	0.3000	0.3400	1.6000	0.0490	14.0000	132.0000	0.9940	3.3000	0.4900	9.5000	6.0000
8.1000	0.2800	0.4000	6.9000	0.0500	30.0000	97.0000	0.9951	3.2600	0.4400	10.1000	6.0000
7.2000	0.2300	0.3200	8.5000	0.0580	47.0000	186.0000	0.9956	3.1900	0.4000	9.9000	6.0000
7.2000	0.2300	0.3200	8.5000	0.0580	47.0000	186.0000	0.9956	3.1900	0.4000	9.9000	6.0000

2. 批式/流式数据导出

CSV 格式的批式和流式数据导出（Sink）组件的参数是一样的，CSV 格式导出组件的参数如表 3-5 所示。

表 3–5 CSV 格式导出组件的参数

参数名称	参数说明
filePath	【必填】CSV 格式文件的路径信息
fieldDelimiter	【可选】每行数据各字段的分隔符，默认为逗号，即","
rowDelimiter	【可选】行数据分隔符，默认为换行符，即"\n"
quoteChar	【可选】引号字符，默认为双引号，即"
numFiles	【可选】保存文件的个数，默认值为 1
overwriteSink	【可选】是否覆盖写已有的数据。默认值为 false，即报错并退出

其中，最后两个参数是导出（Sink）操作所特有的。当参数 numFiles 大于 1 时，参数 filePath 对应的就是文件夹，在该文件夹中将数据写到 numFiles 个文件。如果目标数据文件（夹）不存在，则新建该数据表或文件（夹），然后将其整体导出。如果目标数据文件（夹）存在，则需要考虑参数"overwriteSink"，默认值为 false，即报错并退出。如果用户将其设为 true，则会删除已存在的目标数据文件（夹），并新建该数据表或文件（夹），然后将其整体导出，即该导出操作有覆盖写的效果。

我们结合前面介绍的批式/流式数据源（Source）及文件系统，做个综合性的实验。

首先，定义三个目标文件路径，分别位于以下三个文件系统中：本地文件系统、Hadoop 分布式文件系统、阿里云 OSS 文件系统。具体代码如下：

```
OssFileSystem oss =
  new OssFileSystem(
    OSS_END_POINT,
    OSS_BUCKET_NAME,
    OSS_ACCESS_ID,
    OSS_ACCESS_KEY
  );

FilePath[] filePaths = new FilePath[] {
  new FilePath(LOCAL_DIR + "iris.csv"),
  new FilePath(HDFS_URI + "user/yangxu/alink/data/temp/iris.csv"),
  new FilePath(OSS_PREFIX_URI + "alink/data/temp/iris.csv", oss)
};
```

随后，针对批式处理场景设计实验。从网络的 HTTP 数据源中读取 iris 数据，将其导出（Sink）到某个目标文件路径，然后读取该目标文件路径的数据，使用 count 方法统计其记录总条数，验证此新生成的数据文件是否正确（正确的总记录数为 150 条）。对三个不同文件系统的目标文件路径均执行此操作。具体代码如下：

```
for (FilePath filePath : filePaths) {
  new CsvSourceBatchOp()
    .setFilePath(IRIS_HTTP_URL)
    .setSchemaStr(IRIS_SCHEMA_STR)
    .link(
      new CsvSinkBatchOp()
        .setFilePath(filePath)
        .setOverwriteSink(true)
    );
  BatchOperator.execute();

  System.out.println(
    new CsvSourceBatchOp()
      .setFilePath(filePath)
      .setSchemaStr(IRIS_SCHEMA_STR)
```

```
            .count()
    );
}
```

运行结果如下，均为 150，说明在这三个文件系统中的操作都正确完成：

```
150
150
150
```

最后，针对流式场景设计实验。从网络的 HTTP 数据源中流式读取 iris 数据，将其导出（Sink）到某个目标文件路径，然后读取该目标文件路径的数据。因为从该数据源读取的是流式数据，而流式数据无法直接使用 count 方法进行验证，所以这里就选取满足条件"sepal_length < 4.5"的数据，并打印输出，以此来进行验证。对三个不同文件系统的目标文件路径均执行此操作。代码如下：

```
for (FilePath filePath : filePaths) {
    new CsvSourceStreamOp()
        .setFilePath(IRIS_HTTP_URL)
        .setSchemaStr(IRIS_SCHEMA_STR)
        .link(
            new CsvSinkStreamOp()
                .setFilePath(filePath)
                .setOverwriteSink(true)
        );
    StreamOperator.execute();

    new CsvSourceStreamOp()
        .setFilePath(filePath)
        .setSchemaStr(IRIS_SCHEMA_STR)
        .filter("sepal_length < 4.5")
        .print();
    StreamOperator.execute();
}
```

三次运行结果如下。进行对比后可发现，其结果一致：

```
sepal_length|sepal_width|petal_length|petal_width|category
------------|-----------|------------|-----------|--------
4.4000|2.9000|1.4000|0.2000|Iris-setosa
4.4000|3.2000|1.3000|0.2000|Iris-setosa
4.3000|3.0000|1.1000|0.1000|Iris-setosa
4.4000|3.0000|1.3000|0.2000|Iris-setosa
sepal_length|sepal_width|petal_length|petal_width|category
------------|-----------|------------|-----------|--------
4.4000|3.0000|1.3000|0.2000|Iris-setosa
4.4000|2.9000|1.4000|0.2000|Iris-setosa
4.3000|3.0000|1.1000|0.1000|Iris-setosa
4.4000|3.2000|1.3000|0.2000|Iris-setosa
```

```
sepal_length|sepal_width|petal_length|petal_width|category
------------|-----------|------------|-----------|--------
4.4000|2.9000|1.4000|0.2000|Iris-setosa
4.3000|3.0000|1.1000|0.1000|Iris-setosa
4.4000|3.2000|1.3000|0.2000|Iris-setosa
4.4000|3.0000|1.3000|0.2000|Iris-setosa
```

3.2.2 TSV、LibSVM、Text格式

我们先分别介绍这三种格式的定义，随后通过具体的示例进行讲解。

1. TSV格式

TSV（Tab-Separated Values）为"制表符分隔值"格式，其中的每条记录以制表符（对应键盘上的 Tab 键，字符串表示为"\t"）来分隔各字段值，而记录之间使用换行符分隔，即每条记录为一行。也可以使用 CSV 格式组件执行操作，只要将字段分隔符 fieldDelimiter 设置为"\t"即可。为了方便用户使用，Alink 提供了单独的 TSV 格式组件。

2. LibSVM格式

LibSVM 格式就是 LibSVM（参见链接 3-3）使用的数据格式，是机器学习领域中比较常见的一种形式。其格式定义如下：

```
<label> <index1>:<value1> <index2>:<value2> ...
```

第一项<label>是训练集的目标值。对于分类问题，用整数作为类别的标识（对于 2 分类，多用{0,1}或者{-1,1}表示；对于多分类问题，常用连续的整数，比如用{1,2,3}表示 3 分类的各个类别）；对于回归问题，目标值是实数。其后由若干索引<index>和数值<value>对（索引和数值间以冒号":"作为分隔符）构成，各索引/数值对间以空格作为分隔符。索引<index>是从 1 开始的整数，索引可以是不连续的整数；数值<value>为实数。

下面是几条符合 LibSVM 格式的数据：

```
1 1:-0.555556 2:0.5 3:-0.79661 4:-0.916667
1 1:-0.833333 3:-0.864407 4:-0.916667
1 1:-0.444444 2:0.416667 3:-0.830508 4:-0.916667
1 1:-0.611111 2:0.0833333 3:-0.864407 4:-0.916667
2 1:0.5 3:0.254237 4:0.0833333
2 1:0.166667 3:0.186441 4:0.166667
2 1:0.444444 2:-0.0833334 3:0.322034 4:0.166667
```

注意这条数据：

```
2 1:0.5 3:0.254237 4:0.0833333
```

没有索引值为 2 的项，表明第 2 个特征值为 0。

3. Text格式

Text 格式是 Alink 从使用方式的角度命名的。在处理一些文本数据，或者我们暂时无须区分各字段值，而可以将整条记录看作一个字符串时，每条记录只有一个字符串类型字段，记录之间使用换行符分隔。

下面运行 TSV 示例。MovieLens 数据集中就有 TSV 格式的，相应的链接地址为链接 3-4。使用浏览器打开该链接，如图 3-6 所示，这里有 4 个字段，并以制表符分隔各字段值。

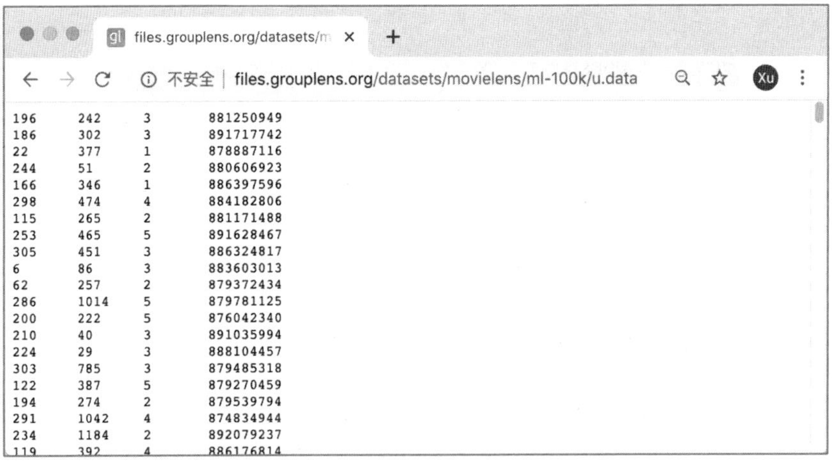

图 3-6

使用 TsvSourceBatchOp 读取该数据的代码如下。除了数据路径，还需要 Schema 信息。

```
new TsvSourceBatchOp()
    .setFilePath(LOCAL_DIR + "u.data")
    .setSchemaStr("user_id long, item_id long, rating float, ts long")
    .firstN(5)
    .print();
```

运行结果如下：

```
user_id|item_id|rating|ts
-------|-------|------|--
196|242|3.0000|881250949
186|302|3.0000|891717742
 22|377|1.0000|878887116
244| 51|2.0000|880606923
166|346|1.0000|886397596
```

第 3 章 文件系统与数据文件

在 LibSVM 网站提供了该格式的数据（参见链接 3-5）。使用浏览器打开该链接，如图 3-7 所示。

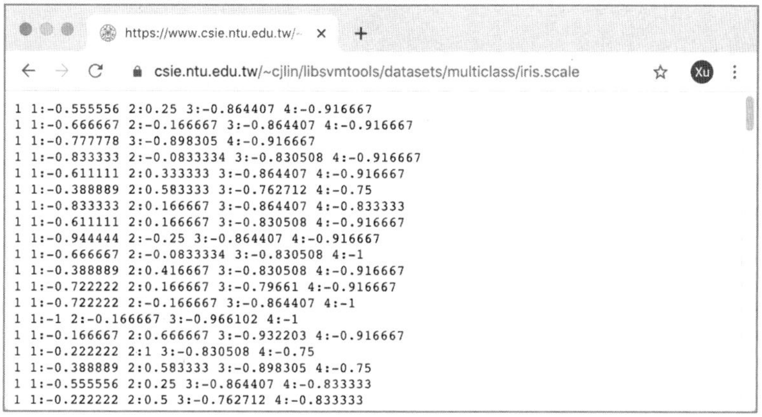

图 3-7

首先，忽略其格式定义，将其当作字符串来读取。使用 TextSourceBatchOp 组件，可以指定输出的字符串列的列名，默认值为"text"。

```
new TextSourceBatchOp()
    .setFilePath(LOCAL_DIR + "iris.scale")
    .firstN(5)
    .print();
```

运行结果如下：

```
text
----
1 1:-0.555556 2:0.25 3:-0.864407 4:-0.916667
1 1:-0.666667 2:-0.166667 3:-0.864407 4:-0.916667
1 1:-0.777778 3:-0.898305 4:-0.916667
1 1:-0.833333 2:-0.0833334 3:-0.830508 4:-0.916667
1 1:-0.611111 2:0.333333 3:-0.864407 4:-0.916667
```

然后，我们使用 LibSvmBatchOp 进行读取，得到两个数据列：标签列（列名被自动命名为 label）和特征列（列名被自动命名为 features）。其中，特征列为向量格式，我们可以接 VectorNormalizeBatchOp（向量正则化）组件，使每个向量的 2-范数为 1。

```
new LibSvmSourceBatchOp()
    .setFilePath(LOCAL_DIR + "iris.scale")
    .firstN(5)
    .lazyPrint(5, "< read by LibSvmSourceBatchOp >")
    .link(
```

```
      new VectorNormalizeBatchOp()
        .setSelectedCol("features")
    )
    .print();
```

运行结果如下：

```
< read by LibSvmSourceBatchOp >
label|features
-----|--------
1.0000|0:-0.555556 1:0.25 2:-0.864407 3:-0.916667
1.0000|0:-0.666667 1:-0.166667 2:-0.864407 3:-0.916667
1.0000|0:-0.777778 2:-0.898305 3:-0.916667
1.0000|0:-0.833333 1:-0.0833334 2:-0.830508 3:-0.916667
1.0000|0:-0.611111 1:0.333333 2:-0.864407 3:-0.916667
label|features
-----|--------
1.0000|0:-0.3969654545518278 1:0.17863431164087318 2:-0.617650997690209 3:-0.6549927141956171
1.0000|0:-0.4645226632894226 1:-0.11613084001826729 2:-0.6023046615566992 3:-0.6387185749250003
1.0000|0:-0.5182689351202653 2:-0.5985815692436788 3:-0.6108170068449973
1.0000|0:-0.5578646805775191 1:-0.05578653500154036 2:-0.5559735185538965 3:-0.6136516172417902
1.0000|0:-0.424541811112608214 1:0.23156807114925168 2:-0.6005077855415192 3:-0.6368130640415773
```

从文件中读取的数据，标签列的类型为 double（因为 LibSVM 格式的数据也可以进行回归计算，标签值可能为浮点值），特征向量的索引是从 0 开始的（Alink 中的稀疏向量索引值都是从 0 开始的）；而在原始数据中，按 LibSVM 的定义，索引值是从 1 开始的。LibSvmBatchOp 组件会在读取和导出时，自动进行索引值的转换。

3.2.3 AK 格式

AK 格式是专门为 Alink 定义的数据格式，在 AK 格式的数据文件中包含了各数据列名称和类型等元数据，并对数据内容进行了压缩。

AK 格式的批式和流式数据源（Source）组件的参数是一样的，都只有一个参数，即 AK 格式文件的路径信息，如表 3-6 所示。

表 3-6 AK 格式数据源组件的参数

参数名称	参数说明
filePath	【必填】AK 格式文件的路径信息

AK 格式的批式和流式数据导出（Sink）组件的参数是一样的，具体参数如表 3-7 所示。

表 3-7 AK 格式导出组件的参数

参数名称	参数说明
filePath	【必填】CSV 格式文件的路径信息
numFiles	【可选】保存文件的个数，默认值为 1
overwriteSink	【可选】是否覆盖写已有的数据，默认值为 False，即报错并退出

其中，最后两个参数是导出（Sink）操作所特有的，其与 CSV 格式的数据导出（Sink）组件的参数含义一样。当参数 numFiles 大于 1 时，参数 filePath 对应的就是文件夹，在该文件夹中指定写到 numFiles 个文件。如果目标数据文件（夹）不存在，则新建该数据表或文件（夹），然后将其整体导出。如果目标数据文件（夹）存在，则需要考虑参数"overwriteSink"，默认值为 false，即报错并退出。如果用户将其设为 true，则会删除已存在的目标数据文件（夹），并新建该数据表或文件（夹），然后将其整体导出，即该导出操作有覆盖写的效果。

与 CSV 格式相比，AK 格式有如下特点：
- 不用额外记录数据的 schemaStr。
- 参数的个数大幅减少。
- 避免了各种数据类型与文本格式转换过程中出现的兼容问题。
- 支持数据压缩，文件占用的空间更小。

示例

本节的实验与对 CSV 格式数据进行的综合实验相似。首先，定义三个目标文件路径，分别位于三个文件系统：本地文件系统、Hadoop 分布式文件系统、阿里云 OSS 文件系统。具体代码如下：

```
OssFileSystem oss =
  new OssFileSystem(
    OSS_END_POINT,
    OSS_BUCKET_NAME,
    OSS_ACCESS_ID,
    OSS_ACCESS_KEY
  );

FilePath[] filePaths = new FilePath[] {
  new FilePath(LOCAL_DIR + "iris.ak"),
  new FilePath(HDFS_URI + "user/yangxu/alink/data/temp/iris.ak"),
  new FilePath(OSS_PREFIX_URI + "alink/data/temp/iris.ak", oss)
};
```

随后，针对批式处理场景设计实验。从网络的 HTTP 数据源中读取 iris 数据，将其导出（Sink）到某个目标文件路径，然后读取该目标文件路径的数据，使用 count 方法统计其记录总条数，验证此新生成的数据文件是否正确（正确的总记录数为 150 条）。对三个不同文件系统的目标文件路径均执行此操作，代码如下：

```
for (FilePath filePath : filePaths) {
    new CsvSourceBatchOp()
        .setFilePath(IRIS_HTTP_URL)
        .setSchemaStr(IRIS_SCHEMA_STR)
        .link(
            new AkSinkBatchOp()
                .setFilePath(filePath)
                .setOverwriteSink(true)
        );
    BatchOperator.execute();

    System.out.println(
        new AkSourceBatchOp()
            .setFilePath(filePath)
            .count()
    );
}
```

运行结果如下：

```
150
150
150
```

最后，针对流式场景设计实验。从网络的 HTTP 数据源中流式读取 iris 数据，将其导出（Sink）到某个目标文件路径，然后读取该目标文件路径的数据。因为从该数据源读取的是流式数据，而流式数据无法直接使用 count 方法进行验证，所以这里就选取满足条件"sepal_length < 4.5"的数据，并打印输出，以此来进行验证。对三个不同文件系统的目标文件路径均执行此操作，代码如下：

```
for (FilePath filePath : filePaths) {
    new CsvSourceStreamOp()
        .setFilePath(IRIS_HTTP_URL)
        .setSchemaStr(IRIS_SCHEMA_STR)
        .link(
            new AkSinkStreamOp()
                .setFilePath(filePath)
                .setOverwriteSink(true)
        );
    StreamOperator.execute();

    new AkSourceStreamOp()
        .setFilePath(filePath)
        .filter("sepal_length < 4.5")
        .print();
    StreamOperator.execute();
}
```

三次运行结果如下。进行对比后可发现,其结果一致。

```
sepal_length|sepal_width|petal_length|petal_width|category
------------|-----------|------------|-----------|--------
4.4000|3.2000|1.3000|0.2000|Iris-setosa
4.4000|3.0000|1.3000|0.2000|Iris-setosa
4.4000|2.9000|1.4000|0.2000|Iris-setosa
4.3000|3.0000|1.1000|0.1000|Iris-setosa
sepal_length|sepal_width|petal_length|petal_width|category
------------|-----------|------------|-----------|--------
4.3000|3.0000|1.1000|0.1000|Iris-setosa
4.4000|3.2000|1.3000|0.2000|Iris-setosa
4.4000|2.9000|1.4000|0.2000|Iris-setosa
4.4000|3.0000|1.3000|0.2000|Iris-setosa
sepal_length|sepal_width|petal_length|petal_width|category
------------|-----------|------------|-----------|--------
4.4000|2.9000|1.4000|0.2000|Iris-setosa
4.4000|3.2000|1.3000|0.2000|Iris-setosa
4.4000|3.0000|1.3000|0.2000|Iris-setosa
4.3000|3.0000|1.1000|0.1000|Iris-setosa
```

4 数据库与数据表

数据库（Database）是重要的数据存储方式，Alink 使用 Flink Catalog 进行数据库和数据表的基本操作。Flink Catalog 提供了元数据，如数据库、表、分区、视图以及访问数据库或其他外部系统中存储的数据所需的函数和信息。同时 Alink 提供了针对数据表的批式/流式数据源（Source）组件及数据导出（Sink）组件。

4.1 简介

4.1.1 Catalog的基本操作

Catalog 允许用户引用数据库系统中现有的元数据，并自动将它们映射到 Flink 的相应元数据中。例如，Flink 可以自动将 JDBC（Java Database Connectivity）表映射到 Flink 表中，用户不必在 Flink 中手动重写 DDL（Data Definition Language）内容。Catalog 极大地简化了用户现有系统开始使用 Flink 所需的步骤，并大大改善了用户体验。关于 Catalog API 的详细介绍，可参见 Flink 官网的说明（参见链接 4-1）。

Catalog 提供了一个统一的 API 来管理元数据，元数据包括数据库（Database）、数据表（Table）、表结构等信息。图 4-1 显示了 Catalog、Database 与 Table 间的层次关系。

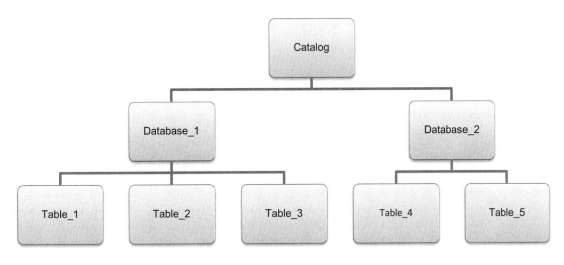

图 4-1

表 4-1 列出了与数据库（Database）和数据表（Table）相关的常用操作。

表 4-1　与数据库和数据表相关的常用操作

与数据库相关的操作	与数据表相关的操作
listDatabases：列出所有的数据库	listTables：列出所有的数据表
databaseExists：判断数据库是否存在	tableExist：判断数据表是否存在
dropDatabases：删除数据库	dropTable：删除数据表
getDefaultDataBase：获取默认的数据库	renameTable：重命名数据表
getDatabase：获取特定的数据库	getTable：获取指定的数据表
createDatabases：创建数据库	createTable：创建数据表
alterDatabases：修改数据库	alterTable：修改数据表

4.1.2　Source和Sink组件

本节所要讨论的数据源（Source）组件与导出（Source）组件都是以单个数据表为基本单位的。这里需要用到Catalog，以及路径信息：

```
public ObjectPath(String databaseName, String objectName)
```

Alink 定义了类 CatalogObject，包括 Catalog 数据源（Source）组件与导出（Source）组件所需的全部元数据，其构造函数有如下两种形式：

```
public CatalogObject(BaseCatalog catalog, ObjectPath objectPath)
public CatalogObject(BaseCatalog catalog, ObjectPath objectPath, Params params)
```

显然，指定 Catalog 和路径信息（数据库→表）是必需的。

Catalog 的批式/流式数据源与导出组件，共有四个，都使用 CatalogObject 类型参数；而且每个组件都只有一个参数，参数的用法也相同，如表 4-2 所示。

表 4-2　数据表的批式/流式数据源与导出组件参数

参数名称	参数类型	参数说明
catalogObject	CatalogObject	包括 Catalog 和路径信息（数据库→表）

四个组件的含义如下：

- 批式数据源组件（CatalogSourceBatchOp）
- 流式数据源组件（CatalogSourceStreamOp）
- 批式数据导出组件（CatalogSinkBatchOp）

  批式数据导出的原则是数据要进行整体导出，不进行追加导出。如果目标数据表不存在，则新建该数据表，然后将其整体导出；如果目标数据表存在，则报错并退出。所以，在执行数据的导出操作前，需要判断目标数据表是否存在，并执行相应的删除操作，以保证数据的导出过程顺利进行。

- 流式数据导出组件（CatalogSinkStreamOp）

  在目标数据表不存在的情况下，流式数据导出与批式数据导出的行为是相似的（新建数据表，然后将数据导出到目标数据表）。需要特别注意的是，在目标数据表存在的情况下，批式导出会报错并退出；而流式导出不会报错，会继续执行下去。流式导出的数据会追加在原有数据上。

4.2　Hive 示例

Hive 是一个基于 Hadoop 的数据仓库平台，在大数据场景中应用广泛。在本节中，我们会演示 Alink 任务如何从 Hive 读取数据，如何将结果写入 Hive。

首先设置常量如下。其中，IRIS_URL 和 IRIS_SCHEMA_STR 用于读取 HTTP 数据，定义 Hive Catalog 中的数据库名称为 DB_NAME，使用批式读/写的数据表名称为 BATCH_TABLE_NAME，使用流式读/写的数据表名称为 STREAM_TABLE_NAME；这里所用的 Hive 版本为 HIVE_VERSION，其配置文件目录为 HIVE_CONF_DIR。

```
static final String IRIS_URL =
    "http://archive.ics.uci.edu/ml/machine-learning-databases/iris/iris.data";
static final String IRIS_SCHEMA_STR =
    "sepal_length double, sepal_width double, petal_length double, petal_width double, category string";
static final String DB_NAME = "test_db";
static final String BATCH_TABLE_NAME = "batch_table";
static final String STREAM_TABLE_NAME = "stream_table";
static final String HIVE_VERSION = "2.3.4";
static final String HIVE_CONF_DIR = ... ;
```

创建 HiveCatalog 对象,创建实验用的数据库 DB_NAME:

```
HiveCatalog hive = new HiveCatalog("hive_catalog",null,HIVE_VERSION,HIVE_CONF_DIR);
hive.open();
hive.createDatabase(DB_NAME, new CatalogDatabaseImpl(new HashMap <>(), ""), true);
```

分别以批式/流式方式读取 iris 数据,并将这些数据以批式/流式方式导出到 Hive 的数据表 BATCH_TABLE_NAME 和 STREAM_TABLE_NAME 中。代码如下:

```
new CsvSourceBatchOp()
    .setFilePath(IRIS_URL)
    .setSchemaStr(IRIS_SCHEMA_STR)
    .lazyPrintStatistics("< origin data >")
    .link(
        new CatalogSinkBatchOp()
            .setCatalogObject(new CatalogObject(hive, new ObjectPath(DB_NAME, BATCH_TABLE_NAME)))
    );
BatchOperator.execute();

new CsvSourceStreamOp()
    .setFilePath(IRIS_URL)
    .setSchemaStr(IRIS_SCHEMA_STR)
    .link(
        new CatalogSinkStreamOp()
            .setCatalogObject(new CatalogObject(hive, new ObjectPath(DB_NAME, STREAM_TABLE_NAME)))
    );
StreamOperator.execute();
```

随后,我们分别以批式/流式方式读取数据内容,并计算统计信息,打印采样数据。详细代码如下:

```
new CatalogSourceBatchOp()
    .setCatalogObject(new CatalogObject(hive, new ObjectPath(DB_NAME, BATCH_TABLE_NAME)))
    .lazyPrintStatistics("< batch catalog source >");
BatchOperator.execute();
```

```
new CatalogSourceStreamOp()
    .setCatalogObject(new CatalogObject(hive, new ObjectPath(DB_NAME, STREAM_TABLE_NAME)))
    .sample(0.02)
    .print();
StreamOperator.execute();
```

运行结果如下：

```
< batch catalog source >
Summary:
|     colName|count|missing|   sum|  mean|variance| min|max|
|------------|-----|-------|------|------|--------|----|---|
|SEPAL_LENGTH|  150|      0| 876.5|5.8433|  0.6857| 4.3|7.9|
| SEPAL_WIDTH|  150|      0| 458.1| 3.054|   0.188|   2|4.4|
|PETAL_LENGTH|  150|      0| 563.8|3.7587|  3.1132|   1|6.9|
| PETAL_WIDTH|  150|      0| 179.8|1.1987|  0.5824| 0.1|2.5|
|    CATEGORY|  150|      0|   NaN|   NaN|     NaN| NaN|NaN|


sepal_length|sepal_width|petal_length|petal_width|category
------------|-----------|------------|-----------|--------
      4.4000|     2.9000|      1.4000|     0.2000|Iris-setosa
      5.5000|     4.2000|      1.4000|     0.2000|Iris-setosa
      6.8000|     3.0000|      5.5000|     2.1000|Iris-virginica
```

最后，演示一下文件操作。下面列出所有的数据表。根据前面的操作，我们知道共有两个表。首先判断数据表 BATCH_TABLE_NAME 是否存在，然后执行删除操作。也可以直接使用 dropTable 方法删除数据表 STREAM_TABLE_NAME。注意将第二个参数设为 true，代表忽略存在的表。最后显示数据表列表。如果操作成功的话，最后的列表应该为空。

```
System.out.println("< tables before drop >");
System.out.println(JsonConverter.toJson(hive.listTables(DB_NAME)));

if (hive.tableExists(new ObjectPath(DB_NAME, BATCH_TABLE_NAME))) {
    hive.dropTable(new ObjectPath(DB_NAME, BATCH_TABLE_NAME), false);
}
hive.dropTable(new ObjectPath(DB_NAME, STREAM_TABLE_NAME), true);

System.out.println("< tables after drop >");
System.out.println(JsonConverter.toJson(hive.listTables(DB_NAME)));
```

运行结果如下。执行删除操作之前，有两个数据表；执行删除操作之后，没有数据表，显然删除操作达到了预期效果：

```
< tables before drop >
["batch_table","stream_table"]
< tables after drop >
[]
```

4.3　Derby 示例

Derby 是完全用 Java 编写的内存数据库，属于 Apache 的一个开源项目，且基于 Apache License 2.0 分发。Derby 的核心部分只有 2MB，非常小巧，适合在单机上运行。

首先需要设置常量如下。这里所用的 Derby 版本为 DERBY_VERSION，其数据所在目录为 DERBY_DIR：

```
static final String DERBY_VERSION = "10.6.1.0";
static final String DERBY_DIR = ... ;
```

创建 DerbyCatalog 对象，创建实验用的数据库 DB_NAME：

```
DerbyCatalog derby = new DerbyCatalog("derby_catalog", null, DERBY_VERSION, DATA_DIR + DERBY_DIR);
derby.open();
derby.createDatabase(DB_NAME, new CatalogDatabaseImpl(new HashMap<>(), ""), true);
```

分别以批式/流式方式读取 iris 数据，并将这些数据以批式/流式方式导出到 Derby 的数据表 BATCH_TABLE_NAME 和 STREAM_TABLE_NAME 中。代码如下：

```
new CsvSourceBatchOp()
    .setFilePath(IRIS_URL)
    .setSchemaStr(IRIS_SCHEMA_STR)
    .lazyPrintStatistics("< origin data >")
    .link(
        new CatalogSinkBatchOp()
            .setCatalogObject(new CatalogObject(derby, new ObjectPath(DB_NAME, BATCH_TABLE_NAME)))
    );
BatchOperator.execute();

new CsvSourceStreamOp()
    .setFilePath(IRIS_URL)
    .setSchemaStr(IRIS_SCHEMA_STR)
    .link(
        new CatalogSinkStreamOp()
            .setCatalogObject(new CatalogObject(derby, new ObjectPath(DB_NAME, STREAM_TABLE_NAME)))
    );
StreamOperator.execute();
```

随后，我们再分别以批式/流式方式读取数据内容，并计算统计信息，打印采样数据。详细代码如下：

```
new CatalogSourceBatchOp()
    .setCatalogObject(new CatalogObject(derby, new ObjectPath(DB_NAME, BATCH_TABLE_NAME)))
    .lazyPrintStatistics("< batch catalog source >");
BatchOperator.execute();
```

```
new CatalogSourceStreamOp()
    .setCatalogObject(new CatalogObject(derby, new ObjectPath(DB_NAME, STREAM_TABLE_NAME)))
    .sample(0.02)
    .print();
StreamOperator.execute();
```

运行结果如下：

```
< batch catalog source >
Summary:
|      colName|count|missing|   sum|  mean|variance|min|max|
|-------------|-----|-------|------|------|--------|---|---|
| SEPAL_LENGTH|  150|      0| 876.5|5.8433|  0.6857|4.3|7.9|
|  SEPAL_WIDTH|  150|      0| 458.1| 3.054|   0.188|  2|4.4|
| PETAL_LENGTH|  150|      0| 563.8|3.7587|  3.1132|  1|6.9|
|  PETAL_WIDTH|  150|      0| 179.8|1.1987|  0.5824|0.1|2.5|
|     CATEGORY|  150|      0|   NaN|   NaN|     NaN|NaN|NaN|

SEPAL_LENGTH|SEPAL_WIDTH|PETAL_LENGTH|PETAL_WIDTH|CATEGORY
------------|-----------|------------|-----------|--------
      5.5000|     3.5000|      1.3000|     0.2000|Iris-setosa
      6.3000|     2.5000|      4.9000|     1.5000|Iris-versicolor
      6.4000|     2.9000|      4.3000|     1.3000|Iris-versicolor
      6.7000|     3.0000|      5.2000|     2.3000|Iris-virginica
```

最后，演示一下文件操作。下面列出所有的数据表。根据前面的操作，我们知道共有两个表。首先判断数据表 BATCH_TABLE_NAME 是否存在，然后执行删除操作。也可以直接使用 dropTable 方法删除数据表 STREAM_TABLE_NAME，注意将第二个参数设为 true，代表忽略存在的表。最后显示数据表列表。如果操作成功的话，最后的列表应该为空。

```
System.out.println("< tables before drop >");
System.out.println(JsonConverter.toJson(derby.listTables(DB_NAME)));

if (derby.tableExists(new ObjectPath(DB_NAME, BATCH_TABLE_NAME))) {
    derby.dropTable(new ObjectPath(DB_NAME, BATCH_TABLE_NAME), false);
}
derby.dropTable(new ObjectPath(DB_NAME, STREAM_TABLE_NAME), true);

System.out.println("< tables after drop >");
System.out.println(JsonConverter.toJson(derby.listTables(DB_NAME)));
```

运行结果如下，显然文件操作达到了预期效果：

```
< tables before drop >
["BATCH_TABLE","STREAM_TABLE"]
< tables after drop >
[]
```

4.4 MySQL示例

MySQL 是最流行的关系型数据库之一。本节将对 MySQL 进行数据源与导出操作，执行显示数据表列表、判断数据表是否存在、删除数据表等操作。

首先设置常量如下。这里所用的 MySQL 版本为 MYSQL_VERSION，其数据服务的链接地址为 MYSQL_URL，端口号为 MYSQL_PORT，用户名和密码分别为 MYSQL_USER_NAME 和 MYSQL_PASSWORD。

```
static final String MYSQL_VERSION = "5.1.27";
static final String MYSQL_URL = ...;
static final String MYSQL_PORT = ...;
static final String MYSQL_USER_NAME = ...;
static final String MYSQL_PASSWORD = ...;
```

创建 MySqlCatalog 对象，创建实验用的数据库 DB_NAME：

```
MySqlCatalog mySql = new MySqlCatalog("mysql_catalog", null, MYSQL_VERSION,
    MYSQL_URL, MYSQL_PORT, MYSQL_USER_NAME, MYSQL_PASSWORD);

mySql.open();
mySql.createDatabase(DB_NAME, new CatalogDatabaseImpl(new HashMap <>(), ""), true);
```

分别以批式/流式方式读取 iris 数据，并将这些数据以批式/流式方式导出到 MySQL 的数据表 BATCH_TABLE_NAME 和 STREAM_TABLE_NAME 中。代码如下：

```
new CsvSourceBatchOp()
    .setFilePath(IRIS_URL)
    .setSchemaStr(IRIS_SCHEMA_STR)
    .lazyPrintStatistics("< origin data >")
    .link(
        new CatalogSinkBatchOp()
            .setCatalogObject(new CatalogObject(mySql, new ObjectPath(DB_NAME, BATCH_TABLE_NAME)))
    );
BatchOperator.execute();

new CsvSourceStreamOp()
    .setFilePath(IRIS_URL)
    .setSchemaStr(IRIS_SCHEMA_STR)
    .link(
        new CatalogSinkStreamOp()
            .setCatalogObject(new CatalogObject(mySql, new ObjectPath(DB_NAME, STREAM_TABLE_NAME)))
    );
StreamOperator.execute();
```

随后，我们再分别以批式/流式方式读取数据内容，并计算统计信息，打印采样数据。详细代码如下：

```
new CatalogSourceBatchOp()
    .setCatalogObject(new CatalogObject(mySql, new ObjectPath(DB_NAME, BATCH_TABLE_NAME)))
    .lazyPrintStatistics("< batch catalog source >");
BatchOperator.execute();

new CatalogSourceStreamOp()
    .setCatalogObject(new CatalogObject(mySql, new ObjectPath(DB_NAME, STREAM_TABLE_NAME)))
    .sample(0.02)
    .print();
StreamOperator.execute();
```

运行结果如下：

```
< batch catalog source >
Summary:
|      colName|count|missing|   sum|   mean|variance|min|max|
|-------------|-----|-------|------|-------|--------|---|---|
| SEPAL_LENGTH|  150|      0| 876.5| 5.8433|  0.6857|4.3|7.9|
|  SEPAL_WIDTH|  150|      0| 458.1|  3.054|   0.188|  2|4.4|
| PETAL_LENGTH|  150|      0| 563.8| 3.7587|  3.1132|  1|6.9|
|  PETAL_WIDTH|  150|      0| 179.8| 1.1987|  0.5824|0.1|2.5|
|     CATEGORY|  150|      0|   NaN|    NaN|     NaN|NaN|NaN|

sepal_length|sepal_width|petal_length|petal_width|category
------------|-----------|------------|-----------|--------
      6.3000|     2.9000|      5.6000|     1.8000|Iris-virginica
      5.0000|     3.6000|      1.4000|     0.2000|Iris-setosa
```

最后，演示一下文件操作。下面列出所有的数据表。根据前面的操作，我们知道共有两个表。首先判断数据表 BATCH_TABLE_NAME 是否存在。然后执行删除操作。也可以直接使用 dropTable 方法删除数据表 STREAM_TABLE_NAME。注意将第二个参数设为 true，代表忽略存在的表。最后显示数据表列表。如果操作成功的话，最后的列表应该为空。

```
System.out.println("< tables before drop >");
System.out.println(JsonConverter.toJson(mySql.listTables(DB_NAME)));

if (mySql.tableExists(new ObjectPath(DB_NAME, BATCH_TABLE_NAME))) {
    mySql.dropTable(new ObjectPath(DB_NAME, BATCH_TABLE_NAME), false);
}
mySql.dropTable(new ObjectPath(DB_NAME, STREAM_TABLE_NAME), true);

System.out.println("< tables after drop >");
System.out.println(JsonConverter.toJson(mySql.listTables(DB_NAME)));
```

运行结果如下,显然文件操作达到了预期效果:

```
< tables before drop >
["batch_table","stream_table"]
< tables after drop >
[]
```

5 支持 Flink SQL

Flink 的 SQL 操作基于实现了 SQL 标准的 Apache Calcite（参见链接 5-1）。Flink SQL 还有一个非常大的优势：对于批式数据表和流式数据表，查询都使用相同的语义，并产生相同的结果。

Alink 批式组件 BatchOperator 的数据格式采用的是批式 Flink Table；同样，Alink 流式组件 StreamOperator 的数据格式采用的是流式 Flink Table。对于 Flink Table 使用 Flink SQL 语句，无须数据格式的转换，只需要在数据表环境（Table Enviroment）中进行相应的注册。这是非常自然和高效的。

Alink 对 Flink SQL 进行了封装，使其和 Alink 组件间的衔接更加自然、方便，而在使用方式上仍延续了 Flink SQL 的风格。

本章以 MovieLens 数据集为例，演示各种 SQL 功能。注意：本书的第 24 章对 MovieLens 数据集有详细的介绍。

5.1 基本操作

5.1.1 注册

SQL 语句中对数据的操作都是通过表名进行的。要对某个 Alink 批式组件 BatchOperator 输出的 Table 类型数据进行操作，需要先为其注册一个名称，具体代码如下：

```
BatchOperator <?> ratings = ...
```

```
BatchOperator <?> users = ...
BatchOperator <?> items = ...

ratings.registerTableName("ratings");
items.registerTableName("items");
users.registerTableName("users");
```

5.1.2 运行

Alink 封装了 Flink 的 sqlQuery 方法,将其输出结果包装成了 BatchOperator,这样就可以使用 link 等方法,连接其他 Alink 组件。

我们使用 SQL 语句做一个示例,从 MovieLens ratings 数据表中计算出某视频网站观众的评价次数最多的 10 部电影,并从 items 数据表中根据观众的 ID 匹配出电影名称,具体代码如下:

```
BatchOperator.sqlQuery(
    "SELECT title, cnt, avg_rating"
    + " FROM ( SELECT item_id, COUNT(*) AS cnt, AVG(rating) AS avg_rating"
    + "        FROM ratings "
    + "        GROUP BY item_id "
    + "        ORDER BY cnt DESC LIMIT 10 "
    + "      ) AS t"
    + " JOIN items"
    + " ON t.item_id=items.item_id"
    + " ORDER BY cnt DESC"
).print();
```

输出结果如下,第一列为影片名称,第二列为观众的评论次数,第三列为电影获得的平均评分:

```
title|cnt|avg_rating
-----|---|----------
Star Wars (1977)|583|4.3585
Contact (1997)|509|3.8035
Fargo (1996)|508|4.1555
Return of the Jedi (1983)|507|4.0079
Liar Liar (1997)|485|3.1567
English Patient, The (1996)|481|3.6570
Scream (1996)|478|3.4414
Toy Story (1995)|452|3.8783
Air Force One (1997)|431|3.6311
Independence Day (ID4) (1996)|429|3.4382
```

为了让读者对影片名有更深刻的印象,下面显示前四名的电影海报,如图 5-1 所示。

看到这里,大家可能会有疑问:排在第 2、3 位的电影似乎没有很大的知名度,它们为什么会排名靠前呢?

图 5-1

稍后，我们会通过使用 UDF（用户定义函数），得到观众每个评分所发生的时间段，具体分析一下其中的原因。

下面我们再举一个更有挑战性的 SQL 操作示例，计算一下男士与女士评分差异最大的电影。具体代码如下，其中定义了 m_rating 和 f_rating 来分别记录对于一部影片的男、女平均评分，最后按评分从大到小排列各影片，并获取前 20 部影片的数据：

```
BatchOperator.sqlQuery(
    "SELECT title, cnt, m_rating, f_rating, ABS(m_rating - f_rating) AS diff_rating"
    + " FROM ( SELECT item_id, COUNT(rating) AS cnt, "
    + "               AVG(CASE WHEN gender='M' THEN rating ELSE NULL END) AS m_rating, "
    + "               AVG(CASE WHEN gender='F' THEN rating ELSE NULL END) AS f_rating "
    + "          FROM (SELECT item_id, rating, gender FROM ratings "
    + "                JOIN users ON ratings.user_id=users.user_id)"
    + "          GROUP BY item_id "
    + "     ) AS t"
    + " JOIN items "
    + " ON t.item_id=items.item_id"
    + " ORDER BY diff_rating DESC LIMIT 20"
).print();
```

结果如下：

```
title|cnt|m_rating|f_rating|diff_rating
-----|---|--------|--------|-----------
Delta of Venus (1994)|2|5.0000|1.0000|4.0000
Two or Three Things I Know About Her (1966)|4|4.6667|1.0000|3.6667
Sliding Doors (1998)|4|4.5000|1.0000|3.5000
Paths of Glory (1957)|33|4.4194|1.0000|3.4194
Magic Hour, The (1998)|5|4.2500|1.0000|3.2500
Love and Death on Long Island (1997)|2|1.0000|4.0000|3.0000
Killer (Bulletproof Heart) (1994)|4|4.0000|1.0000|3.0000
Rough Magic (1995)|2|1.0000|4.0000|3.0000
Visitors, The (Visiteurs, Les) (1993)|2|2.0000|5.0000|3.0000
Little City (1998)|2|5.0000|2.0000|3.0000
……
```

我们看到排名靠前的几部影片的男女评分差异的确很大，排在第一位的电影只有 2 个评分，分别是最高分和最低分。若评分的人数非常少，则样本的偏差就很大，也就无法代表各影片间的真实评分差异。所以，需要在统计的过程中加入评分个数的限制，具体代码如下：

```
BatchOperator.sqlQuery(
    "SELECT title, cnt, m_rating, f_rating, ABS(m_rating - f_rating) AS diff_rating"
    + " FROM ( SELECT item_id, COUNT(rating) AS cnt, "
    + "              AVG(CASE WHEN gender='M' THEN rating ELSE NULL END) AS m_rating, "
    + "              AVG(CASE WHEN gender='F' THEN rating ELSE NULL END) AS f_rating "
    + "         FROM (SELECT item_id, rating, gender FROM ratings "
    + "               JOIN users ON ratings.user_id=users.user_id)"
    + "         GROUP BY item_id "
    + "         HAVING COUNT(rating)>=50 "
    + "     ) AS t"
    + " JOIN items "
    + " ON t.item_id=items.item_id"
    + " ORDER BY diff_rating DESC LIMIT 10"
).print();
```

对评分个数的限制语句为"HAVING COUNT(rating)>=50"，得到的结果如下：

```
title|cnt|m_rating|f_rating|diff_rating
-----|---|--------|--------|------------
Ran (1985)|70|4.3214|3.2143|1.1071
Jane Eyre (1996)|63|3.0526|4.1200|1.0674
Postman, The (1997)|58|2.8378|3.9048|1.0669
First Knight (1995)|86|2.7460|3.7826|1.0366
Cook the Thief His Wife & Her Lover, The (1989)|82|3.1940|2.2667|0.9274
Dirty Dancing (1987)|98|2.7742|3.6667|0.8925
Nosferatu (Nosferatu, eine Symphonie des Grauens) (1922)|54|3.6596|2.8571|0.8024
To Wong Foo, Thanks for Everything! Julie Newmar (1995)|57|2.6216|3.4000|0.7784
Daylight (1996)|57|2.5814|3.3571|0.7757
Big Sleep, The (1946)|73|4.2500|3.4762|0.7738
```

这些影片的评分差异明显变小了。我们关注排在前两位的影片，这两部影片分别被男性观众或女性观众看好。这两部影片的海报及其说明如图 5-2 所示。

该影片讲述的是一个虚构的日本战国时期的故事，某家族因自相残杀而走向灭亡。

著名小说《简·爱》的电影版

图 5-2

5.1.3 内置函数

Flink SQL 提供了丰富的内置函数，比如字符串函数、算术函数、逻辑函数、比较函数等。内置函数的功能列表随着 Flink 版本而变化，该列表基本上是不断增加的。建议读者根据自己所用的 Flink 版本，到 Flink 网站寻找相应版本的内置函数说明。

5.1.4 用户定义函数

Flink SQL 虽然提供了丰富的内置函数，但是仍不能覆盖所有应用场景，还需要用户根据具体情况定义函数。我们通过一个简单的示例来串联函数的定义和使用过程。

在前面的示例中，我们计算了 ratings 数据表中某视频网站观众的评价次数最多的 10 部电影，但其中排在第 2、3 位的电影似乎没有很大的知名度，我们自然会有疑问：它们为什么会排名靠前呢？

我们注意到 ratings 数据表中有一个数据列，名称为 ts，记录的是评分的 UNIX 时间戳，该时间戳用自 UTC 的 1970 年 1 月 1 日以来的 UNIX 秒数来表示，为长整型数值。我们看到的是一个非常大的整数，无法将其与一个时间关联起来。接下来，我们会通过使用 UDF（用户定义函数），将 UNIX 时间戳转换为我们容易理解的日期时间格式，之后进一步分析上述电影排名靠前的原因。

定义函数 FromUnixTimestamp 如下：

```
public static class FromUnixTimestamp extends ScalarFunction {
    public java.sql.Timestamp eval(Long ts) {
        return new java.sql.Timestamp(ts * 1000);
    }
}
```

在使用前需要注册函数，相关代码如下：

```
BatchOperator.registerFunction("from_unix_timestamp", new FromUnixTimestamp());
```

然后就可以在 SQL 语句中调用 from_unix_timestamp 了，我们将 UNIX 时间戳类型转换为日期类型，并统计评分的起始时间和结束时间。

```
BatchOperator.sqlQuery(
    "SELECT MIN(dt) AS min_dt, MAX(dt) AS max_dt "
    + " FROM ( SELECT from_unix_timestamp(ts) AS dt, 1 AS grp FROM ratings) "
    + " GROUP BY grp "
).print();
```

输出结果如下：

```
min_dt|max_dt
------|------
1997-09-20 11:05:10.0|1998-04-23 07:10:38.0
```

前面介绍了基本的 SQL 语句用法，我们在实际使用中还有更简便的用法。比如对于上面的功能，我们使用如下代码，也可以得到相同的结果：

```
ratings
  .select("from_unix_timestamp(ts) AS dt, 1 AS grp")
  .groupBy("grp","MIN(dt) AS min_dt, MAX(dt) AS max_dt")
  .print();
```

显然，其在写法上简洁了很多，而且还会省去注册 ratings 的步骤。接下来介绍 SQL 语句的简化操作。

5.2 简化操作

Alink 对 SQL 语句的常用操作进行了包装。BatchOperator 和 StreamOperator 提供了一些方法，可以简化使用 SQL 语句的流程。比如，对单个数据表进行 SELECT 操作，使用 SQL 语句实现的代码如下：

```
ratings.registerTableName("ratings");
BatchOperator ratings_select = BatchOperator.sqlQuery(
    "SELECT user_id, item_id AS movie_id FROM ratings");
```

而使用 Alink 提供的简化方法，一行就可以搞定，代码如下：

```
BatchOperator ratings_select = ratings.select("user_id, item_id AS movie_id");
```

再举一个例子，对两个数据表进行合并操作，合并的过程中不对数据进行去重操作，使用 SQL 语句实现的代码如下：

```
users_1_4.registerTableName("users_1_4");
users_3_6.registerTableName("users_3_6");
BatchOperator.sqlQuery(
    "SELECT * FROM"
    + " ( (SELECT * FROM users_1_4) "
    + "    UNION ALL "
    + "   (SELECT * FROM users_3_6)"
    + " )"
).print();
```

使用 Alink 提供的简化方法，同样一行就可以搞定，代码如下：

```
new UnionAllBatchOp().linkFrom(users_1_4, users_3_6).print();
```

与标准的 SQL 操作相比，Alink 的简化方法省去了注册数据表的步骤，略去了一些标准样式，只需输入关键语句即可。下面将详细介绍 Alink 封装的一些简化方法。

5.2.1 单表操作

下面将介绍几个针对单个数据表的常用操作。

1. 选择（select）

其类似于 SQL SELECT 语句。执行选择操作，返回的结果还是 Alink 的批式组件。

```
BatchOperator ratings_select = ratings.select("user_id, item_id AS movie_id");
```

也可以直接使用 Alink 提供的简化方法来打印输出结果：

```
ratings.select("user_id, item_id AS movie_id").firstN(5).print();
```

输出结果如下：

```
user_id|movie_id
-------|--------
196|242
186|302
22|377
244|51
166|346
```

注意：可以使用星号（*）充当通配符，选择表中的所有列。

```
BatchOperator ratings_select = ratings.select("*");
```

2. 重命名（as）

重命名字段：

```
ratings.as("f1,f2,f3,f4").firstN(5).print();
```

运行结果如下，可以看到第一行数据列的名称都改变了：

```
f1|f2|f3|f4
--|--|--|--
196|242|3.0000|881250949
186|302|3.0000|891717742
```

```
22|377|1.0000|878887116
244|51|2.0000|880606923
166|346|1.0000|886397596
```

3. 过滤（filter/where）

filter/where 方法类似于 SQL WHERE 子句，可筛选出满足条件的行。

```
ratings.filter("rating > 3").firstN(5).print();
```

或者

```
ratings.where("rating > 3").firstN(5).print();
```

运行结果如下：

```
user_id|item_id|rating|ts
-------|-------|------|--
298|474|4.0000|884182806
253|465|5.0000|891628467
286|1014|5.0000|879781125
200|222|5.0000|876042340
122|387|5.0000|879270459
```

4. 不同记录集（distinct）

distinct 方法类似于 SQL DISTINCT 子句，可返回具有不同值组合的记录。

```
users.select("gender").distinct().print();
```

输出结果如下：

```
gender
------
F
M
```

5. 分组聚合（groupBy）

groupBy 方法类似于 SQL GROUP BY 子句。其指定包含分组键值的列，将数据集的所有行按键值划分为若干个组；然后对每组数据执行聚合运算（求总数、最大值、平均值等）；最后汇合各组的结果进行输出。groupBy 方法共有两个参数：第一个参数用来指定包含分组键值的列，第二个参数定义使用的聚合函数及输出列名。示例代码如下：

```
users.groupBy("gender","gender, COUNT(*) AS cnt").print();
```

输出结果如下：

```
gender|cnt
------|---
F     |273
M     |670
```

6. 排序并输出片段（orderBy）

orderBy 方法类似于 SQL ORDER BY 子句。必须输入的参数为用来排序的数据列名称。在选取输出数据方面，提供了两种方式：

（1）输入一个参数，指定选择输出数据集中最前面的多少行。

（2）输入两个参数，表示从某个偏移量开始，选择多少行。

我们通过示例来展示这两种方式的不同，代码如下：

```
users.orderBy("age", 5).print();
users.orderBy("age", 1,3).print();
```

使用 Java 编辑器可以清晰地显示各参数的含义，如图 5-3 所示。

```
users.orderBy( fieldName: "age",  limit: 5).print();
users.orderBy( fieldName: "age",  offset: 1, fetch: 3).print();
```

图 5-3

运行结果如下：

```
user_id|age|gender|occupation|zip_code
-------|---|------|----------|--------
30     |7  |M     |student   |55436
471    |10 |M     |student   |77459
289    |11 |M     |none      |94619
142    |13 |M     |other     |48118
880    |13 |M     |student   |83702

user_id|age|gender|occupation|zip_code
-------|---|------|----------|--------
471    |10 |M     |student   |77459
289    |11 |M     |none      |94619
142    |13 |M     |other     |48118
```

这里共显示了 2 个结果数据：第 1 个结果数据为"age"从小到大排序的前 5 名；第 2 个结果数据仍以"age"从小到大排序，但偏移了 1 个位置，从第 2 位开始，取了 3 行。

在前面的例子中，数据都是从小到大排序的。如果我们想要查找最年长的几位用户的数据，该怎么处理呢？

orderBy 方法还提供了一个布尔类型的参数：是否按升序排列，其默认值为 true，即按升序排列——从小到大排列。该参数放在方法的最后，下面有两个示例：

第 5 章　支持 Flink SQL

```
users.orderBy("age", 5, false).print();
users.orderBy("age", 1,3, false).print();
```

使用 Java 编辑器可以清晰地显示各参数的含义，如图 5-4 所示。

```
users.orderBy( fieldName: "age",  limit: 5,  isAscending: false).print();
users.orderBy( fieldName: "age",  offset: 1, fetch: 3,  isAscending: false).print();
```

图 5-4

运行结果分别如下：

```
user_id|age|gender|occupation|zip_code
-------|---|------|----------|--------
481|73|M|retired|37771
860|70|F|retired|48322
803|70|M|administrator|78212
767|70|M|engineer|00000
559|69|M|executive|10022

user_id|age|gender|occupation|zip_code
-------|---|------|----------|--------
860|70|F|retired|48322
803|70|M|administrator|78212
767|70|M|engineer|00000
```

这里共显示了 2 个结果数据：第 1 个结果数据为"age"从大到小排序的前 5 名；第 2 个结果数据仍以"age"从大到小排序，但偏移了 1 个位置，从第 2 位开始，取了 3 行。

如果读者想更深入地了解各参数组合之间的关系，可以参见其 Java 接口的代码。

```java
public BatchOperator orderBy(String fieldName, int limit) {
    return orderBy(fieldName, limit, true);
}

public BatchOperator orderBy(String fieldName, int limit, boolean isAscending) {
    // ... ...
}

public BatchOperator orderBy(String fieldName, int offset, int fetch) {
    return orderBy(fieldName, offset, fetch, true);
}

public BatchOperator orderBy(String fieldName, int offset, int fetch,  boolean isAscending) {
    // ... ...
}
```

5.2.2 两表的连接（JOIN）操作

两表的连接（JOIN）操作细分为以下 4 种情况：
- INNER JOIN：如果左表和右表中的数据都匹配，则返回匹配的行。其经常被简写为 "JOIN"。
- LEFT OUTER JOIN：即使右表中没有匹配的数据，也从左表中返回所有的行。
- RIGHT OUTER JOIN：即使左表中没有匹配的数据，也从右表中返回所有的行。
- FULL OUTER JOIN：为 LEFT OUTER JOIN 与 RIGHT OUTER JOIN 结果的并集。

我们从 ratings 中过滤出一些数据，保留少量的 item_id 和 user_id。从 items 数据表中选取少量数据，为了让生成的两个数据集中的 item_id 互不包含，且数量较少，便于查看，我们选取了奇数编号的数据。还有一点，JOIN 类组件在描述匹配关系的时候，由于不能像 SQL 语句那样加数据表名称的前缀，两个数据集中列的名称不应相同，因此，我们要将第 2 个数据集中的列名 "item_id" 转换为 "movie_id"。具体代码如下：

```
BatchOperator left_ratings
    = ratings
    .filter("user_id<3 AND item_id<4")
    .select("user_id, item_id, rating");

BatchOperator right_movies
    = items
    .select("item_id AS movie_id, title")
    .filter("movie_id < 6 AND MOD(movie_id, 2) = 1");
```

我们将这两个数据表打印出来，以便观察其特点。

```
System.out.println("# left_ratings #");
left_ratings.print();
System.out.println("\n# right_movies #");
right_movies.print();
```

显示结果如下：

```
# left_ratings #
user_id|item_id|rating
-------|-------|------
1|2|3.0000
2|1|4.0000
1|1|5.0000
1|3|4.0000

# right_movies #
movie_id|title
--------|-----
```

```
1|Toy Story (1995)
3|Four Rooms (1995)
5|Copycat (1995)
```

在左表 left_ratings 中含有 item_id 的值为{1,2,3}，在右表 right_movies 中含有 movie_id 的值为{1,3,5}，两边有交叉，且不互相包含，该示例可以用来演示出 JOIN 操作的 4 种情况。

下面 4 个示例，都是针对数据表 left_ratings 和 right_movies 进行的操作，匹配条件是"item_id = movie_id"，选择左表 left_ratings 的全部列，以及右表的 title 列，即将左表的数据集加上电影名称。具体代码如下：

```
System.out.println("# JOIN #");
new JoinBatchOp()
    .setJoinPredicate("item_id = movie_id")
    .setSelectClause("user_id, item_id, title, rating")
    .linkFrom(left_ratings, right_movies)
    .print();

System.out.println("\n# LEFT OUTER JOIN #");
new LeftOuterJoinBatchOp()
    .setJoinPredicate("item_id = movie_id")
    .setSelectClause("user_id, item_id, title, rating")
    .linkFrom(left_ratings, right_movies)
    .print();

System.out.println("\n# RIGHT OUTER JOIN #");
new RightOuterJoinBatchOp()
    .setJoinPredicate("item_id = movie_id")
    .setSelectClause("user_id, item_id, title, rating")
    .linkFrom(left_ratings, right_movies)
    .print();

System.out.println("\n# FULL OUTER JOIN #");
new FullOuterJoinBatchOp()
    .setJoinPredicate("item_id = movie_id")
    .setSelectClause("user_id, item_id, title, rating")
    .linkFrom(left_ratings, right_movies)
    .print();
```

打印结果如下：

```
# JOIN #
user_id|item_id|title|rating
-------|-------|-----|-------
2|1|Toy Story (1995)|4.0000
1|1|Toy Story (1995)|5.0000
1|3|Four Rooms (1995)|4.0000

# LEFT OUTER JOIN #
user_id|item_id|title|rating
```

```
-------|-------|-----|------
1|1|Toy Story (1995)|5.0000
2|1|Toy Story (1995)|4.0000
1|2|null|3.0000
1|3|Four Rooms (1995)|4.0000

# RIGHT OUTER JOIN #
user_id|item_id|title|rating
-------|-------|-----|------
2|1|Toy Story (1995)|4.0000
1|1|Toy Story (1995)|5.0000
1|3|Four Rooms (1995)|4.0000
null|null|Copycat (1995)|null

# FULL OUTER JOIN #
user_id|item_id|title|rating
-------|-------|-----|------
1|1|Toy Story (1995)|5.0000
2|1|Toy Story (1995)|4.0000
1|2|null|3.0000
1|3|Four Rooms (1995)|4.0000
null|null|Copycat (1995)|null
```

在此可以看到：

- 执行 JOIN 操作，返回的是左右表中能匹配上的数据。
- 执行 LEFT OUTER JOIN 操作，除了返回左右表中能匹配上的数据，还输出了左表中没有匹配上的数据。对于没有匹配上的电影名称，赋为 null。
- 执行 RIGHT OUTER JOIN 操作，除了返回左右表中能匹配上的数据，还输出了右表中没有匹配上的数据。对于没有匹配上的电影名称，赋为 null。
- 执行 FULL OUTER JOIN 操作，包括了三部分内容：左右表中能匹配上的数据、左表中没有匹配上的数据，以及右表中没有匹配上的数据。

5.2.3 两表的集合操作

以数据表的行作为基本单位，将行看作集合的元素。集合支持的操作可以分为下面两类：

（1）严格的集合操作。集合中没有重复元素，数据表中的相同元素会被看作同一个元素。严格的集合操作有三种：合并（UNION）、相交（INTERSECT）及减（MINUS）操作。

（2）扩展的集合操作。考虑到元素重复的个数，在相应操作名称的后面加上"ALL"，以便与严格的集合操作有所区分。扩展的集合操作也有 3 种：全体合并（UNION ALL）、全体相交（INTERSECT ALL）及全体减（MINUS ALL）操作。

第 5 章 支持 Flink SQL

为了更好地演示下面的操作,我们先生成 2 个小数据表,从 users 中分别取 user_id 为 1~4 号的数据,及 user_id 为 3~6 号的数据,具体代码如下:

```
BatchOperator users_1_4 = users.filter("user_id<5");
System.out.println("# users_1_4 #");
users_1_4.print();

BatchOperator users_3_6 = users.filter("user_id>2 AND user_id<7");
System.out.println("\n# users_3_6 #");
users_3_6.print();
```

输出结果如下:

```
# users_1_4 #
user_id|age|gender|occupation|zip_code
-------|---|------|----------|--------
1|24|M|technician|85711
2|53|F|other|94043
3|23|M|writer|32067
4|24|M|technician|43537

# users_3_6 #
user_id|age|gender|occupation|zip_code
-------|---|------|----------|--------
3|23|M|writer|32067
4|24|M|technician|43537
5|33|F|other|15213
6|42|M|executive|98101
```

这 2 个数据集各有 4 行数据,并且都包含 user_id 为 3 和 4 的两行数据。

1. 全体合并(UNION ALL)

执行 UNION ALL 操作可将两个表合并起来,并要求两个表的列(包括列类型、列顺序)完全一致。执行 UNION ALL 操作只是一种直接的合并,并不会判断两个表是否有相同的行,不会进行去重操作。Alink 提供了两个组件 UnionAllBatchOp 和 UnionAllStreamOp,它们是分别针对批式数据和流式数据设计的。示例代码如下:

```
new UnionAllBatchOp().linkFrom(users_1_4, users_3_6).print();
```

输出结果如下:

```
user_id|age|gender|occupation|zip_code
-------|---|------|----------|--------
1|24|M|technician|85711
2|53|F|other|94043
3|23|M|writer|32067
```

```
4|24|M|technician|43537
3|23|M|writer|32067
4|24|M|technician|43537
5|33|F|other|15213
6|42|M|executive|98101
```

可以看到，user_id 为 3、4 的数据有重复。

2. 合并（UNION）

执行 UNION 操作可将两个表合并起来，并要求两个表的列（包括列类型、列顺序）完全一致。如果两个表有相同的行，会有去重操作。Alink 提供了 UnionBatchOp 组件处理批式数据，示例代码如下：

```
new UnionBatchOp().linkFrom(users_1_4, users_3_6).print();
```

输出结果如下：

```
user_id|age|gender|occupation|zip_code
-------|---|------|----------|--------
1|24|M|technician|85711
2|53|F|other|94043
3|23|M|writer|32067
4|24|M|technician|43537
5|33|F|other|15213
6|42|M|executive|98101
```

在此可以看到，结果中没有相同的数据行，这是去重操作的结果。

3. 相交（INTERSECT）

执行该操作类似于执行 SQL INTERSECT 子句。执行该操作，可返回两个表中都存在的记录。如果一个记录在一个或两个表中存在一次以上，则仅返回一次，即结果表中没有重复的记录。两个表必须具有相同的字段类型。

```
new IntersectBatchOp().linkFrom(users_1_4, users_3_6).print();
```

输出结果如下：

```
user_id|age|gender|occupation|zip_code
-------|---|------|----------|--------
3|23|M|writer|32067
4|24|M|technician|43537
```

这里输出了 users_1_4 和 users_3_6 中都出现的 3 号和 4 号数据。

4. 全体相交（INTERSECT ALL）

执行该操作类似于执行 SQL INTERSECT ALL 子句。执行该操作，可返回两个表中都存在的记录。如果一个记录在两个表中均多次出现，则执行该操作所返回的记录次数，与在两个表中均多次出现该记录时所返回的记录次数一样，即结果表中可能有重复的记录。两个表必须具有相同的字段类型。

我们构造两个带有重复数据的数据集：一个数据集是 users_1_4 的数据重复一次，每个 ID 号（1～4 号）都有两条数据；另外一个数据集是 users_1_4 与 users_3_6 进行 UNION All 操作，该结果集包含 1～6 号的数据，其中 3 号、4 号数据都为两条。然后对这两个数据集进行 INTERSECT ALL 操作，具体代码如下：

```
new IntersectAllBatchOp()
    .linkFrom(
        new UnionAllBatchOp().linkFrom(users_1_4, users_1_4),
        new UnionAllBatchOp().linkFrom(users_1_4, users_3_6)
    )
    .print();
```

输出结果如下：

```
user_id|age|gender|occupation|zip_code
-------|---|------|----------|--------
1|24|M|technician|85711
2|53|F|other|94043
3|23|M|writer|32067
3|23|M|writer|32067
4|24|M|technician|43537
4|24|M|technician|43537
```

两个数据集中的相同数据都会在结果中出现。3 号、4 号数据在原先的两个数据集中均出现了 2 次，所以其在结果中也会出现 2 次。

5. 减（MINUS）

执行该操作类似于执行 SQL EXCEPT 子句。执行减操作，可从左表中返回在右表中不存在的记录。左表中的重复记录仅返回一次，即删除了重复项。两个表必须具有相同的字段类型。

```
new MinusBatchOp().linkFrom(users_1_4, users_3_6).print();
```

输出结果如下：

```
user_id|age|gender|occupation|zip_code
-------|---|------|----------|--------
1|24|M|technician|85711
2|53|F|other|94043
```

在此，左边的数据集 users_1_4，去掉了其与 users_3_6 都有的 3 号、4 号数据。

6. 全体减（MINUS ALL）

执行该操作类似于执行 SQL EXCEPT ALL 子句。执行该操作，可返回右表中不存在的记录。如果某条记录在左表中出现 n 次且在右表中出现 m 次，则将返回 ($n-m$) 次该记录，即，删除与右表中存在的重复项一样多的记录。两个表必须具有相同的字段类型。

采用 INTERSECT ALL 例子中构造的数据集，即 2 个带有重复数据的数据集。其中，一个数据集 users_1_4 的数据重复一次，每个 ID 号（1~4 号）都有两条数据；另外一个数据集是 users_1_4 与 users_3_6 进行 UNION All 操作，该结果集包含 1~6 号的数据，其中 3 号、4 号数据都为两条。然后对这两个数据集进行 MINUS ALL 操作，具体代码如下：

```
new MinusAllBatchOp()
    .linkFrom(
        new UnionAllBatchOp().linkFrom(users_1_4, users_1_4),
        new UnionAllBatchOp().linkFrom(users_1_4, users_3_6)
    )
    .print();
```

输出结果如下：

```
user_id|age|gender|occupation|zip_code
-------|---|------|----------|--------
1|24|M|technician|85711
2|53|F|other|94043
```

> 注意：1 号、2 号数据在左边的数据集中出现了 2 次，在右边的数据集中出现了 1 次；3 号、4 号数据在左右两个数据集中出现的次数相同，均为 2 次。所以，最终的结果为 1 号和 2 号数据各出现了 1 次。

5.3　深入介绍Table Environment

Flink SQL 语句是通过 Batch Table Environment 和 Stream Table Environment 执行的。Alink 的运行环境中包括了 Batch/Stream Table Environment，并提供了接口。

- 获取 Alink 默认使用的 Batch Table Environment，使用 Java 代码：

```
MLEnvironmentFactory.getDefault().getBatchTableEnvironment()
```

- 获取 Alink 默认使用的 Stream Table Environment，使用 Java 代码：

```
MLEnvironmentFactory.getDefault().getStreamTableEnvironment()
```

通过下面的代码可获取 Alink 默认的 Batch Table Environment，并打印出当前 Table Environment 中所有注册的数据表名称列表。

```
String[] tableNames = MLEnvironmentFactory.getDefault().getBatchTableEnvironment().listTables();
System.out.println("Table Names : ");
for (String name : tableNames) {
    System.out.println(name);
}
```

同样的语句，将 getBatchTableEnvironment 换成 getStreamTableEnvironment，就可对流式数据表进行同样的操作。

Flink Table Environment 提供了两种执行方法，即 sqlQuery 和 sqlUpdate：

- sqlQuery 方法会对已在 TableEnvironment 中注册的数据表进行 SQL 查询，结果为数据表。
- sqlUpdate 方法可用来执行 SQL 语句，如 INSERT、UPDATE 或 DELETE；或执行 DDL 语句。

简单来说，sqlQuery 会返回数据表的结果，sqlUpdate 没有返回值。

Alink 对 sqlQuery 进行了简单的封装，定义了静态方法 batchSQL，具体代码如下：

```
/**
 * Evaluates a SQL query on registered tables and retrieves the result as a
 * <code>BatchOperator</code>.
 *
 * @param query The SQL query to evaluate.
 * @return The result of the query as <code>BatchOperator</code>.
 */
public BatchOperator <?> batchSQL(String query) {
    return new TableSourceBatchOp(getBatchTableEnvironment().sqlQuery(query));
}
```

Alink 将其输出结果包装成 BatchOperator，这样就可以使用 link 等方法，连接其他 Alink 组件了。

5.3.1　注册数据表名

Alink BatchOperator 的 registerTableName 方法，可在 Alink 默认的 Batch Table Environment 中，将批式组件的主输出注册为相应的名称。

```
BatchOperator <?> ratings = ...
BatchOperator <?> users = ...
BatchOperator <?> items = ...

ratings.registerTableName("ratings");
```

```
items.registerTableName("items");
users.registerTableName("users");
```

检查一下 Batch Table Environment 的数据表名称列表，代码如下：

```
String[] tableNames
    = MLEnvironmentFactory.getDefault().getBatchTableEnvironment().listTables();
System.out.println("Table Names : ");
for (String name : tableNames) {
    System.out.println(name);
}
```

执行结果如下，前面设置过的 3 个数据表名都显示了出来，说明其已经可以在 SQL 语句中使用了。

```
Table Names : 
ratings
items
users
```

5.3.2 撤销数据表名

与数据表名的注册功能相对应，撤销数据表名可以使用如下操作：

```
BatchTableEnvironment batchTableEnvironment
    = MLEnvironmentFactory.getDefault().getBatchTableEnvironment();

System.out.println("Table Names : ");
for (String name : batchTableEnvironment.listTables()) {
    System.out.println(name);
}

batchTableEnvironment.sqlUpdate("DROP TABLE IF EXISTS users");

System.out.println("\nTable Names After DROP : ");
for (String name : batchTableEnvironment.listTables()) {
    System.out.println(name);
}
```

执行结果如下：

```
Table Names : 
ratings
items
users

Table Names After DROP : 
ratings
items
```

5.3.3 扫描已注册的表

前面介绍的注册操作，指的是将数据表注册到 Table Environment 中；而扫描操作则是反向的操作，可通过扫描的方式来获取 Table Environment 中已注册的表。示例如下：

```
BatchOperator ratings_scan
    = BatchOperator.fromTable(batchTableEnvironment.scan("ratings"));
ratings_scan.firstN(5).print();
```

其中，batchTableEnvironment.scan("ratings")得到的是 Flink Table 类型对象。还需要使用 BatchOperator.fromTable()转化为 Alink 批式组件。

运行结果如下：

```
user_id|item_id|rating|ts
-------|-------|------|--
196|242|3.0000|881250949
186|302|3.0000|891717742
22|377|1.0000|878887116
244|51|2.0000|880606923
166|346|1.0000|886397596
```

6 用户定义函数（UDF/UDTF）

虽然 Alink 提供了丰富的算法组件，也支持 Flink SQL 的丰富操作，但对复杂的实际场景，用户还需要自己能够灵活地定义函数功能，处理批式/流式数据。

用户定义函数分为标量函数（Scalar-Valued Function）和表值函数(Table-Valued Function)，两者的关系如表 6-1 所示。

- 在标量函数中将零个、一个或多个标量值作为输入参数，函数的输出结果为单个标量值。
- 与标量函数相似，在表值函数中也将零个、一个或多个标量值作为输入参数。但是，其输出结果与标量函数有很大差异。该函数可以返回任意数量的行作为输出，而不是单个值。其返回的行可能包含一列或多列数据。

表 6-1　标量函数和表值函数对比表

	输入参数	输出结果
标量函数	零个、一个或多个标量值	单个标量值
表值函数	零个、一个或多个标量值	返回任意数量的行作为输出，返回的行可能包含一列或多列数据

6.1 用户定义标量函数（UDF）

用户定义标量函数（User Defined scalar-valued Function），简称 UDF。自定义标量函数是用户最常用的方式。下面将演示如何定义 UDF，并应用到批式处理和流式处理场景中。

6.1.1　示例数据及问题

我们选择 MovieLens 的一个数据集（参见链接 6-1），该数据集包含了用户对电影评分的具体数据。该数据集包含 943 个用户对 1682 个物品（电影）进行的 100 000 次评分数据；每个用户至少针对 20 部电影进行了评分；用户和物品（电影）都是从 1 开始编号的。图 6-1 是数据文件的部分内容。

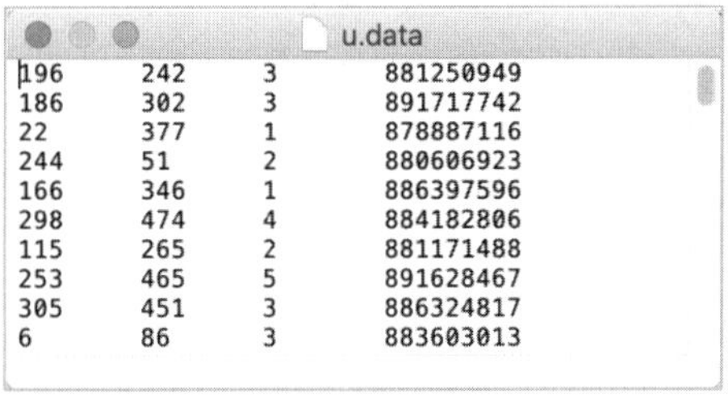
u.data

196	242	3	881250949
186	302	3	891717742
22	377	1	878887116
244	51	2	880606923
166	346	1	886397596
298	474	4	884182806
115	265	2	881171488
253	465	5	891628467
305	451	3	886324817
6	86	3	883603013

图 6-1

在该文件中，每个属性之间用制表符\t 分隔。该文件共包含 4 个属性：

- user_id：用户 ID。
- item_id：物品（电影）ID。
- rating：评分。
- ts：UNIX 时间戳，是自 UTC 的 1970 年 1 月 1 日以来的 UNIX 秒数。

在此可以看到，第 4 列 ts 显示的 UNIX 秒数虽然精确地表示了时间，但是我们却很难从这一串数字中知道其具体表示哪年哪月。下面我们就用自定义方式实现一个这样的函数：将 UNIX 秒数转换为标准格式的时间信息。

6.1.2　UDF的定义

定义 Java 版本标量函数需要注意下面三点：

- 必须继承基类：

```
org.apache.flink.table.functions.ScalarFunction
```

- 必须实现 eval 方法，标量函数的行为是由 eval 方法定义的。
- 可以选择是否重载 getResultType 方法。一般情况下，不必重载该方法。在如下定义中去掉重载，得到的计算结果也是一样的。

示例定义如下：

```
import org.apache.flink.api.common.typeinfo.TypeInformation;
import org.apache.flink.api.common.typeinfo.Types;
import org.apache.flink.table.functions.ScalarFunction;
... ...
public static class FromUnixTimestamp extends ScalarFunction {
...
    public java.sql.Timestamp eval(Long ts) {
        return new java.sql.Timestamp(ts * 1000);
    }

    @Override
    public TypeInformation <?> getResultType(Class <?>[] signature) {
        return Types.SQL_TIMESTAMP;
    }
}
```

6.1.3 使用UDF处理批式数据

定义好 UDF 后，可以通过 UDFBatchOp 组件调用 UDF，该组件需要通过 setFunc 方法指定 UDF，并设置输入和输出的数据列名称参数。注意，如果我们将输出名称与输入名称设置成一样的，则数据结果会替换掉输入列的原始数据。

```
ratings
    .link(
        new UDFBatchOp()
            .setFunc(new FromUnixTimestamp())
            .setSelectedCols("ts")
            .setOutputCol("ts")
    )
    .firstN(5)
    .print();
```

运行结果如下：

```
user_id|item_id|rating|ts
-------|-------|------|--
196|242|3.0000|1997-12-04 23:55:49.0
186|302|3.0000|1998-04-05 03:22:22.0
22|377|1.0000|1997-11-07 15:18:36.0
244|51|2.0000|1997-11-27 13:02:03.0
166|346|1.0000|1998-02-02 13:33:16.0
```

进一步，我们注册此 UDF，注册代码如下：

```
BatchOperator.registerFunction("from_unix_timestamp", new FromUnixTimestamp());
```

这样就可以在 SQL 语句中调用此函数了。一个简单的示例如下：

```
ratings
    .select("user_id, item_id, rating, from_unix_timestamp(ts) AS ts")
    .firstN(5)
    .print();
```

运行结果如下：

```
user_id|item_id|rating|ts
-------|-------|------|--
196|242|3.0000|1997-12-04 23:55:49.0
186|302|3.0000|1998-04-05 03:22:22.0
22|377|1.0000|1997-11-07 15:18:36.0
244|51|2.0000|1997-11-27 13:02:03.0
166|346|1.0000|1998-02-02 13:33:16.0
```

与前面使用 UDFBatchOp 组件调用 UDF 的结果相同。我们还可以在 sqlQuery() 中使用 UDF，相关代码如下，得到的结果与使用 UDFBatchOp 组件得到的结果仍然一致。需要注意的是，使用 SQL 语句前，要先注册数据表 ratings。

```
ratings.registerTableName("ratings");

BatchOperator
    .sqlQuery("SELECT user_id, item_id, rating, from_unix_timestamp(ts) AS ts FROM ratings")
    .firstN(5)
    .print();
```

6.1.4 使用UDF处理流式数据

为了便于打印显示结果，我们先对流式数据进行过滤，减少数据条数，代码如下：

```
ratings = ratings.filter("user_id=1 AND item_id<5");

ratings.print();

StreamOperator.execute();
```

运行结果如下，只有 4 条数据：

```
user_id|item_id|rating|ts
-------|-------|------|--
1|1|5.0000|874965758
1|3|4.0000|878542960
1|2|3.0000|876893171
1|4|3.0000|876893119
```

与批式处理的情况类似,可以通过 UDFStreamOp 组件调用 UDF,该组件需要通过 setFunc 方法指定 UDF,并设置输入和输出的数据列名称参数。注意,在这个示例中,我们仍然将输出名称与输入名称设置成一样的,数据结果会替换掉输入列的原始数据。

```
ratings
  .link(
    new UDFStreamOp()
      .setFunc(new FromUnixTimestamp())
      .setSelectedCols("ts")
      .setOutputCol("ts")
  )
  .print();

StreamOperator.execute();
```

运行结果如下:

```
user_id|item_id|rating|ts
-------|-------|------|--
1|1|5.0000|1997-09-23 06:02:38.0
1|3|4.0000|1997-11-03 15:42:40.0
1|2|3.0000|1997-10-15 13:26:11.0
1|4|3.0000|1997-10-15 13:25:19.0
```

进一步,我们注册此 UDF,注册代码如下。注意,这里需要使用的是 StreamOperator 的 registerFunction 方法。

```
StreamOperator.registerFunction("from_unix_timestamp", new FromUnixTimestamp());
```

随后,就可以在 SQL 语句中调用此函数了。一个简单的示例如下:

```
ratings
  .select("user_id, item_id, rating, from_unix_timestamp(ts) AS ts")
  .print();

StreamOperator.execute();
```

运行结果如下:

```
user_id|item_id|rating|ts
-------|-------|------|--
1|1|5.0000|1997-09-23 06:02:38.0
1|3|4.0000|1997-11-03 15:42:40.0
1|2|3.0000|1997-10-15 13:26:11.0
1|4|3.0000|1997-10-15 13:25:19.0
```

与前面使用 UDFStreamOp 组件调用 UDF 的结果相同。我们还可以在 sqlQuery() 中使用 UDF，相关代码如下，得到的结果与使用 UDFStreamOp 组件得到的结果仍然一致。需要注意的是，使用 SQL 语句前，要先注册 ratings。

```
ratings.registerTableName("ratings");
StreamOperator
    .sqlQuery("SELECT user_id, item_id, rating, from_unix_timestamp(ts) AS ts FROM ratings")
    .print();

StreamOperator.execute();
```

6.2 用户定义表值函数（UDTF）

用户定义表值函数（User Defined Table-valued Function，UDTF），可以用来解决输入一行、输出多行的问题。

6.2.1 示例数据及问题

我们仍然使用 MovieLens 数据，使用影片信息数据集并选择 item_id 和 title 两个字段。在此选择前 10 条数据并打印：

```
item_id|title
-------|-----
1|Toy Story (1995)
2|GoldenEye (1995)
3|Four Rooms (1995)
4|Get Shorty (1995)
5|Copycat (1995)
6|Shanghai Triad (Yao a yao yao dao waipo qiao) (1995)
7|Twelve Monkeys (1995)
8|Babe (1995)
9|Dead Man Walking (1995)
10|Richard III (1995)
```

接下来，我们统计标题（title）字段的词个数，该操作需要两步：首先将标题（title）按词切割开并记录各词在当前标题中出现的个数。然后，进行 groupBy 操作，汇总出每个单词出现的总次数。第一步操作会将一条数据按其包含单词的情况分为多条数据，这不符合前面讲的标量函数（UDF）的输出形式要求，必须通过用户定义表值函数（UDTF）进行处理。

6.2.2　UDTF的定义

定义 Java 版本的表值函数（UDTF）需要注意下面三点：
- 必须继承基类：

```
org.apache.flink.table.functions.TableFunction
```

- 必须实现 eval 方法。与标量函数不同的是，eval() 没有返回值，但是在运行过程中，其可以多次调用 TableFunction 的 collect 方法，将所要输出的数据逐条提交给 collect 方法进行输出。
- 必须重载 getResultType 方法，详细定义函数输出结果的各数据列类型。

具体代码如下：

```java
public static class WordCount extends TableFunction <Row> {
   private HashMap <String, Integer> map = new HashMap <>();

   public void eval(String str) {
      if (null == str || str.isEmpty()) {
         return;
      }
      for (String s : str.split(" ")) {
         if (map.containsKey(s)) {
            map.put(s, 1 + map.get(s));
         } else {
            map.put(s, 1);
         }
      }
      for (Entry <String, Integer> entry : map.entrySet()) {
         collect(Row.of(entry.getKey(), entry.getValue()));
      }
      map.clear();
   }

   @Override
   public TypeInformation <Row> getResultType() {
      return Types.ROW(Types.STRING, Types.INT);
   }
}
```

6.2.3　使用UDTF处理批式数据

定义好 UDTF 后，可以通过 UDTFBatchOp 组件调用 UDTF，该组件需要通过 setFunc 方法指定 UDTF，并设置输入和输出的数据列名称参数。具体代码如下：

第 6 章 用户定义函数（UDF/UDTF）

```
BatchOperator <?> words = items.link(
    new UDTFBatchOp()
        .setFunc(new WordCount())
        .setSelectedCols("title")
        .setOutputCols("word", "cnt")
        .setReservedCols("item_id")
);

words.lazyPrint(10, "<- after word count ->");
```

运行结果如下，可以看到每条数据用空格分隔单词，年份和外面的括号也被看作"单词"进行了个数统计。读者可能会有点疑问：第一条数据为"Toy Story (1995)"，但为什么单词输出的顺序却是反过来的？这是因为在实现过程中使用 HashMap 进行判重和计数操作，HashMap 是按字母升序方式获取结果的。

```
item_id|word|cnt
-------|----|---
1|(1995)|1
1|Story|1
1|Toy|1
2|(1995)|1
2|GoldenEye|1
3|(1995)|1
3|Four|1
3|Rooms|1
4|Shorty|1
4|(1995)|1
4|Get|1
5|(1995)|1
5|Copycat|1
6|a|1
6|(1995)|1
6|dao|1
6|waipo|1
6|Shanghai|1
6|yao|2
6|qiao|1
```

接下来，使用 SQL 语句进行汇总统计。具体代码如下：

```
words.groupBy("word", "word, SUM(cnt) AS cnt")
    .orderBy("cnt", 20, false)
    .print();
```

结果如下面左栏所示，右栏是对统计结果的一些说明。

```
word|cnt
----|---
```

查看单词出现次数的排名，很容易发现：
- 出现次数最多的是"The"。"The"大多数时候出现在片名的开头，有 356 次。

97

```
The|356
(1996)|298
(1995)|296
(1994)|237
(1997)|235
the|145
of|137
(1993)|130
and|66
in|63
(1998)|53
A|50
(1992)|40
a|31
to|31
Love|27
(1991)|24
(1990)|24
Man|23
My|22
```

- 紧随其后的是片名中出现的年份，而年份中又以 1996 年居多。（1996）出现了 298 次。
- 之后是小写的"the"。"the"出现了 145 次，若加上前面首字母大写的"The"，该单词（大小写形式均包括在内）出现的总次数要远超其他单词。
- 介词"of"出现了 137 次。
- 连词"and"出现的次数为 66 次。
- 除以上这些年份、介词、连词、冠词等外，出现次数最多的是"Love"。"Love"出现了 27 次。
- "Man"和"My"出现的次数也不少，其分别排在第 19 位和第 20 位。

进一步，我们换一种 UDTF 的调用方式。注册 UDTF，注册代码如下：

```
BatchOperator.registerFunction("word_count", new WordCount());
```

注册的函数 word_count 可以在 sqlQuery()中使用，相关代码如下，得到的结果与前一种调用方式所得结果仍然一致。需要注意的是，在使用 SQL 语句前，还要先注册数据表 ratings。具体代码如下：

```
items.registerTableName("items");

BatchOperator
    .sqlQuery("SELECT item_id, word, cnt FROM items, "
        + "LATERAL TABLE(word_count(title)) as T(word, cnt)")
    .firstN(10)
    .print();
```

运行结果如下，其与使用 UDTFBatchOp 所调用的结果相同：

```
item_id|word|cnt
-------|----|---
1|(1995)|1
1|Story|1
1|Toy|1
2|(1995)|1
2|GoldenEye|1
3|(1995)|1
3|Four|1
3|Rooms|1
```

```
4|Shorty|1
4|(1995)|1
4|Get|1
5|(1995)|1
5|Copycat|1
6|a|1
6|(1995)|1
6|dao|1
6|waipo|1
6|Shanghai|1
6|yao|2
6|qiao)|1
```

6.2.4 使用UDTF处理流式数据

为了便于打印显示结果,我们先对流式数据进行过滤,减少数据条数。具体代码如下:

```
items = items.select("item_id, title").filter("item_id<4");

items.print();
StreamOperator.execute();
```

运行结果如下,只有 3 条数据:

```
item_id|title
-------|-----
1|Toy Story (1995)
2|GoldenEye (1995)
3|Four Rooms (1995)
```

与批式处理情况类似,通过 UDTFStreamOp 组件调用 UDTF,该组件需要通过 setFunc 方法指定 UDTF,并设置输入和输出的数据列名称参数。具体代码如下:

```
StreamOperator <?> words = items.link(
   new UDTFStreamOp()
     .setFunc(new WordCount())
     .setSelectedCols("title")
     .setOutputCols("word", "cnt")
     .setReservedCols("item_id")
);
words.print();
StreamOperator.execute();
```

运行结果如下:

```
item_id|word|cnt
-------|----|---
```

```
1|(1995)|1
1|Story|1
1|Toy|1
2|(1995)|1
2|GoldenEye|1
3|(1995)|1
3|Four|1
3|Rooms|1
```

在此可以看到每条数据用空格分隔单词，年份和外面的括号也被看作"单词"进行了个数统计。

下面，我们换一种 UDTF 的调用方式。注册此 UDTF，注册代码如下。注意，这里需要使用的是 StreamOperator 的 registerFunction 方法。

```
StreamOperator.registerFunction("word_count", new WordCount());
```

随后，就可以在 sqlQuery()中调用注册的函数 word_count 了，具体代码如下。需要注意的是，使用 SQL 语句前，要先注册数据表 items。

```
items.registerTableName("items");

StreamOperator.sqlQuery("SELECT item_id, word, cnt FROM items, "
    + "LATERAL TABLE(word_count(title)) as T(word, cnt)")
    .print();

StreamOperator.execute();
```

运行结果如下，与前面使用 UDTFStreamOp 所调用 UDTF 的结果相同：

```
item_id|word|cnt
-------|----|---
1|(1995)|1
1|Story|1
1|Toy|1
2|(1995)|1
2|GoldenEye|1
3|(1995)|1
3|Four|1
3|Rooms|1
```

7 基本数据处理

Alink 集成了强大的数据处理能力，包括以下几个方面：
- 支持 Flink SQL 操作（本书第 5 章专门进行了介绍），比如，选取列、数据过滤、类型转换、基本数值计算、判断逻辑、分组聚合、表连接、表合并以及字符串操作等。
- 支持用户定义函数 UDF/UDTF（本书第 6 章专门进行了介绍）。
- 针对机器学习领域一些常用的基本操作，Alink 提供的数据处理组件。本章将重点介绍这些组件。

7.1 采样

本节会先介绍一个和采样比较相似的常用操作：取"前"几条数据。该操作可以被看作不考虑各个数据的获取概率的"采样"。

三种常用的采样方式包括随机采样、加权采样和分层采样，如表 7-1 所示。

表 7-1 常用的采样方式

名称	描述
随机采样	所有数据样本被选中的概率是一样的
加权采样	每个数据样本都有一个权值，该权值与样本被选中的概率成正比
分层采样	每个数据样本都有一个"标签"，可以以据此将数据进行"分层"。可为每个标签（即其对应的一"层"数据）分别设置采样的比例

7.1.1 取"前"N个数据

如果我们只想看几条数据，以便对数据集有一个直观的认识，就可以指定取"前"几条数据（注意："前"是由双引号括起来的，在多并发的情况下，Flink 任务不保证数据的顺序，不能严格地说我们所取出的一定是数据文件的头几条数据）。Alink 提供了 FirstNBatchOp 组件，可通过设置参数 size 控制所取数据的数量。也可以使用包装的方法 firstN(int size)，设置所取数据的数量。这两种方式是等价的。下面的示例展示了这两种使用方式：

```
source
    .link(
        new FirstNBatchOp()
            .setSize(5)
    )
    .print();

source.firstN(5).print();
```

运行结果如下。两次均显示了 5 条数据，但数据并不一致。大家在使用时如果需要确定的顺序，建议采用 SQL Orderby 的方式。

```
sepal_length|sepal_width|petal_length|petal_width|category
------------|-----------|------------|-----------|--------
5.8000|2.8000|5.1000|2.4000|Iris-virginica
6.4000|3.2000|5.3000|2.3000|Iris-virginica
6.5000|3.0000|5.5000|1.8000|Iris-virginica
7.7000|3.8000|6.7000|2.2000|Iris-virginica
7.7000|2.6000|6.9000|2.3000|Iris-virginica
sepal_length|sepal_width|petal_length|petal_width|category
------------|-----------|------------|-----------|--------
5.0000|2.0000|3.5000|1.0000|Iris-versicolor
5.9000|3.0000|4.2000|1.5000|Iris-versicolor
6.0000|2.2000|4.0000|1.0000|Iris-versicolor
6.1000|2.9000|4.7000|1.4000|Iris-versicolor
5.6000|2.9000|3.6000|1.3000|Iris-versicolor
```

7.1.2 随机采样

Alink 提供了随机采样组件 SampleBatchOp 和 SampleWithSizeBatchOp，但是实际应用中经常使用其包装的方法 sample 或者 sampleWithSize。从名称上很容易分辨出两者的区别，前者通过指定采样的比例来确定采样数据的个数，而后者则直接指定确定的采样数据个数。示例代码如下：

```
source
  .sampleWithSize(50)
  .lazyPrintStatistics("< after sample with size 50 >")
  .sample(0.1)
  .print();
```

运行结果如下。从统计结果上能够看出，sampleWithSize(50)的运行结果的确只有 50 条数据。在此基础上的10%采样，实际获得了4条数据。

```
< after sample with size 50 >
|    colName|count|missing|   sum| mean|variance|min|max|
|-----------|-----|-------|------|-----|--------|---|---|
|sepal_length|   50|      0|291.4|5.828|  0.6347|4.4|7.7|
|sepal_width |   50|      0|151.7|3.034|   0.208|2.2|4.4|
|petal_length|   50|      0|186.9|3.738|  2.8991|  1|6.7|
|petal_width |   50|      0| 57.7|1.154|  0.4597|0.2|2.5|
|    category|   50|      0|  NaN|  NaN|     NaN|NaN|NaN|

sepal_length|sepal_width|petal_length|petal_width|category
------------|-----------|------------|-----------|--------
4.8000|3.4000|1.9000|0.2000|Iris-setosa
5.8000|2.7000|5.1000|1.9000|Iris-virginica
5.0000|3.6000|1.4000|0.2000|Iris-setosa
6.1000|2.6000|5.6000|1.4000|Iris-virginica
```

在随机采样时可以选择是否"放回采样"。放回采样指的是逐个抽取样本时，选中的样本还会被放回总体样本中，随后该样本可能被多次选中。若不放回采样，则可保证每个样本最多被选中一次。Alink 的随机采样默认选择"不放回采样"的方式。示例代码如下：

```
source
  .lazyPrintStatistics("< origin data >")
  .sampleWithSize(150, true)
  .lazyPrintStatistics("< after sample with size 150 >")
  .sample(0.03, true)
  .print();
```

结果如下，原始数据为 150 条，选择"放回采样"方式采样时也得到了 150 条数据。虽然我们没有逐条比对，但从统计结果上看，原始数据和采样数据的求和、均值等指标不一致，表明这两个数据表虽然同为 150 条数据，但数据内容不完全一致。其原因是选择"放回采样"方式采样导致个别数据被选到多次。数据的总条数不变，说明有的数据还没被选到，从而导致了统计指标的变化。后面在 150 条数据的基础上，按3%的采样率，得到 3 条数据。由于该采样率较低，因此不容易出现相同的数据。

```
< origin data >
|    colName|count|missing|  sum| mean|variance|min|max|
|-----------|-----|-------|-----|-----|--------|---|---|
```

```
|sepal_length|   150|      0|876.5|5.8433|  0.6857|4.3|7.9|
| sepal_width|   150|      0|458.1|3.054 |  0.188 |  2|4.4|
|petal_length|   150|      0|563.8|3.7587|  3.1132|  1|6.9|
| petal_width|   150|      0|179.8|1.1987|  0.5824|0.1|2.5|
|    category|   150|      0|  NaN|   NaN|     NaN|NaN|NaN|

< after sample with size 150 >
|     colName|count|missing|  sum|  mean|variance|min|max|
|------------|-----|-------|-----|------|--------|---|---|
|sepal_length|   150|      0|863.1| 5.754|  0.6455|4.3|7.7|
| sepal_width|   150|      0|461.6|3.0773|  0.2295|  2|4.4|
|petal_length|   150|      0|537.1|3.5807|  3.1585|  1|6.9|
| petal_width|   150|      0|169.8| 1.132|  0.6119|0.1|2.5|
|    category|   150|      0|  NaN|   NaN|     NaN|NaN|NaN|

sepal_length|sepal_width|petal_length|petal_width|category
------------|-----------|------------|-----------|--------
5.5000|2.4000|3.7000|1.0000|Iris-versicolor
4.5000|2.3000|1.3000|0.3000|Iris-setosa
6.5000|3.2000|5.1000|2.0000|Iris-virginica
```

对于流式数据，Alink 同样提供了随机采样组件，但只支持采用按比例采样的方式。示例代码如下：

```
source_stream.sample(0.1).print();

StreamOperator.execute();
```

运行结果如下：

```
sepal_length|sepal_width|petal_length|petal_width|category
------------|-----------|------------|-----------|--------
5.7000|3.8000|1.7000|0.3000|Iris-setosa
6.1000|2.8000|4.7000|1.2000|Iris-versicolor
6.2000|2.9000|4.3000|1.3000|Iris-versicolor
4.9000|2.5000|4.5000|1.7000|Iris-virginica
4.9000|3.1000|1.5000|0.1000|Iris-setosa
4.4000|3.0000|1.3000|0.2000|Iris-setosa
5.0000|3.5000|1.3000|0.3000|Iris-setosa
5.4000|3.0000|4.5000|1.5000|Iris-versicolor
```

7.1.3 加权采样

加权采样指的是每个数据样本都有一个权值，该权值与该样本被选中的概率成正比。比如，对于样本 1 和样本 2，权值分别为 50 和 10，则样本 1 被选中的概率是样本 2 的 5 倍。Alink 的加权采样组件为 WeightSampleBatchOp。与随机采样相比，其多了一个必选的参数：权重列的名称 WeightCol。

示例代码如下。在采样前使用 SQL 语句新建了权重列，根据花的类别不同而赋予不同的权重：Iris-versicolor 类别的权重为 1，Iris-setosa 类别的权重为 2，Iris-virginica 类别的权重为 4。在采样结束后，统计花中各个类别样本的个数。在理想状态下，这三类花采样后的比例为 1∶2∶4。

```
source
  .select("*, CASE category WHEN 'Iris-versicolor' THEN 1 "
    + "WHEN 'Iris-setosa' THEN 2 ELSE 4 END AS weight")
  .link(
    new WeightSampleBatchOp()
      .setRatio(0.4)
      .setWeightCol("weight")
  )
  .groupBy("category", "category, COUNT(*) AS cnt")
  .print();
```

运行结果如下。其中，Iris-versicolor 类别的数据量最少，Iris-setosa 类别的数据量居中，Iris-virginica 类别的数据量最多。这三个类别的采样数据量之比为 13∶17∶30，接近理想状态的 1∶2∶4。

```
category|cnt
--------|---
Iris-setosa|17
Iris-virginica|30
Iris-versicolor|13
```

7.1.4 分层采样

分层采样，对于含有"分层列"的数据，各"层"数据可以设置不同的采样比例。在处理分类问题时，对于正负样本数量相差悬殊的情况，常用该采样方法，并对拥有数量较多"标签"的样本设置较小的采样比例。Alink 的加权采样组件为 StratifiedSampleBatchOp。需要设定分层列 StrataCol，并设置分层列中各属性值的采样比例。

示例代码如下。其中，根据花的类别不同而赋予不同的采样比例，Iris-versicolor 类别的比例为 0.2，Iris-setosa 类别的比例为 0.4，Iris-virginica 类别的比例为 0.8。在采样结束后，会统计花中各个类别样本的个数。花中三个类别的原始数据量都为 50，理想的采样结果数据量之比为 10∶20∶40。

```
source
  .link(
    new StratifiedSampleBatchOp()
      .setStrataCol("category")
      .setStrataRatios("Iris-versicolor:0.2,Iris-setosa:0.4,Iris-virginica:0.8")
```

```
)
.groupBy("category", "category, COUNT(*) AS cnt")
.print();
```

运行结果如下。其中，Iris-versicolor 类别的数据量最少，Iris-setosa 类别的数据量居中，Iris-virginica 类别的数据量最多。这三个类别的采样数据量之比为 8∶19∶43，接近理想状态的 10∶20∶40。

```
category|cnt
--------|---
Iris-setosa|19
Iris-virginica|43
Iris-versicolor|8
```

对于流式数据，同样可以使用分层采样方式。对应的组件为 StratifiedSampleStreamOp，其参数设置与批式任务相同。示例代码如下：

```
source_stream
   .link(
      new StratifiedSampleStreamOp()
         .setStrataCol("category")
         .setStrataRatios("Iris-versicolor:0.2,Iris-setosa:0.4,Iris-virginica:0.8")
   )
   .print();

StreamOperator.execute();
```

该示例打印的数据较多，这里就不进行展示了。从显示的采样数据中，能够感觉到 Iris-versicolor 类别的数据量较少，Iris-virginica 类别的数据量较多。

7.2 数据划分

我们在做机器学习实验的时候，经常要把原始数据分为训练集和测试集。数据划分组件（SplitBatchOp）就会对应着两个输出：主输出为训练集；侧输出（Side Output）只有一个，输出的是测试集。

首先构建 SplitBatchOp 的实例 spliter。设置其切分比例为 0.9，即主输出占总样本的 90%；侧输出为其余的数据，占总样本的 10%。将原始数据 source 连接到 spliter。具体代码如下：

```
SplitBatchOp spliter = new SplitBatchOp().setFraction(0.9);

source.link(spliter);
```

下面获取一些组件输出的相关信息。直接使用 getSchema 方法，即可获得组件的主输出的

Schema。通过 getSideOutputCount() 获得该组件侧输出的个数,随后便可使用 getSideOutput(index) 方法通过索引号来获取具体的侧输出。每个侧输出是 BatchOperator 或者 StreamOperator。仍然可以继续使用 getSchema 方法。具体代码如下:

```
System.out.println("schema of spliter's main output:");
System.out.println(spliter.getSchema());

System.out.println("count of spliter's side outputs:");
System.out.println(spliter.getSideOutputCount());

System.out.println("schema of spliter's side output :");
System.out.println(spliter.getSideOutput(0).getSchema());
```

输出结果如下。其中,侧输出只有一个;主输出的 Schema 与侧输出的 Schema 相同。

```
schema of spliter's main output:
root
|-- sepal_length: DOUBLE
|-- sepal_width: DOUBLE
|-- petal_length: DOUBLE
|-- petal_width: DOUBLE
|-- category: STRING

count of spliter's side outputs:
1
schema of spliter's side output :
root
|-- sepal_length: DOUBLE
|-- sepal_width: DOUBLE
|-- petal_length: DOUBLE
|-- petal_width: DOUBLE
|-- category: STRING
```

接下来,我们将分别针对主输出和侧输出数据进行操作,使用 lazyPrintStatistics 方法打印当前数据统计结果,并将数据保存到数据文件中。具体代码如下:

```
spliter
  .lazyPrintStatistics("< Main Output >")
  .link(
    new AkSinkBatchOp()
      .setFilePath(DATA_DIR + TRAIN_FILE)
      .setOverwriteSink(true)
  );

spliter.getSideOutput(0)
  .lazyPrintStatistics("< Side Output >")
  .link(
    new AkSinkBatchOp()
      .setFilePath(DATA_DIR + TEST_FILE)
```

```
        .setOverwriteSink(true)
);

BatchOperator.execute();
```

结果打印如下。在此看到主输出有 135 条数据，侧输出有 15 条数据。我们在检查本地文件夹时可发现新生成的两个数据文件，这表明操作成功。

```
< Main Output >
|      colName|count|missing|   sum|  mean|variance|min|max|
|-------------|-----|-------|------|------|--------|---|---|
| sepal_length|  135|      0| 790.5|5.8556|   0.662|4.3|7.9|
|  sepal_width|  135|      0| 413.6|3.0637|  0.1923|2.2|4.4|
| petal_length|  135|      0| 508.3|3.7652|  3.1132|  1|6.9|
|  petal_width|  135|      0|   163|1.2074|  0.5931|0.1|2.5|
|     category|  135|      0|   NaN|   NaN|     NaN|NaN|NaN|

< Side Output >
|      colName|count|missing|   sum|  mean|variance|min|max|
|-------------|-----|-------|------|------|--------|---|---|
| sepal_length|   15|      0|    86|5.7333|  0.9467|4.4|7.3|
|  sepal_width|   15|      0|  44.5|2.9667|   0.151|  2|3.5|
| petal_length|   15|      0|  55.5|   3.7|  3.3314|1.3|6.3|
|  petal_width|   15|      0|  16.8|  1.12|  0.5146|0.2|  2|
|     category|   15|      0|   NaN|   NaN|     NaN|NaN|NaN|
```

7.3 数值尺度变换

数值尺度变换是围绕统计量进行的。我们先回顾几个统计概念，然后再介绍各个变换。对于数据 $X_1, X_2, \cdots, X_n$，有如下定义。

均值（Mean）

$$\bar{X} = \frac{1}{n}\sum_{i=1}^{n} X_i$$

可用均值来描述数据取值的平均位置。

方差（Sample Variance）

$$S_n^2 = \frac{1}{n-1}\sum_{i=1}^{n}(X_i - \bar{X})^2$$

标准差（Standard Deviation）

$$S_n = \sqrt{\frac{1}{n-1}\sum_{i=1}^{n}(X_i - \bar{X})^2}$$

标准差在数值上等于方差的开方，二者都是用来反映数据取值的离散（变异）程度的。标准差的量纲与数据的量纲相同。

最大值（Maximum）

$$X_{\max} = \max(X_1, X_2, \cdots, X_n)$$

最小值（Minimum）

$$X_{\min} = \min(X_1, X_2, \cdots, X_n)$$

7.3.1 标准化

标准化操作的定义很简单，一个表的某一列数据就是$X_1, X_2, \cdots, X_n$，对其进行变换

$$X_i' = \frac{X_i - \bar{X}}{S_n}$$

得到的数据$X_1', X_2', \cdots, X_n'$就是标准化处理的结果。将此结果数据存入结果表中。

标准化操作的结果，也被称作标准分数（Standard Score）或Z分数（Z Score）。

对一个服从一般正态分布的随机变量$X \sim N(\mu, \sigma^2)$，使用标准化操作转换成

$$Z = \frac{X - \mu}{\sigma}$$

则Z为一个服从标准正态分布的随机变量，即$Z \sim N(0,1)$。

标准正态分布的密度分布函数如图 7-1 所示。这里有一些特点值得我们注意：

（1）密度函数相对于平均值所在的直线对称。平均值与它的众数（statistical mode）以及中位数（median）为同一数值，标准差为 1。

（2）68.268949%的面积在平均数左右的一个标准差范围内，即区间为[−1, 1]。

（3）95.449974%的面积在平均数左右的两个标准差范围内，即区间为[−2, 2]。

（4）99.730020%的面积在平均数左右的三个标准差范围内，即区间为[−3, 3]。

（5）99.993666%的面积在平均数左右的四个标准差范围内，即区间为[−4, 4]。

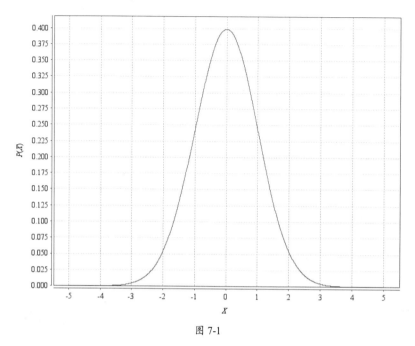

图 7-1

我们遇到的很多数据都近似服从正态分布，那些 Z Score 的绝对值很大时，其是异常点的可能性也很大。

示例代码如下。这里使用了 StandardScaler 组件，通过 fit 方式，计算所需的统计量，得到模型。然后，使用 transform 方法进行预测。对执行标准化操作前后的数据分别打印统计结果，以便对照、理解数据的变化。

```
source.lazyPrintStatistics("< Origin data >");

StandardScaler scaler = new StandardScaler().setSelectedCols(FEATURE_COL_NAMES);

scaler
    .fit(source)
    .transform(source)
    .lazyPrintStatistics("< after Standard Scale >");

BatchOperator.execute();
```

结果如下。在执行标准化操作后，四个数值列的均值都为 0，方差都是 1。

```
< Origin data >
| colName|count|missing| sum |mean|variance|min|max|
|--------|-----|-------|-----|----|--------|---|---|
```

```
|sepal_length|  150|      0|876.5|5.8433|   0.6857|4.3|7.9|
| sepal_width|  150|      0|458.1| 3.054|    0.188|  2|4.4|
|petal_length|  150|      0|563.8|3.7587|   3.1132|  1|6.9|
| petal_width|  150|      0|179.8|1.1987|   0.5824|0.1|2.5|
|    category|  150|      0|  NaN|   NaN|      NaN|NaN|NaN|
< after Standard Scale >
|     colName|count|missing|sum|mean|variance|     min|   max|
|------------|-----|-------|---|----|--------|--------|------|
|sepal_length|  150|      0| -0|  -0|       1| -1.8638|2.4837|
| sepal_width|  150|      0|  0|   0|       1| -2.4308|3.1043|
|petal_length|  150|      0|  0|   0|       1| -1.5635|1.7804|
| petal_width|  150|      0|  0|   0|       1| -1.4396|1.7052|
|    category|  150|      0|NaN| NaN|     NaN|     NaN|   NaN|
```

7.3.2 MinMaxScale

将最小值映射为 0，最大值映射为 1，然后按比例将所有数据映射到区间[0,1]。即对数据 $X_1, X_2, \cdots, X_n$，进行如下变换：

$$X_i' = \frac{X_i - X_{\min}}{X_{\max} - X_{\min}}$$

由最大值、最小值的定义，可知

$$0 \leqslant X_i - X_{\min} \leqslant X_{\max} - X_{\min}$$

所以，$0 \leqslant X_i' \leqslant 1$，即得到的数据 $X_1', X_2', \cdots, X_n'$ 都在 0 与 1 之间。这就是 "归一" 化处理的结果。再将此结果数据存入结果表中。

示例代码如下。这里使用了 MinMaxScaler，通过 fit 方式，计算所需的统计量，得到模型。然后，使用 transform 方法进行预测。对执行该操作前后的数据分别打印统计结果，以便对照、理解数据的变化。

```
source.lazyPrintStatistics("< Origin data >");

MinMaxScaler scaler = new MinMaxScaler().setSelectedCols(FEATURE_COL_NAMES);

scaler
   .fit(source)
   .transform(source)
   .lazyPrintStatistics("< after MinMax Scale >");

BatchOperator.execute();
```

结果如下。在执行 MinMaxScale 操作后，四个数值列的最小值都为 0，最大值都是 1。

```
< Origin data >
|     colName|count|missing|   sum|  mean|variance|min|max|
|------------|-----|-------|------|------|--------|---|---|
|sepal_length|  150|      0| 876.5|5.8433|  0.6857|4.3|7.9|
| sepal_width|  150|      0| 458.1| 3.054|   0.188|  2|4.4|
|petal_length|  150|      0| 563.8|3.7587|  3.1132|  1|6.9|
| petal_width|  150|      0| 179.8|1.1987|  0.5824|0.1|2.5|
|    category|  150|      0|   NaN|   NaN|     NaN|NaN|NaN|

< after MinMax Scale >
|     colName|count|missing|    sum|  mean|variance|min|max|
|------------|-----|-------|-------|------|--------|---|---|
|sepal_length|  150|      0|64.3056|0.4287|  0.0529|  0|  1|
| sepal_width|  150|      0| 65.875|0.4392|  0.0326|  0|  1|
|petal_length|  150|      0|70.1356|0.4676|  0.0894|  0|  1|
| petal_width|  150|      0|68.6667|0.4578|  0.1011|  0|  1|
|    category|  150|      0|    NaN|   NaN|     NaN|NaN|NaN|
```

7.3.3 MaxAbsScale

以 0 为中心,将数据缩放至区间[-1,1],在整个变化过程中,数值 0 不会改变。
定义所有X_i的绝对值的最大值为X_{maxabs},即

$$X_{maxabs} = \max(|X_1|, |X_2|, \cdots, |X_n|)$$

如果已经计算出了X_{min}和X_{max},也可将X_{maxabs}的计算简化为

$$X_{maxabs} = \max(|X_{min}|, |X_{max}|)$$

绝对值最大值的伸缩变换为

$$X'_i = \frac{X_i}{X_{maxabs}}$$

示例代码如下。这里使用了 MaxAbsScaler,通过 fit 方式,计算所需的统计量,得到模型。然后,使用 transform 方法进行预测,并对执行该操作前后的数据分别打印统计结果。

```
source.lazyPrintStatistics("< Origin data >");

MaxAbsScaler scaler = new MaxAbsScaler().setSelectedCols(FEATURE_COL_NAMES);

scaler
    .fit(source)
```

```
.transform(source)
.lazyPrintStatistics("< after MaxAbs Scale >");

BatchOperator.execute();
```

结果如下。在执行 MaxAbsScaler 操作后，四个数值列的最大值都变换为 1，各列的最小值也相应地按比例变化。

```
< Origin data >
|      colName|count|missing|   sum|  mean|variance|min|max|
|-------------|-----|-------|------|------|--------|---|---|
| sepal_length|  150|      0| 876.5|5.8433|  0.6857|4.3|7.9|
|  sepal_width|  150|      0| 458.1| 3.054|   0.188|  2|4.4|
| petal_length|  150|      0| 563.8|3.7587|  3.1132|  1|6.9|
|  petal_width|  150|      0| 179.8|1.1987|  0.5824|0.1|2.5|
|     category|  150|      0|   NaN|   NaN|     NaN|NaN|NaN|

< after MaxAbs Scale >
|      colName|count|missing|     sum|  mean|variance|   min|max|
|-------------|-----|-------|--------|------|--------|------|---|
| sepal_length|  150|      0|110.9494|0.7397|   0.011|0.5443|  1|
|  sepal_width|  150|      0|104.1136|0.6941|  0.0097|0.4545|  1|
| petal_length|  150|      0| 81.7101|0.5447|  0.0654|0.1449|  1|
|  petal_width|  150|      0|   71.92|0.4795|  0.0932|  0.04|  1|
|     category|  150|      0|     NaN|   NaN|     NaN|   NaN|NaN|
```

7.4 向量的尺度变换

本节将介绍四种关于向量的尺度变换：StandardScale、MinMaxScale、MaxAbsScale 和 Normalize。前三种变换（StandardScale、MinMaxScale、MaxAbsScale）在确定变换尺度的时候，都考虑到了整个向量集合，并针对每个向量分量计算整体的变换尺度。但 Normalize（规范化）操作是对于一个向量来说的，其无须考虑其他向量。

7.4.1 StandardScale、MinMaxScale、MaxAbsScale

StandardScale、MinMaxScale、MaxAbsScale 这三个变换均分别对向量中的每个分量进行变换。单个分量的变换可直接参考 7.3 节的相应内容，这里不再赘述。

我们构造一个示例，对原始数据分别做这三种变换。从变换前后的统计量变化上，可以看出其变换效果。仍然使用前面所用的 iris 数据集，但需要将多列的数据转化为一个向量。使用向量组装组件 VectorAssemblerBatchOp，选择将所有特征列组装成向量，并保留标签列。使用 VectorSummarizerBatchOp 计算向量中各个分量的统计指标。先对原始数据进行统计，随后分别对向量列进行 StandardScale、MinMaxScale、MaxAbsScale 操作，并分别统计操作结果。

```java
new CsvSourceBatchOp()
    .setFilePath(DATA_DIR + ORIGIN_FILE)
    .setSchemaStr(SCHEMA_STRING)
    .link(
        new VectorAssemblerBatchOp()
            .setSelectedCols(FEATURE_COL_NAMES)
            .setOutputCol(VECTOR_COL_NAME)
            .setReservedCols(LABEL_COL_NAME)
    );

source.link(
    new VectorSummarizerBatchOp()
        .setSelectedCol(VECTOR_COL_NAME)
        .lazyPrintVectorSummary("< Origin data >")
);

new VectorStandardScaler()
    .setSelectedCol(VECTOR_COL_NAME)
    .fit(source)
    .transform(source)
    .link(
        new VectorSummarizerBatchOp()
            .setSelectedCol(VECTOR_COL_NAME)
            .lazyPrintVectorSummary("< after Vector Standard Scale >")
    );

new VectorMinMaxScaler()
    .setSelectedCol(VECTOR_COL_NAME)
    .fit(source)
    .transform(source)
    .link(
        new VectorSummarizerBatchOp()
            .setSelectedCol(VECTOR_COL_NAME)
            .lazyPrintVectorSummary("< after Vector MinMax Scale >")
    );

new VectorMaxAbsScaler()
    .setSelectedCol(VECTOR_COL_NAME)
    .fit(source)
    .transform(source)
    .link(
        new VectorSummarizerBatchOp()
            .setSelectedCol(VECTOR_COL_NAME)
            .lazyPrintVectorSummary("< after Vector MaxAbs Scale >")
    );

BatchOperator.execute();
```

运行结果如下。执行 StandardScale 操作后，向量的四个分量的均值都为 0，方差都是 1；执行 MinMaxScale 操作后，四个分量的最小值都为 0，最大值都是 1；执行 MaxAbsScaler 操作后，四个数值列的最大值都变换为 1，各列的最小值也相应地按比例变化。

```
< Origin data >
------------------------------ Summary ------------------------------

id|count|sum|mean|variance|standardDeviation|min|max|normL1|normL2
--|-----|---|----|--------|-----------------|---|---|------|------
0|150|876.5000|5.8433|0.6857|0.8281|4.3000|7.9000|876.5000|72.2762
1|150|458.1000|3.0540|0.1880|0.4336|2.0000|4.4000|458.1000|37.7763
2|150|563.8000|3.7587|3.1132|1.7644|1.0000|6.9000|563.8000|50.8232
3|150|179.8000|1.1987|0.5824|0.7632|0.1000|2.5000|179.8000|17.3868
< after Vector Standard Scale >
------------------------------ Summary ------------------------------

id|count|sum|mean|variance|standardDeviation|min|max|normL1|normL2
--|-----|---|----|--------|-----------------|---|---|------|------
0|150|-0.0000|-0.0000|1.0000|1.0000|-1.8638|2.4837|124.5472|12.2066
1|150|-0.0000|-0.0000|1.0000|1.0000|-2.4308|3.1043|115.2321|12.2066
2|150|0.0000|0.0000|1.0000|1.0000|-1.5635|1.7804|132.7847|12.2066
3|150|-0.0000|-0.0000|1.0000|1.0000|-1.4396|1.7052|129.5140|12.2066
< after Vector MinMax Scale >
------------------------------ Summary ------------------------------

id|count|sum|mean|variance|standardDeviation|min|max|normL1|normL2
--|-----|---|----|--------|-----------------|---|---|------|------
0|150|64.3056|0.4287|0.0529|0.2300|0.0000|1.0000|64.3056|5.9541
1|150|65.8750|0.4392|0.0326|0.1807|0.0000|1.0000|65.8750|5.8132
2|150|70.1356|0.4676|0.0894|0.2991|0.0000|1.0000|70.1356|6.7911
3|150|68.6667|0.4578|0.1011|0.3180|0.0000|1.0000|68.6667|6.8191
< after Vector MaxAbs Scale >
------------------------------ Summary ------------------------------

id|count|sum|mean|variance|standardDeviation|min|max|normL1|normL2
--|-----|---|----|--------|-----------------|---|---|------|------
0|150|110.9494|0.7397|0.0110|0.1048|0.5443|1.0000|110.9494|9.1489
1|150|104.1136|0.6941|0.0097|0.0985|0.4545|1.0000|104.1136|8.5855
2|150|81.7101|0.5447|0.0654|0.2557|0.1449|1.0000|81.7101|7.3657
3|150|71.9200|0.4795|0.0932|0.3053|0.0400|1.0000|71.9200|6.9547
```

7.4.2　正则化

对于向量 $x = (x_1, x_2, \cdots, x_m)^T$，$p$-范数的定义如下：

$$\|x\|_p = (|x_1|^p + |x_2|^p + \cdots + |x_m|^p)^{\frac{1}{p}}$$

当 p 取 1、2、∞ 时，分别对应以下三种最简单的情形：

- 1-范数：$\|x\|_1 = |x_1| + |x_2| + \cdots + |x_m|$
- 2-范数：$\|x\|_2 = (x_1^2 + x_2^2 + \cdots + x_m^2)^{\frac{1}{2}}$
- ∞-范数：$\|x\|_\infty = \max(|x_1|, |x_2|, \cdots, |x_m|)$

正则化指的是根据用户指定的 p 值，让每个变换后的向量的 p-范数都为 1。显然，经过正则化操作，原先向量的 0 值分量经过变换后还是 0，并没有破坏向量的稀疏性。

为了方便检验结果，我们选择了 1-范数，然后选择 5 条数据进行打印。示例代码如下：

```
source
    .link(
        new VectorNormalizeBatchOp()
            .setSelectedCol(VECTOR_COL_NAME)
            .setP(1.0)
    )
    .firstN(5)
    .print();
```

运行结果如下。在此可以看出四个分量的和为 1，符合我们选择 1-范数进行正则化的预期。

```
category|vec
--------|---
Iris-setosa|0.5 0.3431372549019608 0.13725490196078433 0.019607843137254905
Iris-setosa|0.5157894736842106 0.3157894736842105 0.14736842105263157 0.021052631578947368
Iris-setosa|0.5 0.3404255319148936 0.13829787234042554 0.02127659574468085
Iris-setosa|0.4893617021276596 0.3297872340425532 0.15957446808510642 0.021276595744680854
Iris-setosa|0.4901960784313726 0.35294117647058826 0.13725490196078433 0.019607843137254905
```

7.5 缺失值填充

缺失值填充指的是将空值或者一个指定的值替换为最大值、最小值、均值或者一个自定义的值。缺失值填充组件的参数如表 7-2 所示。可以使用如下填充方法填充值：

- 通过给定一个缺失值的配置列表，实现将输入表的缺失值用指定的值来填充的目的。
- 将数值型的空值替换为最大值、最小值、均值或者一个自定义的值。
- 将字符型的空值、空字符串或其他指定值替换为一个自定义的值。
- 缺失值若选择空字符，则填充的目标列应是 string 类型的值。

标准化操作及归一化操作需要所处理数据列的整体统计信息，标准化操作需要均值和方差，归一化操作需要最大值和最小值；在进行缺失值填充时，如果需要使用数据的最大值、最小值或均值，也需要事先计算出统计量。

表 7-2　缺失值填充组件的参数

名称	描述
strategy	【可选】缺失值填充的规则 缺失值填充的规则，支持使用 mean、max、min 或者 value 等值进行填充。选择 value 时，需要读取 fillValue 的值。其默认值为 mean
fillValue	【可选】填充缺失值 自定义的填充值。当参数 strategy 为 value 时，读取 fillValue 的值
selectedCols	【必填】选择的列名 计算列对应的列名列表
outputCols	【可选】输出结果列的列名数组 输出结果列的列名数组，默认值为 null，即在原数据列上直接填充

下面通过示例进行说明，代码如下。首先构造一个包含 5 条数据的数据集，最后一条数据的各项都是 null。第一列是字符串类型的，选择指定填充值为 "e"；第二列和第三列都选择使用均值进行填充。

```
Row[] rows = new Row[] {
  Row.of("a", 10.0, 100),
  Row.of("b", -2.5, 9),
  Row.of("c", 100.2, 1),
  Row.of("d", -99.9, 100),
  Row.of(null, null, null)
};

MemSourceBatchOp source
  = new MemSourceBatchOp(rows, new String[] {"col1", "col2", "col3"});

source.lazyPrint(-1, "< origin data >");

Pipeline pipeline = new Pipeline()
  .add(
    new Imputer()
      .setSelectedCols("col1")
      .setStrategy(Strategy.VALUE)
      .setFillValue("e")
  )
  .add(
    new Imputer()
      .setSelectedCols("col2", "col3")
```

```
    .setStrategy(Strategy.MEAN)
);
pipeline.fit(source)
    .transform(source)
    .print();
```

在此可以看到运行情况，原始数据如表 7-3 所示。第 5 行的数据都为 null，需要被填充。

表 7-3　带缺失值的原始数据

col1	col2	col3
a	10.0000	100
b	-2.5000	9
c	100.2000	1
d	-99.9000	100
null	null	null

填充结果如表 7-4 所示。

表 7-4　缺失值填充后的结果

col1	col2	col3
a	10.0000	100
b	-2.5000	9
c	100.2000	1
d	-99.9000	100
e	1.9500	52

第一列正如组件所设定，填充了"e"；第二列为双精度浮点类型的数值，填充值为均值 1.9500；第三列为整型的数值，其均值为 210/4，按整数计算规则得到 52，故填充值为 52。

8 线性二分类模型

本章将介绍纸钞的真假判断示例。根据给定钞票照片的 4 个度量值，预测其是真钞还是假钞。从机器学习的角度来看，这是典型的分类问题，而且分类目标为两个，即二分类问题。这里使用的数据特征已经被很好地数字化，可以直接套用常用的逻辑回归、线性 SVM 模型进行训练、预测。

本章的开头会介绍一些理论知识。从线性模型的角度，可以将逻辑回归模型和线性 SVM 模型（线性支持向量机模型）统一起来，这样更容易理解它们之间的关联与差异。有了二分类模型后，还需要知道如何评估该模型的效果，8.2 节会具体介绍相关的指标和曲线。

从 8.3 节开始，通过纸钞的真假判断示例，展示数据探索、训练线性模型、使用模型进行预测的完整过程。通过对比评估指标，我们可以看到调整模型参数、特征处理对模型效果的影响。最后介绍线性模型的一个扩展：因子分解机（FM）模型。

8.1 线性模型的基础知识

本节首先从如何评估预测值与真实值入手，介绍线性模型，再说明其常用的两个具体形式：逻辑回归模型与线性 SVM 模型。

8.1.1 损失函数

损失函数（Loss Function）或代价函数（Cost Function）用来度量预测值 $f(x)$ 与真实值 y 的不一致程度，将预测误差量化为非负实数，记作 $L(y, f(x))$。

分类损失函数的情况比较复杂，为了方便描述问题，我们只考虑分类标签值为±1的情况，$y = \pm 1$。常用的分类损失函数如下：

- 0-1 损失函数（0-1 Loss Function）

$$L(y, f(x)) = \begin{cases} 1, & y \cdot f(x) < 0 \\ 0, & y \cdot f(x) \geqslant 0 \end{cases}$$

- 对数损失函数（Logarithmic Loss Function）

$$L(y, f(x)) = \log_2(1 + \exp(-y \cdot f(x)))$$

注意：这里是以 2 为底的对数，当 $y \cdot f(x) = 0$ 时，可以保证 $L(y, f(x))$ 的值为 1，与大部分损失函数在 0 点的值相同。

- 指数损失函数（Exponential Loss Function）

$$L(y, f(x)) = \exp\{-y \cdot f(x)\}$$

- 合页损失函数（Hinge Loss Function）

$$L(y, f(x)) = \max(0, 1 - y \cdot f(x))$$

当 $f(x)$ 的符号与 y 相同，且 $|f(x)| \geqslant 1$ 时，损失函数的值为 0。当 $y = +1$，且 $f(x) \geqslant 1$ 时，对应的损失函数 $L(y, f(x)) = 0$；当 $y = -1$，且 $f(x) \leqslant -1$ 时，对应的损失函数 $L(y, f(x)) = 0$。

- 感知损失函数（Perceptron Loss Function）

$$L(y, f(x)) = \max(0, -y \cdot f(x))$$

该函数可以被看作合页损失函数（Hinge Loss Function）的变种。如果 $f(x)$ 的符号与 y 相同，损失函数的值为 0。

我们可以看到，在前面定义的分类损失函数中，y 和 $f(x)$ 往往是作为一个整体 $y \cdot f(x)$ 出现的。$y \cdot f(x)$ 为正，意味着预测正确；$y \cdot f(x)$ 为负，则意味着预测错误。负值越大，意味着模型在当前数据中的预测结果越差。

我们将 $y \cdot f(x)$ 整体看作一个自变量，则可以得到各损失函数的函数曲线，如图 8-1 所示。感兴趣的读者可以自己尝试作图，访问链接 8-1，在函数输入框中填如下表达式：

```
0.5*(1+sign(-x)),log(1+exp(-x))/log(2),exp(-x),max(0,1-x),max(0,-x)
```

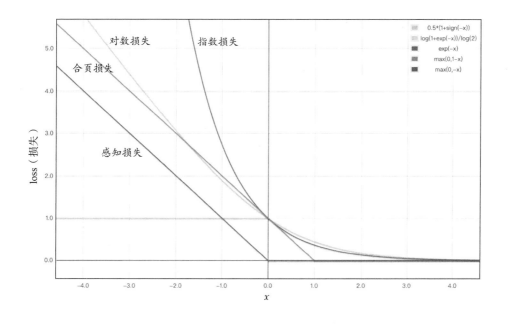

图 8-1

从图 8-1 中看到，在 0 点右侧，各曲线迅速趋近 X 轴，即损失值趋近 0。而在 0 点左侧，各损失函数的增长情况明显不同；损失值增长速度的大小关系如下：

$$\text{指数损失} > \text{对数损失} > \text{合页损失} > \text{感知损失} > \text{0-1 损失}$$

我们常用的逻辑回归算法使用的是对数损失函数，线性支持向量机算法使用的是合页损失函数，二者都是增长速度位于中位的损失函数。

8.1.2 经验风险与结构风险

下面先介绍一些理论上的定义。风险函数（Risk Function），又称期望损失（Expected Loss）。顾名思义，"期望损失"就是损失函数计算出来的损失值的数学期望，即

$$\int_{x \times y} L(y, f(x)) P(x, y) \, \mathrm{d}x \mathrm{d}y$$

其中，$P(x,y)$ 为属性变量 x 与标签变量 y 的联合概率。但实际上联合概率 $P(x,y)$ 是未知的（若已知，则可以直接求出条件概率 $P(y|x)$，也无须再训练模型 $f(x)$ 了）。这就需要一个更实用、

更可操作的指标。

设样本集有N个样本，即$\{(x_1, y_1), (x_2, y_2), (x_3, y_3), \cdots, (x_n, y_n)\}$，则模型$f(x)$关于此样本集的平均损失被称为经验风险（Empirical Risk）或经验损失（Empirical Loss），即

$$L(f) = \frac{1}{n}\sum_{i=1}^{n} L(y_i, f(x_i))$$

模型的经验风险是一个量化的指标，可以将经验风险最小的模型看作最优的模型，即模型寻优的策略是经验风险最小的。

引入正则（Regularizer）项或者罚（Penalty）项的概念，记作$\Omega(f)$，用来表示模型的复杂度。比如，在线性场景下，$\Omega(f)$为模型中各权重参数构成向量的范数。

结构风险，记为$J(f)$，是经验风险与正则项的加权和。即

$$J(f) = L(f) + \lambda \cdot \Omega(f)$$

其中λ被称为超参数（Hyperparameter），可用来平衡经验风险$L(f)$和正则项$\Omega(f)$对于目标函数的影响。这里$\lambda \geq 0$；λ增大，则正则项$\Omega(f)$的重要程度增加；λ越趋近0，则正则项$\Omega(f)$的重要程度越弱。特别地，当$\lambda = 0$时，则可以完全忽略正则项的影响。

8.1.3 线性模型与损失函数

考虑线性模型的情形。设训练集有m个特征$x = \{x_1, \cdots, x_m\}$，分类变量y的取值范围为$\{0, 1\}$，对于权重参数$w = \{w_0, w_1, \cdots, w_m\}$，线性函数为

$$\eta(w, x) = w_0 + w_1 x_1 + \cdots + w_m x_m$$

所以，线性模型的预测值函数为

$$f(x^{(i)}) = \eta(w, x^{(i)}) = w_0 + w_1 x_1^{(i)} + \cdots + w_m x_m^{(i)}$$

即，线性模型的预测值函数是由参数w定义的，前面介绍的不少概念定义都可以被转化为与参数w的关系，如下所示：

- 损失函数$L(y_i, f(x_i)) = L(w, x^{(i)}, y^{(i)})$
- 经验损失函数$L(f) = L(w)$
- 正则项$\Omega(f) = \Omega(w)$
- 结构风险$J(f) = J(w)$

为了表述起来更简单、直接，我们使用如下定义：

经验损失函数为

$$L(w) = \frac{1}{n}\sum_{i=1}^{n} L(w, x^{(i)}, y^{(i)})$$

其结构风险函数为

$$J(w) = L(w) + \lambda \cdot \Omega(w)$$

$\Omega(w)$为衡量模型复杂程度的正则项，经常使用w的 L1 范数和 L2 范数。

- L1 范数：

$$\Omega(w) = \|w\|_1 = |w_0| + |w_1| + |w_2| + \cdots + |w_m|$$

- L2 范数：

$$\Omega(w) = \|w\|_2^2 = w_0^2 + w_1^2 + w_2^2 + \cdots + w_m^2$$

下面将针对两个常用的线性分类模型（逻辑回归模型、线性支持向量机模型），详细介绍其损失函数。

8.1.4 逻辑回归与线性支持向量机（Linear SVM）

逻辑回归（Logistic Regression）算法使用的是对数损失函数$\log_2(1 + e^{-x})$，分类标签取值$y = \pm 1$时，可得

$$L(w, x^{(i)}, y^{(i)}) = \log_2\left(1 + \exp\left(-y^{(i)} \cdot \eta(w, x^{(i)})\right)\right)$$

逻辑回归模型的计算输出是分类值为 1 的概率：

$$P(1|x) = \frac{1}{1 + \exp\{-\eta(w, x)\}}$$

最终的模型预测结果，会根据概率值是否大于 0.5，输出标签值$+1$或者-1。

注意：在一些介绍逻辑回归算法的资料里，也有另外的表示方法，其本质是对取值范围的假设不同。

当标签取值$y = \{0,1\}$时，由上面的正负例对应的损失值，可对应到正负例为$y = 1$和$y = 0$的情形：

- $y = 1$，为正例时，损失值为$\log(1 + \exp(-\eta(w, x)))$。
- $y = 0$，为负例时，损失值为$\log(1 + \exp(\eta(w, x)))$。

可以用一个表达式，将这两种情况表示出来：

$$y \cdot \log(1 + \exp(-\eta(w, x))) + (1 - y) \log(1 + \exp(\eta(w, x)))$$

线性支持向量机（Linear SVM）算法使用的是合页损失函数$\max(0, 1 - x)$，分类标签取值$y = \pm 1$时

$$L(w, x^{(i)}, y^{(i)}) = \max\left(0, 1 - y^{(i)} \cdot \eta(w, x^{(i)})\right)$$

线性支持向量机模型的计算输出为权重与特征值的乘积和：

$$f(x) = \eta(w, x) = w_0 + w_1 x_1 + \cdots + w_m x_m$$

最终的模型预测结果，会根据计算值是否大于 0，输出标签值+1或者−1。

特别地，取线性支持向量机的正则项为 L2 范数正则项，则线性支持向量机的结构风险为

$$J(w) = L(w) + \lambda \cdot \|w\|_2^2$$

如果数据集正好可以被完全区分开，则可以对应一个较直观的描述。
我们将$x^{(i)}$对应到 $m + 1$ 维空间的点$z^{(i)} = (1, x_1^{(i)}, \cdots, x_m^{(i)})$，则

$$f(x^{(i)}) = \eta(w, x^{(i)}) = w_0 + w_1 x_1^{(i)} + \cdots + w_m x_m^{(i)} = w \cdot z^{(i)}$$

数据集正好可以被完全区分开。这就意味着，可以找到w，使下列情形成立：
- 对于所有$y^{(i)} = -1$，对应的$x^{(i)}$满足$f(x^{(i)}) \leqslant -1$。
- 对于所有$y^{(i)} = +1$，对应的$x^{(i)}$满足$f(x^{(i)}) \geqslant 1$。

对应到空间的点上，就是
- 对于所有$y^{(i)} = -1$，对应的$z^{(i)}$满足$w \cdot z^{(i)} \leqslant -1$。
- 对于所有$y^{(i)} = +1$，对应的$z^{(i)}$满足$w \cdot z^{(i)} \geqslant 1$。

即，$\{z^{(1)}, \ldots, z^{(n)}\}$被超平面$w \cdot z = 0$分隔开。更确切地说，所有满足$y^{(i)} = -1$的点都在$w \cdot z = -1$的下方；所有满足$y^{(i)} = +1$的点都在$w \cdot z = 1$的上方。超平面$w \cdot z = -1$与$w \cdot z = 1$之间的距离越大，则模型的分隔效果越好。

我们注意到，$\frac{w}{\|w\|_2}$的 L2 范数为 1，且是与超平面垂直的向量，并且

$$w \cdot \left(z + \frac{w}{\|w\|_2} \cdot \frac{1}{\|w\|_2}\right) = w \cdot z + \frac{w \cdot w}{\|w\|_2^2} = w \cdot z + 1$$

有了这个等式，我们就很容易理解超平面$w \cdot z = -1$与$w \cdot z = 1$之间的距离为

$$\frac{2}{\|w\|_2}$$

要使距离最大，就是要让$\frac{2}{\|w\|_2}$最大，这等价于求$\|w\|_2$的最小值。

我们回头再看线性支持向量机的结构风险：

$$J(w) = L(w) + \lambda \cdot \|w\|_2^2$$

在数据集能被完全区分开的情况下，在迭代求解的过程中，会找到适合的一系列w，使得损失函数为0，即$L(w) = 0$，随后的迭代求解最小$J(w)$就相当于在求$\|w\|_2$的最小值，从而使分隔超平面的间隔最大化。

在实际的分类问题中，标签值往往为"是/否""通过/不通过"等。Alink的逻辑回归组件和线性SVM组件都会在训练前、预测后自动进行实际标签值与±1的转换，用户使用时无须关注标签值是否为±1形式的问题。

逻辑回归使用的对数损失函数$\log_2(1 + e^{-x})$与线性支持向量机使用的合页损失函数$\max(0, 1 - x)$的函数对比图像如图8-2所示。在图8-2的左图中，可以看到在0点附近两个函数的差异；图8-2的右图则体现了在自变量x为更大的负值时，损失函数间的差异。

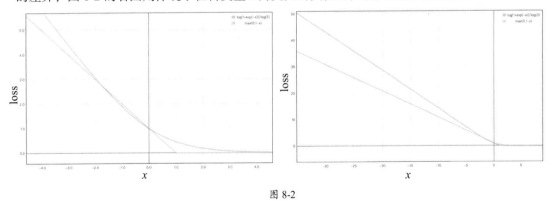

图 8-2

8.2 二分类评估方法

对于二分类问题，即实际分类值为"是"（Yes）或者"否"（No），机器学习模型预测会给出阳性（Positive）或阴性（Negative）的判断，从而产生四种不同的情况：真阳性（True Positive，

TP）、假阳性（False Positive，FP）、真阴性（True Negative，TN）和假阴性（False Negative，FN），如表 8-1 所示。

表 8-1 实际值与预测结果

预测结果	实际值	
	是（Yes）	否（No）
阳性（Positive）	真阳性（TP）	假阳性（FP）
阴性（Negative）	假阴性（FN）	真阴性（TN）

Yes 实例指的是实际值为 Yes 的实例；No 实例指的是实际值为 No 的实例。考虑到实例可能被预测为阳性或阴性，结合表 8-1 的分类，Yes 实例是真阳性或假阴性的实例；No 实例是假阳性或真阴性的实例。

8.2.1 基本指标

由基本概念的定义，很容易理解下面这些基本指标。

真阳性率（True Positive Rate，TPR），也被称为灵敏度（Sensitivity）、召回率（Recall），计算公式为

$$TPR = \frac{TP}{TP + FN}$$

描述的是分类器所识别出阳性的 Yes 实例占所有 Yes 实例的比例。

真阴性率（True Negative Rate，TNR），也被称为特异度（Specificity），计算公式为

$$TNR = \frac{TN}{FP + TN}$$

假阳性率（False Positive Rate，FPR），也被称为 1−特异度（1−Specificity），其计算公式为

$$FPR = \frac{FP}{FP + TN} = 1 - \frac{TN}{FP + TN} = 1 - TNR$$

假阳性率计算的是分类器错认为阳性的 No 实例占所有 No 实例的比例。

准确率（Precision），或被称为 Positive Predictive Value（PPV），定义如下：

$$PPV = \frac{TP}{TP + FP}$$

Negative Predictive Value（NPV）的定义如下：

$$\text{NPV} = \frac{\text{TN}}{\text{TN} + \text{FN}}$$

False Discovery Rate（FDR）的定义如下：

$$\text{FDR} = \frac{\text{FP}}{\text{FP} + \text{TP}} = 1 - \text{PPV}$$

Miss Rate 或 False Negative Rate（FNR）的定义如下：

$$\text{FNR} = \frac{\text{FN}}{\text{FN} + \text{TN}}$$

将上面的定义汇总到表 8-2 中（注意，有几个上面未涉及的使用频次较低的定义也被放到了表 8-2 中，便于读者对比、清晰地了解各指标间的关系）。

表 8-2 指标汇总

预测结果	实际值			
	是（Yes） TP + FN	否（No） FP + TN		
Positive （阳性） TP + FP	True Positive（TP） （真阳性） TP	False Positive（FP） （假阳性） FP	Positive Predictive Value（PPV） 准确率（Precision） $\text{PPV} = \frac{\text{TP}}{\text{TP} + \text{FP}}$	False Discovery Rate（FDR） $\text{FDR} = \frac{\text{FP}}{\text{TP} + \text{FP}}$
Negative （阴性） FN + TN	False Negative（FN） （假阴性） FN	True Negative（TN） （真阴性） TN	False Omission Rate（FOR） $\text{FOR} = \frac{\text{FN}}{\text{FN} + \text{TN}}$	Negative Predictive Value（NPV） $\text{NPV} = \frac{\text{TN}}{\text{TN} + \text{FN}}$
	True Positive Rate（TPR） 灵敏度（Sensitivity） 召回率（Recall） $\text{TPR} = \frac{\text{TP}}{\text{TP} + \text{FN}}$	False Positive Rate（FPR） 1-特异度（1-Specificity） $\text{FPR} = \frac{\text{FP}}{\text{FP} + \text{TN}}$	Positive Likelihood Ratio（LR+） $\text{LR+} = \frac{\text{TPR}}{\text{FPR}}$	Diagnostic Odds Ratio（DOR） $\text{DOR} = \frac{\text{LR+}}{\text{LR-}}$
	False Negative Rate（FNR） 丢失率（Miss Rate） $\text{FNR} = \frac{\text{FN}}{\text{TP} + \text{FN}}$	True Negative Rate（TNR） 特异度（Specificity） $\text{TNR} = \frac{\text{TN}}{\text{FP} + \text{TN}}$	Negative Likelihood Ratio（LR−） $\text{LR-} = \frac{\text{FNR}}{\text{TNR}}$	

在医学上，灵敏度（Sensitivity）指的是在实际的患病人群中，被诊断为阳性的比例；特异度（Specificity）指的是在实际没有患病的人群中，被诊断为阴性的比例。在信息检索领域，召回率（Recall）指的是检索到的相关文件数量与相关文件总数量的比值；准确率（Precision）指的是检索到的相关文件数量与检索到的文件总数量的比值。

8.2.2 综合指标

1. AUC指标

AUC（Area Under the Curve）指的是 ROC 曲线下方的面积。ROC 曲线在 8.2.3 节中会具体介绍。随机分类的 AUC 为 0.5，而完美分类的 AUC 等于 1；实际上大多数分类模型的 AUC 介于 0.5 和 1 之间。

2. 精确度（Accuracy）

精确度（Accuracy，缩写为 ACC）被定义为预测正确的样本在整个样本中的比例，即

$$ACC = \frac{TP + TN}{TP + FN + FP + TN}$$

注意：精确度是一个很容易理解，并且常用的指标。读者要了解其局限性，避免在某些场景下误导自己的判断。比如，这里以选择 100 个样本为例，表 8-3 所示的模型预测情况如下：我们看到数据集中的负样本（实际值为 No 的样本）在数量上占优，有 95 个，达到了样本总数的 95%。分类模型只是简单地将所有样本都预测为阴性（Negative），精确度 Accuracy 就可以达到 95%，但对我们更为关注的正样本（实际值为 Yes 的样本），计算召回率为 0%。显然，这个模型并不是我们想要的。

表 8-3 模型预测情况示例

单位：样本个数

预测结果	实际值		合计
	是（Yes）	否（No）	
阳性（Positive）	0	0	0
阴性（Negative）	5	95	100
合计	5	95	100

3. F_1-Score

下面先介绍一个更泛化的概念，F-Score（F_a）又被称为 F-Measure，可用来拟合准确率（Precision）与召回率（Recall），即将准确率与召回率的加权调和平均作为指标：

$$F_a = \frac{1}{\frac{1}{(a^2+1)} \times \frac{1}{\text{Precision}} + \frac{a^2}{(a^2+1)} \times \frac{1}{\text{Recall}}} = \frac{(a^2+1) \times \text{Precision} \times \text{Recall}}{a^2 \times \text{Precision} + \text{Recall}}$$

当参数 $a = 1$ 时,就是最常见的 F_1-Score,即

$$F_1 = \frac{2 \times \text{Precision} \times \text{Recall}}{\text{Precision} + \text{Recall}}$$

由于 $\text{Precision} = \frac{\text{TP}}{\text{TP+FP}}$,$\text{Recall} = \frac{\text{TP}}{\text{TP+FN}}$

因此,又可以写作

$$F_1 = \frac{2\text{TP}}{2\text{TP} + \text{FP} + \text{FN}}$$

F_1 指标(其也常常被称作 F1)指的是准确率与召回率的调和平均值(Harmonic Mean)。调和平均值又被称为倒数平均数,是总体各统计变量倒数的算术平均数的倒数。

我们做一下类比,以便轻松理解 F_1 与准确率和召回率的关系。小学数学中的有些应用题,实际就是调和平均值的实际应用示例。例题如下:A、B 两地分别位于河流的上下游,路程为 S,由于水流的关系,A 地到 B 地的速度为 v_1,B 地到 A 地的速度为 v_2,求 A 地到 B 地再返回 A 地的平均速度 x。

由

$$\frac{2S}{x} = \frac{S}{v_1} + \frac{S}{v_2}$$

可得:

$$x = \frac{2}{\frac{1}{v_1} + \frac{1}{v_2}}$$

所以,平均速度 x 为速度为 v_1 和 v_2 的调和平均值。

4. Kappa系数

1960 年 Cohen 等人提出用 Kappa 系数(用希腊字母 κ 表示)作为评判诊断试验一致性程度的指标。

对于二分类评估,可以使用 Kappa 系数来测量实际分类情况与预测结果的一致性程度。Kappa 系数定义为

$$\kappa = \frac{P_a - P_e}{1 - P_e}$$

其中 P_a 为精确度(Accuracy),表明实际分类情况与预测分类结果的实际一致率,即

$$P_a = \frac{TP + TN}{TP + FN + FP + TN}$$

P_e 为实际分类情况与预测分类结果的期望一致率，定义如下：

$$P_e = \frac{(TP + FN) \times (TP + FP) + (FP + TN) \times (FN + TN)}{(TP + FN + FP + TN) \times (TP + FN + FP + TN)}$$

Kappa 系数 κ 与精确度 P_a 的关系如下：

$$\kappa \leq P_a$$

证明过程很简单：

$$P_a - \kappa = P_a - \frac{P_a - P_e}{1 - P_e} = \frac{P_e(1 - P_a)}{1 - P_e}$$

由于 $0 \leq P_a \leq 1$，$0 \leq P_e \leq 1$，因此

$$\kappa \leq P_a$$

使用 Kappa 系数的建议参考标准如下：

- κ 的值域为 $[-1, +1]$。
- 若 $0.75 \leq \kappa$，说明一致性较好。
- 若 $0.4 \leq \kappa < 0.75$，说明一致性一般。
- 若 $\kappa < 0.4$，说明一致性较差。

前面介绍精确度（Accuracy）时曾讲过一个例子，如表 8-3 所示。将所有 100 个样本都预测为阴性（Negative），虽然计算出的精确度（Accuracy）很高，有 95%，但是我们认为其分类模型很差。现在计算该模型的 Kappa 系数。

计算

$$P_a = \frac{95+0}{100} = 0.95, \quad P_e = \frac{95 \times 100 + 5 \times 0}{100 \times 100} = 0.95$$

所以，Kappa 系数为

$$\kappa = \frac{P_a - P_e}{1 - P_e} = \frac{0.95 - 0.95}{1 - 0.95} = 0$$

我们看到，在这种情形下，模型的 Kappa 系数为 0。这说明精确度（Accuracy）和 Kappa 系数结合起来，对分类模型的评估更准确。

5. LogLoss

该指标与逻辑回归算法的经验损失函数相关。设分类标签取值 $y = \{0, 1\}$，$P^{(i)}$ 指的是，逻辑回归模型的计算输出是分类值为 1 的概率：

$$P^{(i)} = P(1|x^{(i)}) = \frac{1}{1 + \exp\{-\eta(w, x^{(i)})\}}$$

则，损失函数

$$\begin{aligned}&L(w, x^{(i)}, y^{(i)}) \\ &= y^{(i)} \cdot \log\left(1 + \exp\left(-\eta(w, x^{(i)})\right)\right) + (1 - y^{(i)}) \log\left(1 + \exp\left(\eta(w, x^{(i)})\right)\right) \\ &= -y^{(i)} \cdot \log\left(\frac{1}{1 + \exp(-\eta(w, x^{(i)}))}\right) - (1 - y^{(i)}) \log\left(1 - \frac{1}{1 + \exp(-\eta(w, x^{(i)}))}\right) \\ &= -y^{(i)} \cdot \log(P^{(i)}) - (1 - y^{(i)}) \log(1 - P^{(i)})\end{aligned}$$

所以，经验损失函数为

$$L(w) = \frac{1}{n} \sum_{i=1}^{n} L(w, x^{(i)}, y^{(i)}) = \frac{1}{n} \sum_{i=1}^{n} \left[-y^{(i)} \cdot \log(P^{(i)}) - (1 - y^{(i)}) \log(1 - P^{(i)}) \right]$$

此经验损失函数值就是 LogLoss 指标，即

$$\text{LogLoss} = \frac{1}{n} \sum_{i=1}^{n} \left[-y^{(i)} \cdot \log(P^{(i)}) - (1 - y^{(i)}) \log(1 - P^{(i)}) \right]$$

注意：对于分类标签 y 为其他取值的情况，在 Alink 二分类评估组件中可以输入参数 PositiveLabelValueString，即指定 $y^{(i)} = 1$ 所对应的标签值；在 Alink 的预测详情信息中包含了正、负分类标签值及其概率。

8.2.3 评估曲线

对于二分类问题（分类值为"是"（Yes）或者"否"（No）），使用机器学习模型可以给每个实例预测出一个数值（不同分类模型的返回值有差异，比如，使用逻辑回归模型返回的"是"（Yes）的概率，值域为[0, 1]；使用线性支持向量机模型得到的线性变换的值，值域为 $(-\infty, +\infty)$）。假设已确定一个阈值，比如 0.5，大于这个值的实例被预测为阳性，小于这个值

的实例则被预测为阴性,由公式可以计算出 FPR、TPR。如果增加阈值(如增加到 0.8),则会减少识别出的 Yes 实例,也就是降低了识别出的 Yes 实例占所有 Yes 实例的比例,即降低了 TPR;但同时也会有更少的 No 实例被预测为阳性,即降低了 FPR。如果减小阈值(如减少到 0.4),虽然能识别出更多的 Yes 实例,也就是提高了 TPR,但同时也将更多的 No 实例预测为阳性,即提高了 FPR。

根据我们的需求,如何确定最优的阈值?如何评估各种分类方法的优劣?

下面即将介绍的 ROC 曲线等提供了形象化的图形展示,便于我们做出决策。

1. ROC(Receiver Operating Characteristic)曲线

ROC(Receiver Operating Characteristic)曲线最初用来评价雷达性能,被称为接收者操作特性曲线。ROC 曲线指的是,根据一系列不同的阈值,得到相应的一系列二分类预测方式,每一个预测结果以一个点表示,每个点以真阳性率(TPR)为纵坐标,以假阳性率(FPR)为横坐标,再将所有的点连接起来,绘制成曲线。真阳性率代表获利能力,其值越高,获得的利益越多。假阳性率可以被看作成本,其值越高,成本就越高。在 ROC 曲线中,斜率较高的一段可获得较大的利益,同时付出较小的成本。

ROC 曲线的范围是由(0,1)、(1,1)、(1,0)、(0,0)四点构成的单位正方形区域。最好的可能预测方式在左上角,即(0,1)点,这代表 100% 灵敏(没有假阴性)和 100% 特异(没有假阳性)。一个完全随机预测会得到一条从左下到右上对角线(也叫无识别率线)上的一个点。一个最直观的采用随机预测的方式做决定的例子就是抛硬币。这条斜线将 ROC 空间划分为两个区域,在这条线以上的点代表一个好的分类结果,而在这条线以下的点代表差的分类结果。

为了将 ROC 曲线概括成单一的数量,可以选用曲线下方的面积(Area Under Curve, AUC)。一般来说,AUC 值越大,分类预测效果越好。AUC 也可以被看作,分类预测将任意抽取的 Yes 实例排列在任意抽取的 No 实例之前的概率。

为了便于理解该算法,下面引用一个例子(参见链接 8-2)。表 8-4 是一个逻辑回归预测结果,它将得到的实数值按从大到小的顺序划分成 10 个个数大致相同的部分。

表 8-4 逻辑回归预测结果

百分位数	实例数	Yes 实例数	(1-特异度)/%	敏感度/%
10	6180	4879	2.73	34.64
20	6180	2804	9.80	54.55
30	6180	2165	18.22	69.92
40	6180	1506	28.01	80.62
50	6180	987	38.90	87.62

续表

百分位数	实例数	Yes 实例数	(1-特异度)/%	敏感度/%
60	6180	529	50.74	91.38
70	6180	365	62.93	93.97
80	6180	294	75.26	96.06
90	6180	297	87.59	98.17
100	6177	258	100.00	100.00

将逻辑回归得到的结果按从大到小的顺序排列，若以前 10% 的数值作为阈值，即将前 10% 的实例都预测为阳性，共 6180 个。其中，Yes 实例的个数为 4879 个，占所有 Yes 实例的百分比如下：4879/14084×100%=34.64%（即敏感度）。另外，可得 6180−4879=1301 个 No 实例被错误地预测为阳性，占所有 No 实例的百分比如下：1301/47713×100%=2.73%（即 1-特异度）。以这两组值分别作为 x 值和 y 值，在 Excel 中制作散点图，得到 ROC 曲线，横轴为 1-特异度（%），纵轴为敏感度（%），如图 8-3 所示。

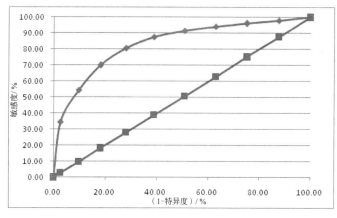

图 8-3

2. KS（Kolmogorov–Smirnov）曲线

对于每个阈值点 α_i，其真阳性率 $\text{TPR}_i = \frac{\text{TP}_i}{\text{TP}_i + \text{FN}_i}$ 和假阳性率 $\text{FPR}_i = \frac{\text{FP}_i}{\text{FP}_i + \text{TN}_i}$ 可以被看作正样本集合与负样本集合中，被预测为阳性的比例。如果采用随机采样模型，则正负样本集合中被采样到（即预测为阳性）的比例是相同的；这两个比例的差异越大，模型的效率就越高。

KS 曲线的横轴为 α_i 的值，纵轴为真阳性率和假阳性率的差值，即 $\text{TPR}_i - \text{FPR}_i$。

该曲线纵轴绝对值的最大值，就等于 KS（Kolmogorov-Smirnov）检验的统计量，所以，该曲线被命名为 KS 曲线。进一步，可以使用 KS 检验判断，在正样本集合与负样本集合中，样本评分的分布是否有显著差异。

3. PR（Precision Recall）曲线

PR 曲线很好理解，就如其名称，纵坐标和横坐标分别对应准确率（Precision）和召回率（Recall）。

对于每个阈值点 α_i，计算曲线上的点。

- 横坐标为召回率（Recall）：

$$\text{Recall}_i = \frac{TP_i}{TP_i + FN_i}$$

- 纵坐标为准确率（Precision）：

$$\text{Precision}_i = \frac{TP_i}{TP_i + FP_i}$$

很多算法会用指标 Precision 和 Recall 来评估自身的好坏，但事实上这两者在某些情况下是此消彼长的关系。可以通过绘制 Precision-Recall 曲线来分析、获取适合的阈值。

4. Lift（提升）图

如果我们从训练样本中随机采样，则得到的样本为阳性的概率，就是训练样本中的正样本的概率值（$\pi_+ = \frac{TP+FN}{TP+FN+FP+TN}$）。现在我们使用模型的预测效果应该更好，即预测为阳性的样本中是正样本的概率值（即 $\text{Precision} = \text{PPV} = \frac{TP}{TP+FP}$）会更大。Lift 就是这两个概率值的比值，表示提升的倍数，即

$$\text{Lift} = \frac{\text{Precision}}{\pi_+} = \frac{\frac{TP}{TP + FP}}{\frac{TP + FN}{TP + FN + FP + TN}}$$

显然，Lift 值越大，模型的效果就越好。当 Lift 值为 1 时，表示效果没有提升，依靠模型预测的效果与随机采样的效果是一样的。

5. CAP（Cumulative Accuracy Profile）曲线

CAP 曲线上各点坐标的定义如下：

- X 轴代表预测结果为阳性的样本占总样本的比例：

$$PR_i = \frac{TP_i + FP_i}{TP_i + FN_i + FP_i + TN_i}$$

- Y 轴代表正样本中被预测为阳性的比例，即真阳性率（TPR）、召回率（Recall）：

$$\text{Recall}_i = \text{TPR}_i = \frac{\text{TP}_i}{\text{TP}_i + \text{FN}_i}$$

对于每个阈值点 α_i，可以得到相应的 $\{\text{TP}_i, \text{FN}_i, \text{FP}_i, \text{TN}_i\}$。下面将 5 种曲线所需要的指标总结如下。

（1）真阳性率（TPR）、召回率（Recall）：

$$\text{Recall}_i = \text{TPR}_i = \frac{\text{TP}_i}{\text{TP}_i + \text{FN}_i}$$

（2）假阳性率（FPR）：

$$\text{FPR}_i = \frac{\text{FP}_i}{\text{FP}_i + \text{TN}_i}$$

（3）准确率（Precision）、Positive Predictive Value（PPV）：

$$\text{Precision}_i = \text{PPV}_i = \frac{\text{TP}_i}{\text{TP}_i + \text{FP}_i}$$

（4）预测结果为阳性的样本占总样本的比例：

$$\text{PR}_i = \frac{\text{TP}_i + \text{FP}_i}{\text{TP}_i + \text{FN}_i + \text{FP}_i + \text{TN}_i}$$

（5）训练样本中正样本的概率值：

$$\pi_+ = \frac{\text{TP}_i + \text{FN}_i}{\text{TP}_i + \text{FN}_i + \text{FP}_i + \text{TN}_i}$$

注意：π_+ 可以被看作一个常量，TP_i 与 FN_i 的和为正样本的总数，是不随阈值点 α_i 而改变的；同样，分母上 4 个数的和为全体样本的总数，也是不随阈值点 α_i 而改变的。

基于上面汇总的指标，可以更清晰地看到 5 种评估曲线的关联与区别，如表 8-5 所示。

表 8-5 5 种评估曲线的对比

名称	横坐标（X 轴）	纵坐标（Y 轴）
ROC 曲线	FPR_i	TPR_i
KS 曲线	α_i	$\text{TPR}_i - \text{FPR}_i$

续表

名称	横坐标（X轴）	纵坐标（Y轴）
PR 曲线	$Recall_i$	$Precision_i$
Lift（提升）图	PR_i	$\dfrac{Precision_i}{\pi_+}$
CAP 曲线	PR_i	$Recall_i$

8.3 数据探索

纸钞认证数据集（Banknote Authentication Data Set）来源于 UC Machine Learning Repository，可从链接 8-3 下载。

其数据是从真钞和假钞的图像样本中提取出来的。每个图像均为 400 像素×400 像素。这些图像是分辨率约为 660dpi 的灰度图像。可通过小波变换从图像中提取 4 个特征。一共有 1372 个样本，每个样本有 4 个输入变量和 1 个输出变量。变量名如下：

（1）variance：小波变换图像的方差。

（2）skewness：小波变换图像的偏度。

（3）kurtosis：小波变换图像的峰度。

（4）entropy：图像熵。

（5）class：类别（0 代表真钞，1 代表假钞）。

我们先下载数据，之后使用文本编辑器即可打开该数据文件，如图 8-4 所示。

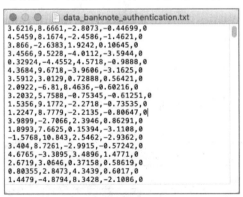

图 8-4

我们看到每行为一条数据，各列数据间以逗号分隔。这是一个典型的 CSV 格式数据，可以使用 CsvSourceBatchOp 组件直接读取。关于 CsvSourceBatchOp 等数据文件相关操作的详细介绍，请参阅 3.2.1 节。

首先，定义相关的一些常量、原始数据及后续处理数据所在的文件夹路径 DATA_DIR（该路径在数据根文件夹下的 banknote 子文件夹中）、原始数据文件的名称、SCHEMA_STRING（CsvSourceBatchOp 组件所需的参数）。将各列的名称和类型使用字符串形式表示出来，此内容在 2.8 节中已有详细介绍，这里不再展开说明。

```
static final String DATA_DIR = Utils.ROOT_DIR + "banknote" + File.separator;
static final String ORIGIN_FILE = "data_banknote_authentication.txt";
static final String SCHEMA_STRING
    = "variance double, skewness double, kurtosis double, entropy double, class int";
```

然后，使用 CsvSourceBatchOp 组件读取数据，并打印显示信息，代码如下：

```
CsvSourceBatchOp source =
    new CsvSourceBatchOp()
        .setFilePath(DATA_DIR + ORIGIN_FILE)
        .setSchemaStr(SCHEMA_STRING);

System.out.println("schema of source:");
System.out.println(source.getSchema());

System.out.println("column names of source:");
System.out.println(ArrayUtils.toString(source.getColNames()));

System.out.println("column types of source:");
System.out.println(ArrayUtils.toString(source.getColTypes()));

source.firstN(5).print();
```

前 4 行代码定义了 CsvSourceBatchOp 的实例 source；随后，展示了可以从 source 中获取 Schema 信息、列名数组信息、列类型数组信息；最后一行的 firstN(5)方法，是从批式组件中取前 5 行数据，然后选择打印方法输出结果。

```
schema of source:
root
 |-- variance: DOUBLE
 |-- skewness: DOUBLE
 |-- kurtosis: DOUBLE
 |-- entropy: DOUBLE
 |-- class: INT

column names of source:
```

```
{variance,skewness,kurtosis,entropy,class}
column types of source:
{Double,Double,Double,Double,Integer}
variance|skewness|kurtosis|entropy|class
--------|--------|--------|-------|-----
3.6216|8.6661|-2.8073|-0.4470|0
4.5459|8.1674|-2.4586|-1.4621|0
3.8660|-2.6383|1.9242|0.1065|0
3.4566|9.5228|-4.0112|-3.5944|0
0.3292|-4.4552|4.5718|-0.9888|0
```

前面打印输出的 Schema 信息、列名数组信息、列类型数组信息，都和我们对数据的理解相一致，不用进行说明。后面打印的 5 行数据内容，采用的是 Markdown 表格的形式。可以通过 Markdown 编辑器将这些内容转换为一般的表格形式，以便更好地展现数据，如表 8-6 所示。

表 8-6　5 条 iris 数据

variance	skewness	kurtosis	entropy	class
3.6216	8.6661	−2.8073	−0.4470	0
4.5459	8.1674	−2.4586	−1.4621	0
3.8660	−2.6383	1.9242	0.1065	0
3.4566	9.5228	−4.0112	−3.5944	0
0.3292	−4.4552	4.5718	−0.9888	0

8.3.1　基本统计

使用 SummarizerBatchOp 可以轻松计算数据表各列数据的基本统计信息。下列代码给出了 SummarizerBatchOp 的基本使用方式：

```
TableSummary summary = new SummarizerBatchOp().linkFrom(source).collectSummary();
System.out.println("Count of data set : " + summary.count());
System.out.println("Max value of entropy : " + summary.max("entropy"));
System.out.println(summary);
```

运行结果如下，先打印输出记录的总条数，随后输出"entropy"列的最大值，最后执行 println(summary) 操作，将 summary.toString() 的结果打印了出来：

```
Count of data set : 1372
Max value of entropy : 2.4495
|colName|count|missing|    sum|  mean|variance|     min|    max|
|-------|-----|-------|-------|------|--------|--------|-------|
|variance| 1372|      0|595.0848|0.4337|  8.0813| -7.0421| 6.8248|
```

```
|skewness| 1372|        0| 2637.4685|  1.9224| 34.4457|-13.7731|12.9516|
|kurtosis| 1372|        0| 1917.5444|  1.3976| 18.5764| -5.2861|17.9274|
| entropy| 1372|        0|-1634.9527| -1.1917|  4.4143| -8.5482| 2.4495|
|   class| 1372|        0|       610|  0.4446|  0.2471|       0|     1|
```

上述代码通过 collectSummary 方法，触发 SummarizerBatchOp 及其上游关联组件的执行，得到任务运行结果（TableSummary 类的实例 summary），从而可以获取、打印输出任何基本统计量。Alink 提供了 Lazy 机制（详见 2.5 节中的相关内容），SummarizerBatchOp 组件不会马上被触发执行，而会和其他组件共同在一个任务中执行，一起返回结果。这样就节省了多次启动任务的时间，也可以减少重复计算，如此一来极大地提高了我们处理批式任务的效率。使用 lazyCollectSummary 方法的代码如下，通过 Java 的 Consumer 接口，异步接收处理 summary 的返回结果：

```
source
    .link(
        new SummarizerBatchOp()
            .lazyCollectSummary(
                new Consumer <TableSummary>() {
                    @Override
                    public void accept(TableSummary tableSummary) {
                        System.out.println("Count of data set : " + tableSummary.count());
                        System.out.println("Max value of entropy : " + tableSummary.max("entropy"));
                        System.out.println(tableSummary);
                    }
                }
            )
    );
```

上述代码可以获得与前面示例代码一样的输出，这里就不重复显示了。

基本统计的输出表包含了常用的信息，我们还可以简化代码，如下所示：

```
source
    .link(
        new SummarizerBatchOp()
            .lazyPrintSummary()
    );
```

这里显示的就是基本统计的信息，以表格形式呈现。与前面使用 lazyCollectSummary 方法相比，这种写法简单了很多。

基本统计是常用操作。在任务中多次调用基本统计操作，可以帮助我们了解整个流程中发生的变化。基本统计的调用能否再简单一点呢？Alink 为每个批式组件提供了 lazyPrintStatistics 方法，该方法就是对 new SummarizerBatchOp().lazyPrintSummary() 的包装。其用法更加简单，示例代码如下。这里对原始数据进行了统计，然后取前 5 条数据，再对取出的这 5 条数据进行统计，最后打印输出。注意：经过 lazyPrintStatistics 操作，数据没有改变。

```
source
    .lazyPrintStatistics("<- origin data ->")
    .firstN(5)
    .lazyPrintStatistics("<- first 5 data ->")
    .print();
```

运行结果如下。第一次对原始数据进行统计，共有 1372 条数据。执行 firstN(5)后，第二次统计结果显示的是 5 条数据。经过 lazyPrintStatistics 操作，数据没有改变，最后打印输出的仍是 5 条数据。

```
<- origin data ->
| colName|count|missing|      sum|   mean|variance|    min|    max|
|--------|-----|-------|---------|-------|--------|-------|-------|
|variance| 1372|      0| 595.0848| 0.4337|  8.0813|-7.0421| 6.8248|
|skewness| 1372|      0|2637.4685| 1.9224| 34.4457|-13.7731|12.9516|
|kurtosis| 1372|      0|1917.5444| 1.3976| 18.5764|-5.2861|17.9274|
| entropy| 1372|      0|-1634.9527|-1.1917|  4.4143|-8.5482| 2.4495|
|   class| 1372|      0|      610| 0.4446|  0.2471|      0|      1|

<- first 5 data ->
| colName|count|missing|      sum|   mean|variance|    min|    max|
|--------|-----|-------|---------|-------|--------|-------|-------|
|variance|    5|      0|   4.0304| 0.8061|  3.0154|-1.2528| 3.5458|
|skewness|    5|      0|  17.2018| 3.4404| 47.9574|-4.4552|10.2036|
|kurtosis|    5|      0|   7.8499|   1.57| 10.8332|-4.0351| 4.5718|
| entropy|    5|      0| -11.7274|-2.3455|  5.3836|-5.6038|-0.1935|
|   class|    5|      0|        0|      0|       0|      0|      0|

variance|skewness|kurtosis|entropy|class
--------|--------|--------|-------|-----
-1.2528|10.2036|2.1787|-5.6038|0
0.5195|-3.2633|3.0895|-0.9849|0
0.3292|-4.4552|4.5718|-0.9888|0
0.8887|5.3449|2.0450|-0.1936|0
3.5458|9.3718|-4.0351|-3.9564|0
```

8.3.2 相关性

前面，我们通过基本统计对单个数值列有了一定的了解。下面，我们关注各数据列之间的关系，尤其是特征数据列与类别数据列之间的关系。这里先计算和显示一下相关系数，代码如下。使用 collectCorrelation 触发任务执行，返回相关性结果，从中抽取所需信息进行输出。

```java
CorrelationResult correlation = new CorrelationBatchOp().linkFrom(source).collectCorrelation();
String[] colNames = correlation.getColNames();
System.out.print("Correlation of " + colNames[0] + " with " + colNames[1]);
System.out.println(" is " + correlation.getCorrelation()[0][1]);
System.out.println(correlation.getCorrelationMatrix());
```

显示结果如下:

```
Correlation of variance with skewness is 0.26402552997043593
mat[5,5]:
 1.0,0.26402552997043593,-0.3808499720462522,0.2768166960053636,-0.7248431424446056
 0.26402552997043593,1.0,-0.7868952243065793,-0.5263208425437146,-0.44468775759659307
-0.3808499720462522,-0.7868952243065793,1.0,0.31884088768744584,0.15588323600923013
 0.2768166960053636,-0.5263208425437146,0.31884088768744584,1.0,-0.023423678954851614
-0.7248431424446056,-0.44468775759659307,0.15588323600923013,-0.023423678954851614,1.0
```

与基本统计操作类似,我们可以使用 Lazy 方式提高效率。具体代码如下:

```java
source
  .link(
    new CorrelationBatchOp()
      .lazyCollectCorrelation(new Consumer <CorrelationResult>() {
        @Override
        public void accept(CorrelationResult correlationResult) {
          String[] colNames = correlationResult.getColNames();
          System.out.print("Correlation of " + colNames[0] + " with " + colNames[1]);
          System.out.println(" is " + correlationResult.getCorrelation()[0][1]);
          System.out.println(correlationResult.getCorrelationMatrix());
        }
      })
  );
```

运行结果与前面使用 collectCorrelation 方法得到的结果相同。

相关性计算结果可以通过把列名信息与相关性矩阵整合为一张表格来清晰地展现。相关性组件提供的 lazyPrintCorrelation 方法,便可直接打印结果表格,具体 Java 代码如下所示。特别地,我们默认使用 Pearson 相关系数,另外一种 Spearman 相关系数可通过设置 Method 参数进行切换。很明显,使用 lazyPrintCorrelation 方法比使用 lazyCollectCorrelation 方法的代码简化了很多。

```java
source
  .link(
    new CorrelationBatchOp()
      .lazyPrintCorrelation("< Pearson Correlation >")
  );
source.link(
```

```
new CorrelationBatchOp()
    .setMethod(Method.SPEARMAN)
    .lazyPrintCorrelation("< Spearman Correlation >")
);

BatchOperator.execute();
```

运行结果如下：

```
< Pearson Correlation >
------------------------------- Correlation -------------------------------
colName|variance|skewness|kurtosis|entropy|class
-------|--------|--------|--------|-------|-----
variance|1.0000|0.2640|-0.3808|0.2768|-0.7248
skewness|0.2640|1.0000|-0.7869|-0.5263|-0.4447
kurtosis|-0.3808|-0.7869|1.0000|0.3188|0.1559
entropy|0.2768|-0.5263|0.3188|1.0000|-0.0234
class|-0.7248|-0.4447|0.1559|-0.0234|1.0000
< Spearman Correlation >
------------------------------- Correlation -------------------------------
colName|variance|skewness|kurtosis|entropy|class
-------|--------|--------|--------|-------|-----
variance|1.0000|0.2551|-0.3267|0.2415|-0.6404
skewness|0.2551|1.0000|-0.7294|-0.5725|-0.3834
kurtosis|-0.3267|-0.7294|1.0000|0.4333|0.0694
entropy|0.2415|-0.5725|0.4333|1.0000|-0.0139
class|-0.6404|-0.3834|0.0694|-0.0139|1.0000
```

结合两种相关系数可知，类别 class 列与 variance 列的相关性最强，类别 class 列与 skewness 列的相关性次之；类别 skewness 列与 kurtosis 列之间有较强的负相关性。

可使用 Python 的工具函数，使用多变量图来展示出各变量间两两的关系。

如图 8-5 所示，对角线上为单个变量的直方分布图。由于选择了分类值，因此每个柱子是按分类值叠加展示的。在此可以看出每个柱子所代表区间内，不同分类的比例。variance 所对应的直方图在图 8-5 的左上角。不同分类值所对应的分布有显著不同。分类值 1 所对应的灰色部分偏向左侧，而分类值 0 所对应的黑色部分偏向右边。再看与 variance 相关的散点图，灰色部分与黑色部分有交叠，但也有较大的纯色区域。

在前面的散点图中，选择两个特征，比如 variance 和 skewness，数据点间的交叠部分较小。随着更多特征的加入，数据点应该有更好的区分，但是我们没法形象地"看到"其在高维空间如何分布。这就需要借助数据降维与可视化方法，将数据展示在平面或三维空间中。使用 t-SNE 方法得到的可视化图像参见图 8-6。

第 8 章 线性二分类模型

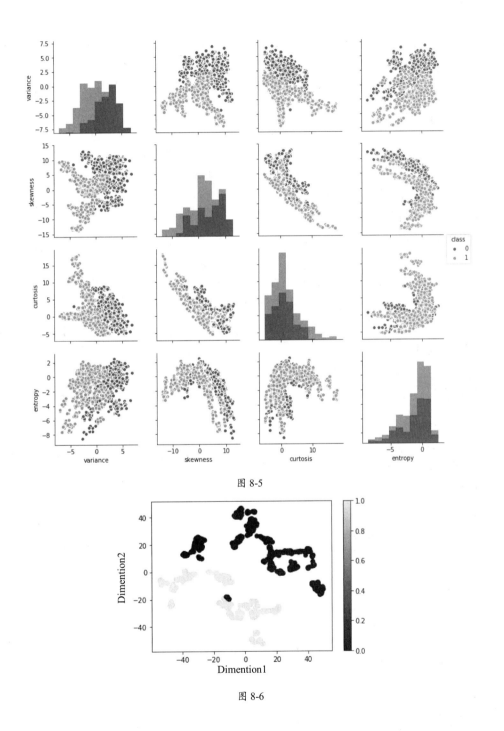

图 8-5

图 8-6

从图 8-6 中可以看到，两个类别间似乎有一个分隔，大部分样本都会被分到正确的类别。这也给了我们很强的信心：后面我们会尝试各种分类方法，争取拿到更好的分类模型。

8.4 训练集和测试集

将原始的数据集分为训练集和测试集两部分。训练集用来建立分类模型，然后，使用分类模型对测试集的数据进行预测，并对比预测值与原始分类标记值，给出模型评估指标。

本节内容涉及两个重要组件：

- 数据划分组件 SplitBatchOp，详见 7.2 节。
- 批式 AK 文件格式导出组件 AkSinkBatchOp，详见 3.2.3 节。

由于数据划分在后面的章节中频繁出现，因此我们在工具类 Utils 中封装定义了相关的函数。具体代码如下。原始数据 source 会被分成两部分，第一部分为 ratio，其余在第二部分。第一部分会被看作训练集，另一部分被看作测试集。我们会直接把划分出来的数据集分别保存为本地数据文件。在数据格式方面选择 AK 格式，以使数据读取操作更加简便。

```java
public static void splitTrainTestIfNotExist(
    BatchOperator <?> source,
    String trainFilePath,
    String testFilePath,
    double ratio
) throws Exception {
    if ((!new File(trainFilePath).exists()) && (!new File(testFilePath).exists())) {
        SplitBatchOp spliter = new SplitBatchOp().setFraction(ratio);

        source.link(spliter);

        spliter
            .link(
                new AkSinkBatchOp()
                    .setFilePath(trainFilePath)
            );

        spliter.getSideOutput(0)
            .link(
                new AkSinkBatchOp()
                    .setFilePath(testFilePath)
            );

        BatchOperator.execute();
    }
}
```

使用如下代码进行数据划分,其中训练集占总数据量的 80%。训练集和测试集的数据文件都以 ".ak" 为扩展名,即均为 AK 格式的数据文件。

```
static final String TRAIN_FILE = "train.ak";
static final String TEST_FILE = "test.ak";

Utils.splitTrainTestIfNotExist(
    source,
    DATA_DIR + TRAIN_FILE,
    DATA_DIR + TEST_FILE,
    0.8
);
```

有了训练集和测试集的数据,每次实验的时候就可以直接从文件路径读取。定义相应的 AkSourceBatchOp,代码如下:

```
AkSourceBatchOp train_data = new AkSourceBatchOp().setFilePath(DATA_DIR + TRAIN_FILE);
AkSourceBatchOp test_data = new AkSourceBatchOp().setFilePath(DATA_DIR + TEST_FILE);
```

8.5 逻辑回归模型

这里主要介绍对于当前的二分类问题,通过逻辑回归训练组件和预测组件来实现建模和预测。

首先设置逻辑回归训练(LogisticRegressionTrainBatchOp)组件和逻辑回归预测(LogisticRegressionPredictBatchOp)组件,代码如下。训练组件需要指定特征列和标签列;预测组件需要指定预测结果列名,即参数 predictionCol,分类预测结果会被放在该列。另外,如果想要获取更多的信息,了解预测过程中对各分类情况的概率值,可以通过设置预测详情列名,即参数 predictionDetailCol 来完成。设置了该列名,预测结果数据集中就会有这样一列:该列为字符串类型值,给出了属于各分类值的概率,可用于二分类评估。

```
LogisticRegressionTrainBatchOp lrTrainer =
    new LogisticRegressionTrainBatchOp()
        .setFeatureCols(FEATURE_COL_NAMES)
        .setLabelCol(LABEL_COL_NAME);

LogisticRegressionPredictBatchOp lrPredictor =
    new LogisticRegressionPredictBatchOp()
        .setPredictionCol(PREDICTION_COL_NAME)
        .setPredictionDetailCol(PRED_DETAIL_COL_NAME);
```

然后,通过 link 方法,建立起整个实验流程,代码如下。我们通过这两行代码的描述,可以想象出整个流程:训练数据源 train_data 连接逻辑回归训练组件 lrTrainer;预测过程需要模型,也需要待预测的数据,所以逻辑回归预测组件 lrPredictor 需要使用 linkFrom 方法来同时连接 lrTrainer 输出的模型和待预测的数据源 test_data；lrPredictor 组件的输出即预测结果。

```
train_data.link(lrTrainer);
lrPredictor.linkFrom(lrTrainer, test_data);
```

在训练过程中,我们还希望看到训练信息（TrainInfo）和模型信息（ModelInfo）,以及最终的预测结果。我们希望能看到几条数据,对运行结果有一个直观的了解。以上这些信息最好能运行一次任务就可全部得到。因此,这里选择了 Lazy 方式。具体代码如下:

```
lrTrainer.lazyPrintTrainInfo().lazyPrintModelInfo();
lrPredictor.lazyPrint(5, "< Prediction >");
BatchOperator.execute()
```

训练信息如下:

```
------------------------- train meta info -------------------------
{model name: Logistic Regression, num feature: 4}
------------------------- train importance info -------------------------
|colName|importanceValue| colName|weightValue|
|-------|---------------|--------|-----------|
|skewness|   24.05967904| entropy|-0.68830239|
|kurtosis|   22.17694646|skewness|-4.14845206|
|variance|   21.81129809|kurtosis|-5.20781697|
| entropy|    1.47179204|variance|-7.68311606|
------------------------- train convergence info -------------------------
step:0 loss:0.67538777 gradNorm:0.42586577 learnRate:0.40000000
step:1 loss:0.50807036 gradNorm:0.41072258 learnRate:1.60000000
step:2 loss:0.27567906 gradNorm:0.26515855 learnRate:4.00000000
... ...
step:32 loss:0.02060997 gradNorm:0.00055431 learnRate:4.00000000
step:33 loss:0.02060759 gradNorm:0.00009935 learnRate:4.00000000
step:34 loss:0.02060757 gradNorm:0.00000829 learnRate:4.00000000
```

在训练信息中给出了对于训练数据计算出来的特征重要性（importance）。在此可以看到特征 skewness、kurtosis 和 variance 的重要性数值差别较小,特征 entropy 的重要性数值比前面几个特征的重要性数值要小得多。随后可看到逻辑回归迭代训练中每步的主要指标。在此可看到本次训练是在第 34 轮中结束的,属于满足终止条件、提前结束迭代训练的情况。

模型信息输出如下,在此可以看到各个模型的权重系数:

```
------------------------- model meta info -------------------------
{hasInterception: true, model name: Logistic Regression, num feature: 4}
------------------------- model weight info -------------------------
|intercept|  variance|  skewness|  kurtosis|   entropy|
|---------|----------|----------|----------|----------|
|   6.9651|-7.68311606|-4.14845206|-5.20781697|-0.68830239|
```

预测结果的前 5 项如下所示。预测结果项返回的是预测标签值，详细信息列返回的是各标签值的概率。

```
< Prediction >
variance|skewness|kurtosis|entropy|class|pred|predinfo
--------|--------|--------|-------|-----|----|--------
-2.4953|11.1472|1.9353|-3.4638|0|0|{"0":"0.9999999999991567","1":"8.433254095052689E-13"}
-1.3000|10.2678|-2.9530|-5.8638|0|0|{"0":"0.9980296917461061","1":"0.001970308253893882"}
5.2423|11.0272|-4.3530|-4.1013|0|0|{"0":"1.0","1":"0.0"}
2.0843|6.6258|0.4838|-2.2134|0|0|{"0":"1.0","1":"0.0"}
3.8200|10.9279|-4.0112|-5.0284|0|0|{"0":"1.0","1":"0.0"}
```

从这 5 条数据上看，预测得比较准确。

8.6 线性SVM模型

本节将介绍如何使用线性 SVM 训练组件和线型 SVM 预测组件来解决二分类问题。

首先设置线性 SVM 训练组件（LinearSvmTrainBatchOp）组件和线型 SVM 预测组件（LinearSvmPredictBatchOp），代码如下。与逻辑回归算法相似，线性 SVM 训练组件需要指定特征列和标签列；线性 SVM 预测组件需要指定预测结果列名，在此还可以通过设置预测详情列名，列出属于各分类值的概率，用于后面的二分类评估。

```
LinearSvmTrainBatchOp svmTrainer =
    new LinearSvmTrainBatchOp()
        .setFeatureCols(FEATURE_COL_NAMES)
        .setLabelCol(LABEL_COL_NAME);

LinearSvmPredictBatchOp svmPredictor =
    new LinearSvmPredictBatchOp()
        .setPredictionCol(PREDICTION_COL_NAME)
        .setPredictionDetailCol(PRED_DETAIL_COL_NAME);
```

建立起整个实验流程：训练数据源 train_data 连接线性 SVM 训练组件 svmTrainer；预测过程需要线性 SVM 模型，也需要待预测的数据，所以预测组件 svmPredictor 需要使用 linkFrom 方法来同时连接 svmTrainer 输出的模型和待预测的数据源 test_data；svmPredictor 组件的输出即预测结果。代码如下：

```
train_data.link(svmTrainer);

svmPredictor.linkFrom(svmTrainer, test_data);
```

在训练过程中，我们还仿照前面的逻辑回归算法，使用 Lazy 方式打印输出训练信息（TrainInfo）和模型信息（ModelInfo），以及最终的预测结果。具体代码如下：

```
svmTrainer.lazyPrintTrainInfo().lazyPrintModelInfo();

svmPredictor.lazyPrint(5, "< Prediction >");

BatchOperator.execute();
```

训练信息如下：

```
------------------------- train meta info -------------------------
{model name: Linear SVM, num feature: 4}
------------------------- train importance info -------------------------
| colName|importanceValue| colName|weightValue|
|--------|---------------|--------|-----------|
|skewness|     8.43892212| entropy|-0.22797316|
|kurtosis|     7.74791237|skewness|-1.45506778|
|variance|     7.62486689|kurtosis|-1.81944388|
| entropy|     0.48747337|variance|-2.68588953|
------------------------- train convergence info -------------------------
step:0  loss:0.43263605  gradNorm:0.85130570  learnRate:0.40000000
step:1  loss:0.26491660  gradNorm:0.73496319  learnRate:1.60000000
step:2  loss:0.06516288  gradNorm:0.45340859  learnRate:4.00000000
... ... ... ...
step:24 loss:0.01192161  gradNorm:0.00012197  learnRate:4.00000000
step:25 loss:0.01192159  gradNorm:0.00003354  learnRate:4.00000000
step:26 loss:0.01192159  gradNorm:0.00000841  learnRate:4.00000000
```

在此可从特征重要性的角度来说（参考训练信息的 importanceValue 项），特征 skewness、kurtosis 和 variance 的重要性数值差别较小，特征 entropy 的重要性数值比前面几个特征的重要性数值要小得多。本次训练是在第 26 轮中满足终止条件，提前结束的。在上一个实验中，逻辑回归训练是在第 34 轮结束的。

模型信息如下：

```
------------------------- model meta info -------------------------
{hasInterception: true, model name: Linear SVM, num feature: 4}
------------------------- model weight info -------------------------
|intercept| variance| skewness|kurtosis| entropy|
|---------|---------|---------|--------|--------|
|  2.4171 |-2.68588953|-1.45506778|-1.81944388|-0.22797316|
```

预测结果如下：

```
< Prediction >
variance|skewness|kurtosis|entropy|class|pred|predinfo
```

```
--------|--------|--------|--------|-----|----|---------
0.8887|5.3449|2.0450|-0.1936|0|0|{"0":"0.9999890670812613","1":"1.0932918738659758E-5"}
2.6799|3.1349|0.3407|0.5849|0|0|{"0":"0.9995587398819237","1":"4.12601180762584E-5"}
4.5459|8.1674|-2.4586|-1.4621|0|0|{"0":"0.9999999528621162","1":"4.7137883818493265E-8"}
4.4338|9.8870|-4.6795|-3.7483|0|0|{"0":"0.9999995002912768","1":"4.997087231783937E-7"}
-0.7829|11.3603|-0.3764|-7.0495|0|0|{"0":"0.9999398211526664","1":"6.0178847333558494E-5"}
```

8.7 模型评估

二分类模型评估需要在模型预测结果数据集中包含如下信息：
- 标签列——即原始的分类标记结果，用来验证预测的正确性。
- 预测详情列——该列包含了该条记录预测为各分类值的概率，概率最大的就是分类预测结果。

二分类评估组件 EvalBinaryClassBatchOp 需要两个必选参数，即所要评估的预测结果数据集的标签列名称和预测详情列名称。此外，还有一个可选参数 PositiveLabelValueString（作为正例的标签值，选择不同的标签值作为正例会影响召回率等指标的值。当用户没有输入该参数时，默认会使用标签值降序排列。位于首位的标签值作为正例标签值。注意：为了统一输入类型，这里使用的是正例标签值转为字符串的结果）。

从二分类评估组件 EvalBinaryClassBatchOp 中获取评估指标。这里使用的是该组件的 collectMetrics 方法。在此沿用 Flink 的风格，collect 方法和 print 方法都会触发运行相应的任务。任务结束后返回结果，所以我们无须再调用 execute 方法。

具体代码如下。将前面的逻辑回归预测结果存储于 AK 格式的文件 DATA_DIR + LR_PRED_FILE 中。在此可以先定义评估组件 EvalBinaryClassBatchOp，然后使用 linkFrom 方法，将预测结果数据源与评估组件连接起来。

```
BinaryClassMetrics lr_metrics =
  new EvalBinaryClassBatchOp()
    .setPositiveLabelValueString("1")
    .setLabelCol(LABEL_COL_NAME)
    .setPredictionDetailCol(PRED_DETAIL_COL_NAME)
    .linkFrom(
      new AkSourceBatchOp().setFilePath(DATA_DIR + LR_PRED_FILE)
    )
    .collectMetrics();
```

接下来，从评估结果中抽取我们需要的指标。首先抽取常用的指标值：

```
StringBuilder sbd = new StringBuilder();
sbd.append("< LR >\n")
```

```
    .append("AUC : ").append(lr_metrics.getAuc())
    .append("\t Accuracy : ").append(lr_metrics.getAccuracy())
    .append("\t Precision : ").append(lr_metrics.getPrecision())
    .append("\t Recall : ").append(lr_metrics.getRecall())
    .append("\n");
System.out.println(sbd.toString());
```

运行结果如下:

```
< LR >
AUC : 0.9999457847655191    Accuracy : 0.9963503649635036    Precision : 1.0    Recall : 0.9915966386554622
```

直接对评估结果对象 lr_metrics 进行打印,会显示几个常用的指标,并输出混淆矩阵。代码如下:

```
System.out.println(lr_metrics);
```

运行结果如下。其中,横向为 Real(测试数据中真实的标签值),纵向为 Pred(预测结果的标签值),标签值为 1 和 0,中间交叉的部分即混淆矩阵。

```
-------------------------------- Metrics: --------------------------------
{Accuracy: 0.9964, LogLoss: 0.0089, Precision: 1, Recall: 0.9916, F1: 0.9958, Auc: 0.9999}
|Pred\Real|  1|  0|
|---------|---|---|
|        1|118|  0|
|        0|  1|155|
```

二分类评估中的各种评估曲线也是非常重要的。Alink 提供了将曲线保存为图片的方法,相关代码如下:

```
lr_metrics.saveRocCurveAsImage(DATA_DIR + "lr_roc.jpg", true);
lr_metrics.saveRecallPrecisionCurveAsImage(DATA_DIR + "lr_recallprec.jpg", true);
lr_metrics.saveLiftChartAsImage(DATA_DIR + "lr_lift.jpg", true);
lr_metrics.saveKSAsImage(DATA_DIR + "lr_ks.jpg", true);
```

运行成功后,可以在指定的路径下,看到评估曲线图片。比如,ROC 曲线图像文件打开后,如图 8-7 所示。

至此,我们展示了二分类评估组件的常用功能。在实际使用中,我们除了使用 collectMetrics 方法直接触发任务执行、获取评估指标,也可以采用 Lazy 方式,将评估组件与任务中的其他组件一起执行,提高效率。参考代码如下。以评估 SVM 预测结果为例,其测试集采用了 AK 数据文件格式,保存在 DATA_DIR + SVM_PRED_FILE 中。首先建立预测结果数据源,然后连接二分类评估组件 EvalBinaryClassBatchOp,并设置标签列名称和预测详情列名称。使用 lazyPrintMetrics 方法打印二分类评估常用的指标及混淆矩阵;再使用 lazyCollectMetrics 方法,设置通过 Java 的 Consumer 接口来异步接收处理 BinaryClassMetrics 的返回结果。这里略去了打印输出评估指标的代码。

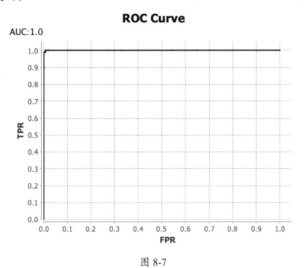

图 8-7

```
new AkSourceBatchOp()
  .setFilePath(DATA_DIR + SVM_PRED_FILE)
  .link(
    new EvalBinaryClassBatchOp()
      .setPositiveLabelValueString("1")
      .setLabelCol(LABEL_COL_NAME)
      .setPredictionDetailCol(PRED_DETAIL_COL_NAME)
      .lazyPrintMetrics()
      .lazyCollectMetrics(new Consumer <BinaryClassMetrics>() {
        @Override
        public void accept(BinaryClassMetrics binaryClassMetrics) {
          try {
            binaryClassMetrics.saveRocCurveAsImage(
              DATA_DIR + "svm_roc.jpg", true);
            binaryClassMetrics.saveRecallPrecisionCurveAsImage(
              DATA_DIR + "svm_recallprec.jpg", true);
            binaryClassMetrics.saveLiftChartAsImage(
              DATA_DIR + "svm_lift.jpg", true);
            binaryClassMetrics.saveKSAsImage(
              DATA_DIR + "svm_ks.jpg", true);
          } catch (IOException e) {
            e.printStackTrace();
          }
        }
      })
  );
```

前面，我们针对逻辑回归和线性 SVM 的预测结果，分别进行了二分类评估。表 8-7 将输出的各种指标及曲线汇总起来，进行对比分析。

表 8-7　逻辑回归与线性 SVM 的二分类评估指标对比

逻辑回归	线性 SVM
AUC:0.9999	AUC:0.9999
Accuracy:0.9964	Accuracy:0.9964
Precision:1	Precision:1
Recall:0.9916	Recall:0.9916
F1:0.9958	F1:0.9958
LogLoss:0.0089	LogLoss:0.0233
混淆矩阵： |Pred\Real|　1|　0| |---------|---|---| |　　　　1|118|　0| |　　　　0|　1|155|	混淆矩阵： |Pred\Real|　1|　0| |---------|---|---| |　　　　1|118|　0| |　　　　0|　1|155|
ROC 曲线（AUC:1.0）	ROC 曲线（AUC:1.0）
PR 曲线（PRC:1.0）	PR 曲线（PRC:1.0）

续表

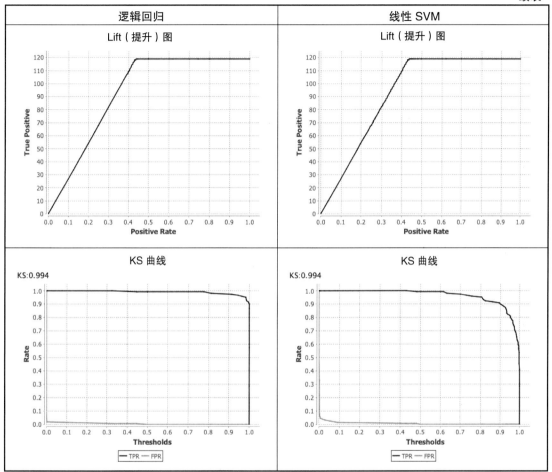

逻辑回归和线性 SVM 算法对当前的问题都非常有效，其测试集中的数据都只有一个预测错误。在表 8-7 中显示的几个指标中，两者的数值都相同；在 ROC 曲线、PR 曲线和 Lift 图上，二者看不出差别；但在 KS 曲线上二者有些差别，逻辑回归算法略好。

8.8 特征的多项式扩展

前面介绍的线性模型形式如下：

$$y = w_0 + \sum_{i=1}^{m} w_i x_i$$

在此基础上，定义二阶多项式模型如下：

$$y = w_0 + \sum_{i=1}^{m} w_i x_i + \sum_{i=1}^{m}\sum_{j=i}^{m} w_{ij} x_i x_j$$

在实际应用中，我们可以将原始特征（$x_1, x_2, \cdots, x_m$），通过特征间交叉相乘的方式加入二阶特征，扩展为（$x_1, x_2, \cdots, x_m, x_1^2, x_1 x_2, \cdots, x_{m-1} x_m, x_m^2$），从而将二阶多项式模型训练转化为关于新特征的线性模型训练。

基于这种想法，使用 Alink 的向量多项式扩展组件 VectorPolynomialExpand 可以实现特征的展开，设置多项式的阶数为 2，即参数 Degree=2。由于该组件需要输入向量形式的数据，因此在运行前需要使用向量组装组件 VectorAssembler，将各数值列合并为向量。具体的调用代码如下：

```
PipelineModel featureExpand = new Pipeline()
  .add(
    new VectorAssembler()
      .setSelectedCols(FEATURE_COL_NAMES)
      .setOutputCol(VEC_COL_NAME + "_0")
  )
  .add(
    new VectorPolynomialExpand()
      .setSelectedCol(VEC_COL_NAME + "_0")
      .setOutputCol(VEC_COL_NAME)
      .setDegree(2)
  )
  .fit(train_data);

train_data = featureExpand.transform(train_data);
test_data = featureExpand.transform(test_data);

train_data.lazyPrint(1);
```

运行结果如下：

```
variance|skewness|kurtosis|entropy|class|vec_0|vec
--------|--------|--------|-------|-----|-----|---
2.5328|7.5280|-0.4193|-2.6478|0|2.5328 7.528 -0.41929 -2.6478|2.5328 6.41507584 7.528 19.0669184 56.67078399999999 -0.41929 -1.061977712 -3.1564151199999997 0.1758041041 -2.6478 -6.70634784 -19.9326384 1.110196062 7.010844840000001
```

每个特征的数值在打印输出时会保留小数点后 4 位，各数值列合并成的向量为（2.5328 7.5280 -0.4193 -2.6478）；二阶扩展后得到的向量为（2.5328 6.41507584 7.528 19.0669184

56.67078399999999 -0.41929 -1.061977712 -3.1564151199999997 0.1758041041 -2.6478 -6.70634784 -19.9326384 1.110196062 7.010844840000001），其中包含了初始的 4 个特征，再加上 10 个二阶项特征，共 14 个特征。

下面，我们对新特征向量试用线性 SVM 算法，看看效果如何。具体代码如下，使用 Pipeline 的机制简化代码，LinearSvm 为 Pipeline 的 Estimator（详细内容可参考 2.4 节）。

```
new LinearSvm()
  .setVectorCol(VEC_COL_NAME)
  .setLabelCol(LABEL_COL_NAME)
  .setPredictionCol(PREDICTION_COL_NAME)
  .setPredictionDetailCol(PRED_DETAIL_COL_NAME)
  .fit(train_data)
  .transform(test_data)
  .link(
    new EvalBinaryClassBatchOp()
      .setPositiveLabelValueString("1")
      .setLabelCol(LABEL_COL_NAME)
      .setPredictionDetailCol(PRED_DETAIL_COL_NAME)
      .lazyPrintMetrics("LinearSVM")
  );
```

模型评估结果如下：

```
LinearSVM
-------------------------------- Metrics: --------------------------------
Auc:1 Accuracy:1 Precision:1    Recall:1   F1:1 LogLoss:0.0066
|Pred\Real|   1|   0|
|---------|---|---|
|        1|119|   0|
|        0|  0|155|
```

测试集的数据均被正确分类！特征的扩展直接提升了模型分类效果。

再看另外一个线性模型——逻辑回归模型，具体代码如下：

```
new LogisticRegression()
  .setVectorCol(VEC_COL_NAME)
  .setLabelCol(LABEL_COL_NAME)
  .setPredictionCol(PREDICTION_COL_NAME)
  .setPredictionDetailCol(PRED_DETAIL_COL_NAME)
  .fit(train_data)
  .transform(test_data)
  .link(
    new EvalBinaryClassBatchOp()
      .setPositiveLabelValueString("1")
      .setLabelCol(LABEL_COL_NAME)
      .setPredictionDetailCol(PRED_DETAIL_COL_NAME)
      .lazyPrintMetrics("LogisticRegression")
  );
```

运行结果如下：

```
LogisticRegression
------------------------------- Metrics: -------------------------------
Auc:0.9967 Accuracy:0.9964 Precision:0.9917    Recall:1   F1:0.9958  LogLoss:0.1261
|Pred\Real|  1|  0|
|---------|---|---|
|        1|119|  1|
|        0|  0|154|
```

这里有一个数据被错分类，其分类效果比线性 SVM 模型稍差。

下面做一个实验，调整一下 LogisticRegression 组件的优化方法选择参数 OptimMethod。逻辑回归算法默认使用 OptimMethod.LBFGS 优化方法，这是一种在内存使用、收敛速度等方面综合表现最优的算法；而 OptimMethod.Newton 方法在收敛性方面表现得更好，只是其内存使用量为特征数的平方量级，不适合特征数很多的场景。这次的特征数很少，只有 14 个，因此可以尝试采用 OptimMethod.Newton 方法，具体代码如下，

```
new LogisticRegression()
    .setOptimMethod(OptimMethod.Newton)
    .setVectorCol(VEC_COL_NAME)
    .setLabelCol(LABEL_COL_NAME)
    .setPredictionCol(PREDICTION_COL_NAME)
    .setPredictionDetailCol(PRED_DETAIL_COL_NAME)
    .fit(train_data)
    .transform(test_data)
    .link(
        new EvalBinaryClassBatchOp()
            .setPositiveLabelValueString("1")
            .setLabelCol(LABEL_COL_NAME)
            .setPredictionDetailCol(PRED_DETAIL_COL_NAME)
            .lazyPrintMetrics("LogisticRegression + OptimMethod.Newton")
    );
```

运行结果如下，这种方法也做到了完全分类：

```
LogisticRegression + OptimMethod.Newton
------------------------------- Metrics: -------------------------------
Auc:1 Accuracy:1 Precision:1    Recall:1   F1:1  LogLoss:0
|Pred\Real|  1|  0|
|---------|---|---|
|        1|119|  0|
|        0|  0|155|
```

从本节内容可以看到，特征扩展直接提升了分类器的效果。在后面的章节中，我们还会陆续介绍特征工程方面的方法。

8.9 因子分解机

前面介绍的二阶多项式模型如下：

$$y = w_0 + \sum_{i=1}^{m} w_i x_i + \sum_{i=1}^{m} \sum_{j=i}^{m} w_{ij} x_i x_j$$

二阶多项式模型比线性模型具有更强的分类能力。可以将原始特征（$x_1, x_2, \cdots, x_m$），加入二阶特征，扩展为（$x_1, x_2, \cdots, x_m, x_1^2, x_1 x_2, \cdots, x_{m-1} x_m, x_m^2$）。

二阶多项式模型中参数的个数为特征数的平方级。当特征数较多的时候，比如有 10 万个特征，则模型参数就要到百亿量级。利用因子分解机（Factorization Machine, FM）算法，可在模型效果和模型规模间取得一个很好的平衡。FM 模型的公式如下：

$$y = w_0 + \sum_{i=1}^{m} w_i x_i + \sum_{i=1}^{m-1} \sum_{j=i+1}^{m} <v^{(i)}, v^{(j)}> x_i x_j$$

即，每个特征 x_i 对应一个向量 $v^{(i)}$，向量的维度为 k，使用向量内积 $<v^{(i)}, v^{(j)}>$ 代替参数 w_{ij}。实际应用中对于千万级特征，向量的维度可以取 100 维左右。FM 模型参数的个数约等于特征数乘以向量的维度。

注意：为了降低计算复杂度，FM 模型公式在实际计算时会采用另外一种形式。在 FM 模型公式定义中，计算量集中在交叉项：

$$\sum_{i=1}^{m-1} \sum_{j=i+1}^{m} <v^{(i)}, v^{(j)}> x_i x_j$$

其计算复杂度为 $O(km^2)$，将此交叉项进行等式变换：

$$\sum_{i=1}^{m-1} \sum_{j=i+1}^{m} <v^{(i)}, v^{(j)}> x_i x_j$$
$$= \sum_{i=1}^{m-1} \sum_{j=i+1}^{m} <x_i \cdot v^{(i)}, x_j \cdot v^{(j)}>$$

$$= \frac{1}{2}\left(<\sum_{i=1}^{m} x_i \cdot \boldsymbol{v}^{(i)}, \sum_{j=1}^{m} x_j \cdot \boldsymbol{v}^{(j)}> - \sum_{i=1}^{m} <x_i \cdot \boldsymbol{v}^{(i)}, x_i \cdot \boldsymbol{v}^{(i)}>\right)$$

$$= \frac{1}{2}\left(\left\|\sum_{i=1}^{m} x_i \cdot \boldsymbol{v}^{(i)}\right\|_2 - \sum_{i=1}^{m} \|x_i \cdot \boldsymbol{v}^{(i)}\|_2\right)$$

可知，变换后表达式的计算复杂度为$O(km)$。

下面将此算法应用到本章的数据，代码如下。设置向量的维度为 2，即参数 NumFactor=2；参数 NumEpochs 表示训练几遍数据，这里设置 NumEpochs=10；并设置学习率 LearnRate=0.5。参数 NumEpochs 和 LearnRate 是需要根据具体数据进行调节的。我们使用了 enableLazyPrintTrainInfo 方法来打印输出训练过程的数据：

```
new FmClassifier()
    .setNumEpochs(10)
    .setLearnRate(0.5)
    .setNumFactor(2)
    .setFeatureCols(FEATURE_COL_NAMES)
    .setLabelCol(LABEL_COL_NAME)
    .setPredictionCol(PREDICTION_COL_NAME)
    .setPredictionDetailCol(PRED_DETAIL_COL_NAME)
    .enableLazyPrintTrainInfo()
    .enableLazyPrintModelInfo()
    .fit(train_data)
    .transform(test_data)
    .link(
        new EvalBinaryClassBatchOp()
            .setPositiveLabelValueString("1")
            .setLabelCol(LABEL_COL_NAME)
            .setPredictionDetailCol(PRED_DETAIL_COL_NAME)
            .lazyPrintMetrics("FM")
    );
```

训练过程的信息如下：

```
------------------------- train meta info -------------------------
{numFeature: 4, numFactor: 2, hasLinearItem: true, hasIntercept: true}
------------------------- train convergence info -------------------------
step: 0 loss: 0.08713016 auc: 0.99632587 accuracy: 0.96083789
step: 1 loss: 0.05184192 auc: 0.99943728 accuracy: 0.98633880
step: 2 loss: 0.04138949 auc: 1.00000000 accuracy: 0.99635701
... ...
step: 7 loss: 0.02687007 auc: 1.00000000 accuracy: 0.99726776
step: 8 loss: 0.02532989 auc: 1.00000000 accuracy: 1.00000000
step: 9 loss: 0.02396747 auc: 1.00000000 accuracy: 1.00000000
```

我们可以看到经过每一个 Epoch 后主要指标 loss、auc 和 accuracy 的变化情况。参考这些信息有助于我们调整参数 NumEpochs 和 LearnRate。

模型信息如下。其中，bias 为模型的 0 阶截距（Intercept）系数，每个特征都对应一个一阶系数（linearItem）和一个向量系数。

```
---------------------------- model meta info ----------------------------
{numFeature: 4, numFactor: 2, bias: 1.8750569560538486, hasLinearItem: true, hasIntercept: true}
---------------------------- model label values ----------------------------
[1, 0]
---------------------------- model info ----------------------------
| colName| linearItem|                       factor|
|--------|-----------|-----------------------------|
|variance|-2.22505520| 0.34821299  0.31662868      |
|skewness|-0.93356183| 0.10569094  0.14378680      |
|kurtosis|-1.53638153|-0.05461301 -0.23339019      |
| entropy| 0.11982146| 0.22809465  0.04826799      |
```

二分类评估结果如下，测试集实现了全部正确分类：

```
---------------------------- Metrics: ----------------------------
Auc:1 Accuracy:1 Precision:1    Recall:1   F1:1  LogLoss:0.0173
|Pred\Real|  1|  0|
|---------|---|---|
|        1|119|  0|
|        0|  0|155|
```

一般来说，FM 的分类效果要优于逻辑回归和线性 SVM，但在使用 FM 时，需要在调节训练参数方面多花费一些时间。

9

朴素贝叶斯模型与决策树模型

前面介绍了线性模型,本章将重点介绍两种常用的非线性模型:朴素贝叶斯模型与决策树模型。

9.1 节和 9.2 节将分别对这两种算法模型进行介绍。从 9.3 节开始,以蘑菇是否有毒的分类为例,介绍如何使用模型、如何查看模型。

9.1 朴素贝叶斯模型

朴素贝叶斯(Naive Bayes)是基于"朴素"的假设(各特征属性相互独立)和贝叶斯定理的分类算法。它是一种简单、常用的分类算法,"朴素"的假设不但使计算量大幅减少,而且对分类效果的影响也不大。对于现实中的一些复杂场景,该算法也表现得很好。

设有 k 个分类值 $\{c_1, c_2, \cdots, c_k\}$,共有 M 个特征属性 $a_1, a_2, \cdots, a_M$,使用朴素贝叶斯算法预测一个待分类项,分为两步:

(1)计算出该属性值对于各分类值的条件概率,即

$$P(c_1|a_1, a_2, \cdots, a_M), P(c_2|a_1, a_2, \cdots, a_M), \cdots, P(c_k|a_1, a_2, \cdots, a_M)$$

(2)最大概率值所对应的分类值即预测结果。

可以看出计算核心是 $P(c_k|a_1, a_2, \cdots, a_M)$。这里需要用到贝叶斯公式及朴素(Naive)的假设,下面将详细解释。

对于概率 $P(c_k|a_1, a_2, \cdots, a_M)$,由贝叶斯公式,可得

$$P(c_k|a_1,a_2,\cdots,a_M) = \frac{P(a_1,a_2,\cdots,a_M|c_k)P(c_k)}{P(a_1,a_2,\cdots,a_M)}$$

再利用"朴素"的假设（各个特征属性是相互独立的），则其联合概率可以表示为更简单的，也更容易计算的形式：

$$P(a_1,a_2,\cdots,a_M|c_k) = P(a_1|c_k)P(a_2|c_k)\cdots P(a_{M-1}|c_k)P(a_M|c_k)$$

所以，

$$P(c_k|a_1,a_2,\cdots,a_M) = \frac{P(a_1|c_k)P(a_2|c_k)\cdots P(a_{M-1}|c_k)P(a_M|c_k)P(c_k)}{P(a_1,a_2,\cdots,a_M)}$$

我们要比较的k个概率：

$$P(c_1|a_1,a_2,\cdots,a_M), P(c_2|a_1,a_2,\cdots,a_M), \cdots, P(c_k|a_1,a_2,\cdots,a_M)$$

都是以$P(a_1,a_2,\cdots,a_M)$为分母的，所以$P(a_1,a_2,\cdots,a_M)$的具体值对k个概率的大小顺序是没有影响的，即

$$P(c_k|a_1,a_2,\cdots,a_M) \propto P(a_1|c_k)P(a_2|c_k)\cdots P(a_{M-1}|c_k)P(a_M|c_k)P(c_k)$$

因此，我们只要计算$P(a_i|c_k)$与$P(c_k)$，带入上面的公式，就可知道哪个分类的概率最大。我们从训练数据中很容易计算出

$$P(c_k) = \frac{\text{分类为}c_k\text{的训练样本数}}{\text{总训练样本数}}$$

在条件概率$P(a_i|c_k)$的计算中，将训练数据按其分类值分为k个集合，对每个子集计算各个特征属性的概率分布。特征属性a_i为连续特征或者离散特征时，计算方法是不一样的，分别介绍如下：

（1）若特征属性a_i为连续特征，则在所有分类为c_k的训练样本上，计算其均值μ_{ik}与标准差σ_{ik}，并假设a_i是服从正态分布的，则概率密度如下：

$$P(a_i|c_k) = \frac{1}{\sigma_{ik}\sqrt{2\pi}}\exp\left(-\frac{(a_i-\mu_{ik})^2}{2\sigma_{ik}^2}\right)$$

（2）若特征属性a_i为离散特征，共有s_i个离散值$a_i^{(1)}, a_i^{(2)}, \cdots, a_i^{(s_i)}$，则关于分类值$c_k$的条件概率的极大似然估计如下：

$$P(a_i^{(j)}|c_k) = \frac{\text{在分类为} c_k \text{的训练样本中，} a_i = a_i^{(j)} \text{的样本数}}{\text{分类为} c_k \text{的训练样本数}}$$

尽管这个表达式很容易理解，但是在实际使用中却会遇到问题。因为在某个分类所对应的数据集中，有的属性的属性值可能不出现。即，可能存在 c_k 与 $a_i^{(j)}$，使得：

在分类为 c_k 的训练样本中，$a_i = a_i^{(j)}$ 的样本数为 0

则对应的

$$P(a_i^{(j)}|c_k) = 0$$

当使用模型预测新的数据时，如果正好遇到预测的数据的属性值为 $a_i^{(j)}$，则因为该项的概率为 0，会使整体的概率为 0。这样无论其他属性的概率有多高，都对分类结果没有影响。为了避免出现这种情况，我们希望给 $P(a_i^{(j)}|c_k)$ 赋个很小的概率。这样既体现了该属性的作用，又可以整体用到其他属性的概率信息。

下面使用离散数据的平滑技术：加法平滑[Additive Smoothing，或称拉普拉斯平滑（Laplace Smoothing）]来解决这个问题，即，定义平滑参数 $\lambda \geq 0$，条件概率如下：

$$P(a_i^{(j)}|c_k) = \frac{\text{在分类为} c_k \text{的训练样本中，} a_i = a_i^{(j)} \text{的样本数} + \lambda}{\text{分类为} c_k \text{的训练样本数} + s_i \cdot \lambda}$$

常取参数 $\lambda = 1$。在表达式的分子中加 λ，分母中加 $s_i \cdot \lambda$，这样可保证条件概率的和为 1，即

$$\sum_{j=1}^{s_i} P(a_i^{(j)}|c_k) = 1$$

在实际应用中，遍历计算所有离散值关于分类值 c_k 的条件概率，并保存到模型中。在预测新的数据时，对于其属性值 a_i，可以通过查表的方式找到 $P(a_i|c_k)$。

9.2 决策树模型

决策树模型再现了人们做决策的过程，该过程由一系列的判断构成，后面的判断基于前面判断的结果，不断缩小范围，最终推出结果。我们先看一个使用决策树模型的简单示例。通过

天气状态判断某一天是否适合打高尔夫球，决策的过程如下。首先看天气的整体情况：晴天？阴天？下雨？如果是阴天的话，可以直接给出结论。对于晴天的情况，需要根据湿度情况进一步判断；对于雨天的情况，需要根据是否刮风，给出最终的结论。整个决策过程如图9-1所示，用树状结构可以形象地展示。

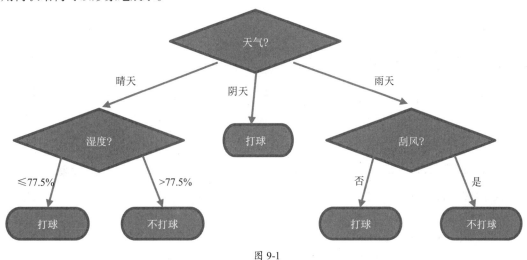

图 9-1

在此可以看到：
- 树的叶节点，给出结论。
- 树的枝节点，做出判断。
  - 作用：产生子节点（可能是枝节点或叶节点），即根据本轮的判断，引申出进一步的判断或得出结论。
  - 方式：针对某个特征（分裂特征），看特征值属于哪种划分方式（分裂方式）。比如，湿度按≤77.5%与>77.5%进行划分；天气按晴天、阴天和雨天进行划分。

在上述例子中，是否适合打高尔夫球的场景，就对应着Golf数据集，如表9-1所示。表9-1记录了天气状态［天气的整体情况（Outlook）、温度（Temperature，本列采用的是华氏度）、湿度（Humidity）、是否刮风（Windy）］及当天是否适合打高尔夫球。注意：本示例Golf数据集中的湿度百分比号省略。

表 9-1　Golf 数据集

序号	Outlook	Temperature	Humidity	Windy	Play
1	sunny	85	85	false	no
2	sunny	80	90	true	no
3	overcast	83	78	false	yes

续表

序号	Outlook	Temperature	Humidity	Windy	Play
4	rainy	70	96	false	yes
5	rainy	68	80	false	yes
6	rainy	65	70	true	no
7	overcast	64	65	true	yes
8	sunny	72	95	false	no
9	sunny	69	70	false	yes
10	rainy	75	80	false	yes
11	sunny	75	70	true	yes
12	overcast	72	90	true	yes
13	overcast	81	75	false	yes
14	rainy	71	80	true	no

决策树的构建可以被看作一个递归的过程，具体过程如下所示，先确定根节点的分裂特征和分裂方式，然后从其最左边的节点开始：

（1）设节点的数据集为D，
- 如果所有数据都属于同一个类别C_j，则将该节点作为叶节点；
- 如果满足某种条件，该数据集不能再划分，则将D中实例数最大的类C_j作为该节点的类别；
- 否则，考虑所有特征及其可能的划分方式。找到使分裂指标最优的划分方式，定为该节点的分裂特征和分裂方式。此节点作为枝节点，每个分裂方式对应一个子节点S_i，并按此分裂特征进行判断，每个子节点S_i均对应一个非空子集D_i。

（2）对子节点S_i（相应的数据集D_i）递归调用（1）。

（3）汇总所有计算出的枝节点和叶节点，生成决策树。

对于 Golf 数据集，我们可以得到如图 9-2 所示的决策树，每个节点的 Instances 项为属于该节点的样本个数，并表明了标签值为 yes 或 no 的样本数量和比例。

注意：图 9-2 由 Alink 决策树组件生成，在 9.2.5 节中有相关的具体示例。

在实际操作中还要考虑其他因素，以控制树的生长，常用下面一些判断：

（1）当决策树达到一定的高度时，就让决策树停止生长。

（2）到达某节点的训练样本具有相同的特征属性，即使这些样本不属于同一类，也可以让决策树停止生长。

（3）当某节点的训练样本个数小于叶节点最低样本数的时候，也可以让决策树停止生长。
（4）如果分裂指标的增量小于某个阈值，也可以让决策树停止生长。

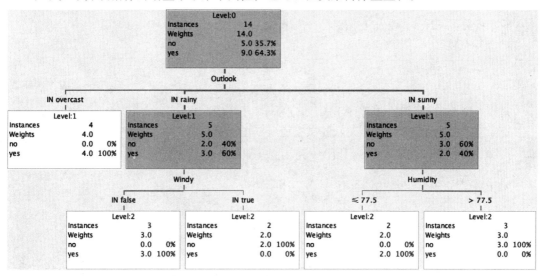

图 9-2

9.2.1 决策树的分裂指标定义

对于离散随机变量X，有k个状态$\{x_1, x_2, \cdots, x_k\}$，其每个状态对应的概率为$\{p_1, p_2, \cdots, p_k\}$，则$X$的信息熵定义如下：

$$H(X) = -\sum_{i=1}^{k} p_i \log_a(p_i)$$

其中a为对数的底，当$a = 2$时，信息熵的单位为比特（bit）；当$a = e$时，信息熵的单位为奈特（nat）。由定义可以看出，离散随机变量X的信息熵为"信息量的加权平均"。

设Y有m个取值$\{y_1, y_2, \cdots, y_m\}$。条件熵表示在已知随机变量$X$的条件下随机变量$Y$的不确定性，其定义为在$X$给定的条件下，$Y$的条件概率分布对$X$的数学期望：

$$H(Y|X) = \sum_{i=1}^{k} P(x_i) H(Y|X = x_i)$$

设训练集为C，用$|\cdot|$表示集合的元素个数，则$|C|$为样本容量，即样本个数。
设训练数据共有k个分类，按分类划分为k个样本集合$\{C_1, C_2, \cdots, C_k\}$，满足：

$$\begin{cases} C_i \cap C_j = \emptyset, \quad \forall i \neq j \\ C = C_1 \cup C_2 \cup \cdots \cup C_k \end{cases}$$

对于各分类的样本个数$|C_i|$，有

$$|C| = \sum_{i=1}^{k} |C_i|$$

对于特征A，按照某种规则，可将C划分为m个子集$\{A_1, A_2, \cdots, A_m\}$，满足：

$$\begin{cases} A_i \cap A_j = \emptyset, \quad \forall i \neq j \\ C = A_1 \cup A_2 \cup \cdots \cup A_m \end{cases}$$

对于各子集的样本个数$|A_i|$，有

$$|C| = \sum_{i=1}^{m} |A_i|$$

信息增益（Information Gain，IG）表示得知特征A的信息而使分类C的信息的熵减少的程度。

$$IG(C, A) = H(C) - H(C|A)$$

由前面介绍的内容可知，信息增益也是特征A与分类C的互信息。

$$IG(C, A) = I(C; A) = H(C) + H(A) - H(C, A)$$

各类别的子集为$\{C_1, C_2, \cdots, C_k\}$，相应的概率为$\left\{\frac{|C_1|}{|C|}, \frac{|C_2|}{|C|}, \cdots, \frac{|C_k|}{|C|}\right\}$。由信息熵和条件熵的定义，数据集$C$的信息熵如下：

$$H(C) = -\sum_{i=1}^{k} \frac{|C_i|}{|C|} \ln \frac{|C_i|}{|C|}$$

根据特征A的取值，将数据集划分为m个子集$\{A_1, A_2, \cdots, A_k\}$，则特征$A$对于数据集$C$的条件熵如下：

$$H(C|A) = -\sum_{i=1}^{m} \frac{|A_i|}{|C|} \sum_{j=1}^{k} \frac{|A_i \cap C_j|}{|A_i|} \ln \frac{|A_i \cap C_j|}{|A_i|}$$

所以，信息增益

$$\text{IG}(C,A) = H(C) - H(C|A) = \sum_{i=1}^{m} \frac{|A_i|}{|C|} \sum_{j=1}^{k} \frac{|A_i \cap C_j|}{|A_i|} \ln \frac{|A_i \cap C_j|}{|A_i|} - \sum_{i=1}^{k} \frac{|C_i|}{|C|} \ln \frac{|C_i|}{|C|}$$

信息增益率（Information Gain Ratio，IGR）为信息增益$\text{IG}(C,A)$与特征A的信息熵的比值，即

$$\text{IGR}(C,A) = \frac{\text{IG}(C,A)}{H(A)}$$

这里，假设数据集C中属于各类别的子集为$\{C_1, C_2, \cdots, C_k\}$，并根据特征$A$的值将数据集划分为$m$个子集$\{A_1, A_2, \cdots, A_k\}$，则$H(A)$和$\text{IG}(C,A)$的计算公式如下：

$$H(A) = -\sum_{i=1}^{k} \frac{|A_i|}{|C|} \ln \frac{|A_i|}{|C|}$$

$$\text{IG}(C,A) = \sum_{i=1}^{m} \frac{|A_i|}{|C|} \sum_{j=1}^{k} \frac{|A_i \cap C_j|}{|A_i|} \ln \frac{|A_i \cap C_j|}{|A_i|} - \sum_{i=1}^{k} \frac{|C_i|}{|C|} \ln \frac{|C_i|}{|C|}$$

样本集合C的基尼指数（Gini Index，或称基尼系数）与其分类情况相关：

$$\text{Gini}(C) = 1 - \sum_{i=1}^{k} \left(\frac{|C_i|}{|C|} \right)^2$$

根据属性特征A，若样本集合C被划分为两个非空子集A_1和A_2，则

$$\text{Gini}(C,A) = \frac{|A_1|}{|C|} \text{Gini}(A_1) + \frac{|A_2|}{|C|} \text{Gini}(A_2)$$

其中，

$$\text{Gini}(A_j) = 1 - \sum_{i=1}^{k} \left(\frac{|A_j \cap C_i|}{|A_j|} \right)^2$$

9.2.2 常用的决策树算法

常用的决策树算法包括 ID3、C4.5 和 CART，各算法的简介如表 9-2 所示。

表 9-2 常用的决策树算法简介

算法	简介	作者	年份
ID3	Iterative Dichotomiser 3 的缩写	Ross Quinlan	1986 年

续表

算法	简介	作者	年份
C4.5	对 ID3 算法的改进	Ross Quinlan	1993 年
CART	Classification And Regression Tree 的缩写	L. Breiman, J. Friedman, R. Olshen, C. Stone	1984 年

决策树对于离散特征和连续特征，处理方法有些差异，下面将分别讨论。

1. 离散特征

对于离散特征，直接套用公式，计算分裂指标即可，参见表 9-3。

表 9-3 离散特征的决策树分裂指标

算法	分裂指标
ID3	使用信息增益（Information Gain，IG） $$IG(C, A^{(i)}) = H(C) - H(C|A^{(i)})$$ 选择使 $IG(C, A^{(i)})$ 最大的特征属性 $A^{(i)}$ 划分方法：将枚举特征属性 $A^{(i)}$ 的每一个取值 $a_j^{(i)}$，分别作为一个划分子集
C4.5	使用信息增益率（Information Gain Ratio，IGR） $$IGR(C, A^{(i)}) = \frac{IG(C, A^{(i)})}{H(A^{(i)})}$$ 选择使 $IGR(C, A^{(i)})$ 最大的特征属性 $A^{(i)}$ 划分方法：将枚举特征属性 $A^{(i)}$ 的每一个取值 $a_j^{(i)}$，分别作为一个划分子集
CART	使用基尼指数（Gini Index） 需要选择特征属性 $A^{(i)}$ 是否取某一个可能值 $a_j^{(i)}$，将数据分割成 $A_1^{(i)}$ 和 $A_2^{(i)}$ 两部分，并计算 $$Gini(C, A^{(i)} = a_j^{(i)})$$ 考虑基尼增量： $$\Delta Gini(C) = Gini(C) - Gini(C, A^{(i)} = a_j^{(i)})$$ 目标是使基尼增量最大，即选择使基尼指数 $Gini(C, A^{(i)} = a_j^{(i)})$ 最小的属性特征 $A^{(i)}$。每次将是否等于 $a_j^{(i)}$ 作为划分条件，得到两个划分子集。

2. 连续特征

处理连续特征的方法如下：确定可能的阈值集合；每个阈值点，可将数据集划分为 2 个部分；计算每个划分的指标（信息增益、信息增益率，或者基尼指数）；找到使指标达到最优的阈值点，并将该阈值点作为分割阈值点。连续特征的决策树分裂指标如表 9-4 所示。

计算分割阈值点的具体步骤如下：

（1）确定可能的阈值集合。

  （a）对特征的取值按升序进行排序。

  （b）将两个特征取值之间的中点作为可能的分裂点。（如果有 n 条训练样本，那么可能有 $n-1$ 个阈值，过多的阈值会对应大量的指标计算。我们可以通过判断两个相邻特征取值所对应的样本的分类信息是否相同来减少计算次数。如果该分类信息相同，则该点不适合作为分割点，可以略过，这样能显著减少阈值的数量。）

（2）通过计算分类指标，确定指标最优的阈值，并将其作为分割阈值点。

  （a）对于阈值 a_j，则 $(-\infty, a_j]$ 和 $(a_j, +\infty)$ 就是该特征的划分子集，计算分类指标。

  （b）找出指标最优的阈值，并将其作为分割阈值点。

表 9-4 连续特征的决策树分裂指标

算法	分裂指标
ID3	使用信息增益（Information Gain，IG） $$IG(C, A^{(i)}) = H(C) - H(C\|A^{(i)})$$ 选择使 $IG(C, A^{(i)})$ 最大的特征属性 $A^{(i)}$ 以及阈值分割点 $a^{(i)}$
C4.5	使用信息增益率（Information Gain Ratio，IGR） $$IGR(C, A^{(i)}) = \frac{IG(C, A^{(i)})}{H(A^{(i)})}$$ 选择使 $IGR(C, A^{(i)})$ 最大的特征属性 $A^{(i)}$ 以及阈值分割点 $a^{(i)}$
CART	使用基尼指数（Gini Index） $$\text{Gini}(C, A^{(i)})$$ 考虑基尼增量： $$\Delta\text{Gini}(C) = \text{Gini}(C) - \text{Gini}\left(C, A^{(i)} = a_j^{(i)}\right)$$ 目标是使基尼增量最大，即选择基尼指数 $\text{Gini}\left(C, A^{(i)} = a_j^{(i)}\right)$ 最小的属性特征 $A^{(i)}$ 以及阈值分割点 $a^{(i)}$

9.2.3 指标计算示例

我们以 Golf 数据集（见表 9-1）为例，目标是根据某一天的天气状态（天气的整体情况、温度、湿度、是否刮风)来判断这一天是否适合打高尔夫球。

分类C共有两个分类值 yes 和 no,对应的分类子集为C_1和C_2,其统计值如表 9-5 所示。

表 9-5 分类值的统计

	个数
yes	9
no	5
总数	14

计算数据集的信息熵如下:

$$H(C) = -\sum_{i=1}^{k} \frac{|C_i|}{|C|} \ln \frac{|C_i|}{|C|} = -\left(\frac{9}{14} \ln \frac{9}{14} + \frac{5}{14} \ln \frac{5}{14}\right) = 0.6517565611726531$$

计算 Outlook 属性的指标,该属性的统计值如表 9-6 所示。

表 9-6 Outlook 属性的统计值

	sunny	overcast	rainy
yes	2	4	3
no	3	0	2
总数	5	4	5

$$H(A_{\text{Windy}}) = -\sum_{i=1}^{k} \frac{|A_i|}{|C|} \ln \frac{|A_i|}{|C|} = 1.0933747175566468$$

$$H(C|A_{\text{Outlook}}) = -\sum_{i=1}^{m} \frac{|A_i|}{|C|} \sum_{j=1}^{k} \frac{|A_i \cap C_j|}{|A_i|} \ln \frac{|A_i \cap C_j|}{|A_i|} = 0.48072261929232607$$

所以,

$$\text{IG}(C, A_{\text{Outlook}}) = H(C) - H(C|A_{\text{Outlook}}) = 0.17103394188032706$$

$$\text{IGR}(C, A_{\text{Outlook}}) = \frac{\text{IG}(C, A_{\text{Outlook}})}{H(A_{\text{Outlook}})} = 0.15642756242111751$$

与计算 Outlook 属性类似,可以计算出 Windy 属性,该属性的统计值如表 9-7 所示。

表 9-7 Windy 属性的统计值

	true	false
yes	3	6
no	3	2
总数	6	8

$$H(A_{\text{Windy}}) = -\sum_{i=1}^{k} \frac{|A_i|}{|C|} \ln \frac{|A_i|}{|C|} = 0.6829081047004717$$

$$H(C|A_{\text{Windy}}) = -\sum_{i=1}^{m} \frac{|A_i|}{|C|} \sum_{j=1}^{k} \frac{|A_i \cap C_j|}{|A_i|} \ln \frac{|A_i \cap C_j|}{|A_i|} = 0.6183974457364384$$

所以，

$$\text{IG}(C, A_{\text{Windy}}) = H(C) - H(C|A_{\text{Windy}}) = 0.033359115436214726$$

$$\text{IGR}(C, A_{\text{windy}}) = \frac{\text{IG}(C, A_{\text{Windy}})}{H(A_{\text{Windy}})} = 0.04884861551152079$$

显然，在信息增益（IG）和信息增益率（IGR）指标上，Outlook 属性均优于 Windy 属性。我们再计算一下连续值属性 Temperature 的指标。

按温度高低对样本进行排序。我们只关注温度与分类列，排序后的结果如下：

64	65	68	69	70	71	72	72	75	75	80	81	83	85
yes	no	yes	yes	yes	no	no	yes	yes	yes	no	yes	yes	no

下面我们先找可能的阈值点。我们可以通过下面的两个判断准则来减少候选阈值点的数量：
（1）阈值点一定位于两个不同的数值之间。比如，阈值点可以在 64 和 65 中间。
（2）对于相邻的不重复的数值，如果它们的分类值一样，则其中点不是候选阈值点。
于是，我们得到候选阈值点的分割显示如下：

64	65	68	69	70	71	72	72	75	75	80	81	83	85
yes	no	yes	yes	yes	no	no	yes	yes	yes	no	yes	yes	no

候选阈值点集合：{64.5, 66.5, 70.5, 71.5, 73.5, 77.5, 80.5, 84}。

以候选阈值点 71.5 为例，计算分类指标，该阈值的统计值如表 9-8 所示。

表 9-8 阈值点 71.5 的统计值

	≤71.5	>71.5
yes	4	5
no	2	3
总数	6	8

$$H(A_{\text{Temperature}} = 71.5) = -\sum_{i=1}^{k} \frac{|A_i|}{|C|} \ln \frac{|A_i|}{|C|} = 0.6829081047004717$$

$$H(C|A_{\text{Temperature}} = 71.5) = -\sum_{i=1}^{m} \frac{|A_i|}{|C|} \sum_{j=1}^{k} \frac{|A_i \cap C_j|}{|A_i|} \ln \frac{|A_i \cap C_j|}{|A_i|} = 0.6508279225023381$$

计算出各个阈值点对应的指标如表 9-9 所示。

表 9-9 各个阈值点对应的指标

θ	$H(A_{\text{Temperature}} = \theta)$	$H(C\|A_{\text{Temperature}} = \theta)$	$IG(C, A_{\text{Temperature}} = \theta)$	$IGR(C, A_{\text{Temperature}} = \theta)$
64.5	0.25731864054383163	0.618687125099342	0.03306943607331114	0.12851550903354822
66.5	0.410116318288409	0.6446045986184031	0.007151962554249991	0.01743886364750906
70.5	0.6517565611726531	0.6203333076476321	0.031423253525021067	0.04821317558887898
71.5	0.6829081047004717	0.6508279225023381	9.286386703150074E-4	0.0013598296226434664
73.5	0.6829081047004717	0.6508279225023381	9.286386703150074E-4	0.0013598296226434664
77.5	0.5982695885852573	0.6343736959134797	0.01738286525917343	0.029055237957655706
80.5	0.5195798391305154	0.651417286985697	3.392741869561178E-4	6.529779668200214E-4
84	0.25731864054383163	0.5731530718924601	0.078603489280193	0.3054714151841778

注意：对于阈值为 64.5 和 84 的情况（表 9-9 中背景为灰色的区域），在划分的集合中，其中一个划分子集只有一个训练样本。考虑到划分后要满足每个叶节点的样本数大于 1，所以这两种阈值的情况不予考虑。

9.2.4 分类树与回归树

决策树分为两大类：分类树、回归树，其可分别解决机器学习领域中的分类和回归问题。分类树预测分类标签值，回归树用于预测数值。

分类树在每次分枝时，会针对当前枝节点的样本，找出最优的分裂特征和分裂方式，从而得到若干新节点；继续分枝，直到所有样本都被分入类别唯一的叶节点，或者满足设定的终止条件（比如，叶节点最低样本数的限制，或者决策树层数的限制）。若最终叶节点中的类别不唯一，则以多数样本的类别作为该叶节点的类别。

回归树的建模流程与分类树类似；回归树在分枝时的分裂指标与分类树不同，回归树一般以均方差为分裂指标；回归树的分枝操作很难做到每个叶节点上的数值都一样，在实际中，以最终叶节点上所有样本的平均数值作为该叶节点的预测数值。

9.2.5 经典的决策树示例

根据天气因素判断能否打高尔夫球是一个经典的决策树示例。该数据集只有 14 条数据，我们使用内存数据源组件 MemSourceBatchOp，可以直接在代码中定义数据记录 Row 类型的数组，并输入相应的列名称。构造函数代码如下。这里省略了中间的一些数据；最后一句为使用 Lazy 方式进行数据的打印输出，参数-1 表示打印全部数据。

```
MemSourceBatchOp source = new MemSourceBatchOp(
    new Row[] {
        Row.of("sunny", 85.0, 85.0, false, "no"),
... ...
        Row.of("overcast", 81.0, 75.0, false, "yes"),
        Row.of("rainy", 71.0, 80.0, true, "no")
    },
    new String[] {"Outlook", "Temperature", "Humidity", "Windy", "Play"}
);

source.lazyPrint(-1);
```

打印输出数据，如表 9-10 所示。前面四列为特征列，最后一列为是否打高尔夫球的判断结果列，即分类问题的标签列。

表 9-10　Golf 数据集

Outlook	Temperature	Humidity	Windy	Play
sunny	85.0000	85.0000	false	no
sunny	80.0000	90.0000	true	no
overcast	83.0000	78.0000	false	yes
rainy	70.0000	96.0000	false	yes
rainy	68.0000	80.0000	false	yes
rainy	65.0000	70.0000	true	no
overcast	64.0000	65.0000	true	yes
sunny	72.0000	95.0000	false	no
sunny	69.0000	70.0000	false	yes

续表

Outlook	Temperature	Humidity	Windy	Play
rainy	75.0000	80.0000	false	yes
sunny	75.0000	70.0000	true	yes
overcast	72.0000	90.0000	true	yes
overcast	81.0000	75.0000	false	yes
rainy	71.0000	80.0000	true	no

下面，将此数据连接到 C45 决策树训练组件，设置特征数据列，设置其中的离散特征列，并设置标签列。这些参数设置与前面的朴素贝叶斯训练组件类似。决策树还有其他的训练参数，这里暂时使用其默认值。打印输出默认的模型信息。如果需要查看树结构的模型展示，则需要使用 lazyCollectModelInfo 方法来获取决策树模型信息。类 DecisionTreeModelInfo 中有 saveTreeAsImage 方法，其第一个参数为图片路径，第二个参数表示是否覆盖已有图片。

```
source
  .link(
    new C45TrainBatchOp()
      .setFeatureCols("Outlook", "Temperature", "Humidity", "Windy")
      .setCategoricalCols("Outlook", "Windy")
      .setLabelCol("Play")
      .lazyPrintModelInfo()
      .lazyCollectModelInfo(new Consumer <DecisionTreeModelInfo>() {
        @Override
        public void accept(DecisionTreeModelInfo decisionTreeModelInfo) {
          try {
            decisionTreeModelInfo.saveTreeAsImage(
              DATA_DIR + "weather_tree_model.png", true);
          } catch (IOException e) {
            e.printStackTrace();
          }
        }
      })
  );
```

打印输出的模型信息如下：

```
Classification trees modelInfo:
Number of trees: 1
Number of features: 4
Number of categorical features: 2
Labels: [no, yes]

Categorical feature info:
|feature|number of categorical value|
|-------|---------------------------|
```

```
|outlook|                               3|
| Windy|                                2|
Table of feature importance:
|    feature|importance|
|-----------|----------|
|   Humidity|    0.4637|
|      Windy|    0.4637|
|    Outlook|    0.0725|
|Temperature|         0|
```

决策树的结构如图 9-3 所示。

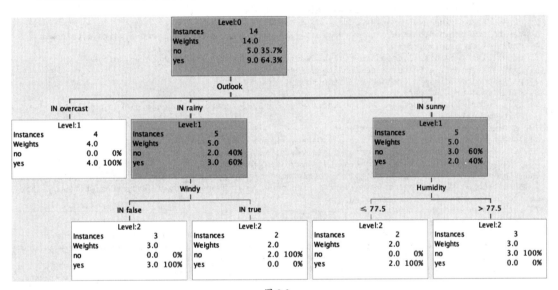

图 9-3

该树结构展示了完整的推断逻辑：

（1）根节点显示的是整个训练集的情况：14 条样本数据。在没有输入权重列参数的情况下，每条样本的权重都是 1.0，并注明了这些样本数据中各个标签值的分布比例。

（2）按 Outlook 特征进行判断。对于取值为 overcast 的样本，直接判断结果为 yes；对于其他取值的情况，若两种标签值都存在，则需要继续判断。

（3）对于 Outlook=sunny 的情况，可根据特征 Humidity 进行判断：该值小于或等于 77.5 时，均判断为 yes；该值大于 77.5 时，均判断为 no。对于 Outlook=rainy 的情况，可根据特征 Windy 进行判断：若值为 true，则判断结果为 no；若值为 false，则判断结果为 yes。

9.3 数据探索

本节使用的是 UCI 的 Mushroom 数据集，可通过链接 9-1 下载。其数据来自 *The Audubon Society Field Guide to North American Mushrooms*（1981 年）。

该数据集有 8124 条数据、23 个变量。其中，变量 class 为分类标签，其有 2 个标签值："e" 表示 edible（可食用），"p" 表示 poisonous（有毒）。该数据集共有 22 个特征变量，各特征名称及缩写字母含义如下所示：

```
1. cap-shape: bell=b,conical=c,convex=x,flat=f, knobbed=k,sunken=s
2. cap-surface: fibrous=f,grooves=g,scaly=y,smooth=s
3. cap-color: brown=n,buff=b,cinnamon=c,gray=g,green=r, pink=p,purple=u,red=e,white=w,yellow=y
4. bruises?: bruises=t,no=f
5. odor: almond=a,anise=l,creosote=c,fishy=y,foul=f, musty=m,none=n,pungent=p,spicy=s
6. gill-attachment: attached=a,descending=d,free=f,notched=n
7. gill-spacing: close=c,crowded=w,distant=d
8. gill-size: broad=b,narrow=n
9. gill-color: black=k,brown=n,buff=b,chocolate=h,gray=g, green=r,orange=o,pink=p,purple=u,red=e,white=w,yellow=y
10. stalk-shape: enlarging=e,tapering=t
11. stalk-root: bulbous=b,club=c,cup=u,equal=e, rhizomorphs=z,rooted=r,missing=?
12. stalk-surface-above-ring: fibrous=f,scaly=y,silky=k,smooth=s
13. stalk-surface-below-ring: fibrous=f,scaly=y,silky=k,smooth=s
14. stalk-color-above-ring: brown=n,buff=b,cinnamon=c,gray=g,orange=o, pink=p,red=e,white=w,yellow=y
15. stalk-color-below-ring: brown=n,buff=b,cinnamon=c,gray=g,orange=o, pink=p,red=e,white=w,yellow=y
16. veil-type: partial=p,universal=u
17. veil-color: brown=n,orange=o,white=w,yellow=y
18. ring-number: none=n,one=o,two=t
19. ring-type: cobwebby=c,evanescent=e,flaring=f,large=l, none=n,pendant=p,sheathing=s,zone=z
20. spore-print-color: black=k,brown=n,buff=b,chocolate=h,green=r, orange=o,purple=u,white=w,yellow=y
21. population: abundant=a,clustered=c,numerous=n, scattered=s,several=v,solitary=y
22. habitat: grasses=g,leaves=l,meadows=m,paths=p, urban=u,waste=w,woods=d
```

这里有几个重要的单词需要说明一下：菌盖(cap)、菌褶(gill)、菌柄(stalk)、菌环(ring)、气味(odor)。图 9-4 说明了蘑菇各个部位的名称。

图 9-4

我们将数据文件下载到本地，使用文本编辑器打开该数据文件，数据显示如图 9-5 所示。

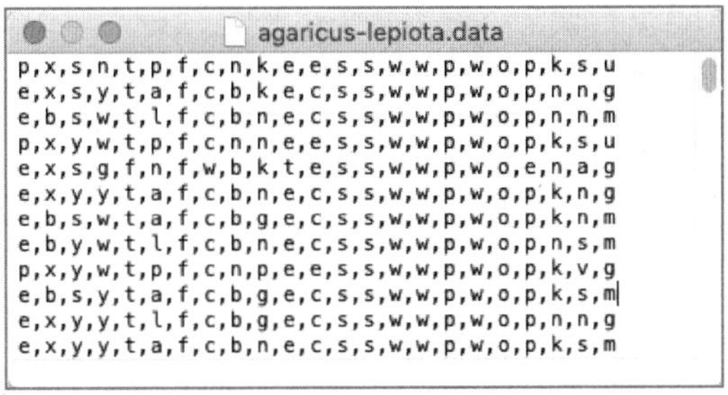

图 9-5

每个特征值都是一个字母，特征值之间以逗号分隔，一行为一条数据。这是典型的 CSV 格式。只需定义好每列的名称，其类型都为 string 类型，便可以使用 CsvSourceBatchOp 组件读取数据了。首先，各列数据的名称和类型如下。标签列为 class 列，除标签列外的各列均为特征列。

```java
private static final String DATA_DIR = Chap01.ROOT_DIR + "mushroom" + File.separator;
private static final String ORIGIN_FILE = "agaricus-lepiota.data";
private static final String[] COL_NAMES = new String[] {
    "class",
    "cap_shape", "cap_surface", "cap_color", "bruises", "odor",
    "gill_attachment", "gill_spacing", "gill_size", "gill_color",
    "stalk_shape", "stalk_root", "stalk_surface_above_ring", "stalk_surface_below_ring",
    "stalk_color_above_ring", "stalk_color_below_ring",
    "veil_type", "veil_color",
    "ring_number", "ring_type", "spore_print_color", "population", "habitat"
};

private static final String[] COL_TYPES = new String[] {
    "string",
    "string", "string", "string", "string", "string",
    "string", "string", "string", "string",
    "string", "string", "string", "string",
    "string", "string",
    "string", "string",
    "string", "string"
};

static final String LABEL_COL_NAME = "class";
static final String[] FEATURE_COL_NAMES = ArrayUtils.removeElement(COL_NAMES, LABEL_COL_NAME);
```

使用 CsvSourceBatchOp 读取并显示前 5 条数据，代码如下：

```java
CsvSourceBatchOp source = new CsvSourceBatchOp()
    .setFilePath(DATA_DIR + ORIGIN_FILE)
    .setSchemaStr(Utils.generateSchemaString(COL_NAMES, COL_TYPES));

source.lazyPrint(5, "< origin data >");
```

打印输出如下。在此，可以看到，第一列 class 为分类标签列，随后是各个离散特征列。

```
< origin data >
class|cap_shape|cap_surface|cap_color|bruises|odor|gill_attachment|gill_spacing|gill_size|gill_color|s
talk_shape|stalk_root|stalk_surface_above_ring|stalk_surface_below_ring|stalk_color_above_ring|stalk_c
olor_below_ring|veil_type|veil_color|ring_number|ring_type|spore_print_color|population|habitat
-----|---------|-----------|---------|-------|----|---------------|------------|---------|----------|-
------|----------|------------------------|------------------------|----------------------|----------
------|---------|----------|-----------|---------|-----------------|----------|-------
p|f|y|y|f|f|f|c|b|p|e|b|k|k|b|p|p|w|o|l|h|y|d
p|x|f|y|f|f|f|c|b|p|e|b|k|k|b|n|p|w|o|l|h|y|g
p|x|y|g|f|f|f|c|b|g|e|b|k|k|n|p|p|w|o|l|h|v|d
p|f|f|g|f|f|f|c|b|g|e|b|k|k|n|b|p|w|o|l|h|y|d
p|f|f|y|f|f|f|c|b|g|e|b|k|p|n|p|w|o|l|h|y|p
```

获得原始数据后，我们还需要将其分为训练集和测试集，并按 9∶1 的比例，分别保存为 AK 格式的文件 TRAIN_FILE 和 TEST_FILE，具体代码如下：

```
private static final String TRAIN_FILE = "train.ak";
private static final String TEST_FILE = "test.ak";
... ...
Utils.splitTrainTestIfNotExist(source, DATA_DIR + TRAIN_FILE, DATA_DIR + TEST_FILE, 0.9);
```

接下来，我们使用卡方特征选择组件 ChiSqSelectorBatchOp，从当前特征列中选出对分类问题最重要的 3 个特征。具体代码如下：

```
new AkSourceBatchOp()
    .setFilePath(DATA_DIR + TRAIN_FILE)
    .link(
        new ChiSqSelectorBatchOp()
            .setSelectorType(SelectorType.NumTopFeatures)
            .setNumTopFeatures(3)
            .setSelectedCols(FEATURE_COL_NAMES)
            .setLabelCol(LABEL_COL_NAME)
            .lazyPrintModelInfo("< Chi-Square Selector >")
    );
```

输出结果如下：

```
< Chi-Square Selector >
---------------------------- ChisqSelectorModelInfo ----------------------------
Number of Selector Features: 3
Number of Features: 22
Type of Selector: NumTopFeatures
Number of Top Features: 3
Selector Indices:
    |          ColName|ChiSquare|PValue| DF|Selected|
    |-----------------|---------|------|---|--------|
    |             odor|6897.9815|     0|  8|    true|
```

```
|        spore_print_color|4197.5627|   0|  8|  true|
|                gill_color|3399.9436|   0| 11|  true|
|                 ring_type|2686.4367|   0|  4| false|
| stalk_surface_above_ring|2542.7242|   0|  3| false|
| stalk_surface_below_ring|2453.9015|   0|  3| false|
|                 gill_size|2161.6473|   0|  1| false|
|    stalk_color_above_ring|2027.4274|   0|  8| false|
|    stalk_color_below_ring|1950.8527|   0|  8| false|
|                   bruises|1874.4286|   0|  1| false|
|                population|1769.4532|   0|  5| false|
|                   habitat|1432.2396|   0|  6| false|
|                stalk_root|1254.6198|   0|  4| false|
|              gill_spacing| 884.8095|   0|  1| false|
|                 cap_shape| 455.8245|   0|  5| false|
|                 cap_color| 350.5556|   0|  9| false|
|               ring_number| 339.0443|   0|  2| false|
|               cap_surface| 276.0812|   0|  3| false|
|                veil_color| 171.2708|   0|  3| false|
|            gill_attachment| 120.5918|   0|  1| false|
|               stalk_shape|  70.8908|   0|  1| false|
|                 veil_type|        0|   1|  0| false|
```

显然，排在前三位的是 odor、spore_print_color 和 gill_color。在后面的示例中，我们会重点观察这三列。我们也注意到，排在最后的特征列的卡方值为 0。下面通过一个简单的操作来了解该特征列。我们单独选择该特征列，然后看其有多少不同的值，具体代码如下：

```
new AkSourceBatchOp()
    .setFilePath(DATA_DIR + TRAIN_FILE)
    .select("veil_type")
    .distinct()
    .lazyPrint(100);
```

运行结果如下：

```
veil_type
---------
p
```

即特征列 veil_type 只有唯一的一个值'p'。显然，该列不会对分类问题有任何贡献，我们可以完全不予考虑。

9.4 使用朴素贝叶斯方法

朴素贝叶斯训练组件（NaiveBayesTrainBatchOp）和朴素贝叶斯预测组件（NaiveBayes-

PredictBatchOp)的参数设置如下面的代码所示。朴素贝叶斯训练组件除了要设置特征列和标签列，还需要指出特征列中的哪些列是离散特征列（对应参数 CategoricalCols）。在当前的分类问题中，所有特征都是离散特征，所以该参数填入所有的特征列 FEATURE_COL_NAMES。朴素贝叶斯预测组件的设置与逻辑回归预测组件和线性 SVM 预测组件一样，需要指定预测结果的列名。朴素贝叶斯预测组件还可以通过设置预测详情的列名，得到属于各分类值的概率，用于后面的二分类评估。

```
NaiveBayesTrainBatchOp trainer =
    new NaiveBayesTrainBatchOp()
        .setFeatureCols(FEATURE_COL_NAMES)
        .setCategoricalCols(FEATURE_COL_NAMES)
        .setLabelCol(LABEL_COL_NAME);
NaiveBayesPredictBatchOp predictor = new NaiveBayesPredictBatchOp()
    .setPredictionCol(PREDICTION_COL_NAME)
    .setPredictionDetailCol(PRED_DETAIL_COL_NAME);
```

之后，通过 link 方法，建立起整个实验流程：训练数据源 train_data 连接朴素贝叶斯训练组件 trainer；预测过程需要模型，也需要待预测的数据，所以朴素贝叶斯预测组件 predictor 需要使用 linkFrom 方法来同时连接 trainer 输出的模型和待预测的数据源 test_data；predictor 组件的输出即预测结果。代码如下：

```
train_data.link(trainer);
predictor.linkFrom(trainer, test_data);
```

为了查看朴素贝叶斯模型的信息，选择训练组件的 lazyPrintModelInfo 方法：

```
trainer.lazyPrintModelInfo();
```

模型信息打印如下，包括特征列的名称、标签值。特别地，对每个标签值给出了其在整个训练集中的比例。标签值'p'（有毒）的比例为 0.4824；标签值'e'（可食用）的比例为 0.5176。然后，给出了各离散特征中每个离散值对应类别'p'和'e'的比例。模型打印出的信息较多，这里只显示了特征 gill_color、spore_print_color 和 odor 的信息，略去了其他特征的信息。

```
---------------------- NaiveBayesTextModelInfo ----------------------

=========================== model meta info ===========================
{label number: 2, feature size: 22, feature col names:
["cap_shape","cap_surface","cap_color","bruises","odor","gill_attachment","gill_spacing","gill_size","gill_color","stalk_shape","stalk_root","stalk_surface_above_ring","stalk_surface_below_ring","stalk_color_above_ring","stalk_color_below_ring","veil_type","veil_color","ring_number","ring_type","spore_print_color","population","habitat"], labels: ["p","e"]}
====================== label proportion information ======================

label info:[p, e]
```

```
proportion:[0.4824, 0.5176]
======================== category information ========================
categorical features: [stalk_surface_above_ring, habitat, gill_spacing, ..., stalk_color_above_ring,
gill_size, ring_type]
gaussian features: []
============== categorical features proportion information ==============

... ...

The features proportion information of gill_color:
| |     p|  r|  b|...|     k|     n|  o|
|---|------|---|---|---|------|------|---|
| p|0.4251|  1|  1|...|0.1432|0.1065|  0|
| e|0.5749|  0|  0|...|0.8568|0.8935|  1|

The features proportion information of spore_print_color:
| |  r|  b|  u|...|     k|     n|  o|
|---|---|---|---|---|------|------|---|
| p|  1|  0|  0|...|0.1136|0.1142|  0|
| e|  0|  1|  1|...|0.8864|0.8858|  1|

The features proportion information of odor:
| |  p|  a|  c|...|  l|     m|     n|
|---|---|---|---|---|---|------|------|
| p|  1|  0|  1|...|  0|     1|0.0339|
| e|  0|  1|  0|...|  1|     0|0.9661|

... ...

============== continuous features mean sigma information ==============

There is no continuous feature.
```

对于特征 gill_color、spore_print_color 和 odor，每个特征的离散值较多，在此无法完全显示，我们只选择打印输出了 6 个离散值。如果读者希望了解这三个特征的全部离散值相对于分类标签值的概率，可以使用 lazyCollectModelInfo 方法，获取到 NaiveBayesModelInfo 对象，选择输出内容，具体代码如下：

```
trainer.lazyCollectModelInfo(new Consumer<NaiveBayesModelInfo>() {
  @Override
  public void accept(NaiveBayesModelInfo naiveBayesModelInfo) {
    StringBuilder sbd = new StringBuilder();
    for (String feature : new String[] {"odor", "spore_print_color", "gill_color"}) {
      HashMap<Object, HashMap<Object, Double>> map2 =
        naiveBayesModelInfo.getCategoryFeatureInfo().get(feature);
      sbd.append("\nfeature:").append(feature);
      for(Entry<Object, HashMap<Object, Double>> entry:map2.entrySet()){
```

```
        sbd.append("\n").append(entry.getKey()).append(" : ")
            .append(entry.getValue().toString());
      }
    }
    System.out.println(sbd.toString());
  }
});
```

运行结果如下。在此可看到很多离散值对应着概率值 1.0，即取到这些值，便可以 100% 判断样本的所属分类。尤其是 odor 特征，当离散值不为'n'的时候，均可以完全判断样本的类别；当离散值为'n'的时候，属于标签值'e'（可食用）的概率值也很大，是 96.6%。

```
feature:odor
p : {p=1.0, s=1.0, c=1.0, f=1.0, y=1.0, m=1.0, n=0.033871478315922764}
e : {a=1.0, l=1.0, n=0.9661285216840773}
feature:spore_print_color
p : {r=1.0, w=0.763585694379935, h=0.9720518064076347, k=0.11355529131985731, n=0.11423747889701745}
e : {b=1.0, u=1.0, w=0.23641430562006502, h=0.02794819359236537, y=1.0, k=0.8864447086801427, n=0.8857625211029826, o=1.0}
feature:gill_color
p : {p=0.42509225092250924, b=1.0, r=1.0, u=0.09610983981693363, w=0.20888468809073724, g=0.668141592920354, h=0.7216338880484114, y=0.25675675675675674, k=0.14323607427055704, n=0.10649627263045794}
e : {p=0.5749077490774908, e=1.0, u=0.9038901601830663, w=0.7911153119092628, g=0.33185840707964603, h=0.2783661119515885, y=0.7432432432432432, k=0.8567639257294429, n=0.8935037273695421, o=1.0}
```

最后，我们打印 10 条预测记录，并根据预测结果集，计算输出朴素贝叶斯二分类模型的评估指标，具体代码如下：

```
predictor.lazyPrint(10, "< Prediction >");

predictor
  .link(
    new EvalBinaryClassBatchOp()
      .setPositiveLabelValueString("p")
      .setLabelCol(LABEL_COL_NAME)
      .setPredictionDetailCol(PRED_DETAIL_COL_NAME)
      .lazyPrintMetrics()
  );

BatchOperator.execute();
```

结果如下，最右面的列为预测详情列，右边第二列是预测结果列。各项评估指标都比较高，但从混淆矩阵中可看到，仍有一例误判的情况：将一个有毒（标签值为 p）的蘑菇判断为可食用（标签值为 e）的蘑菇：

```
< Prediction >
class|cap_shape|cap_surface|cap_color|bruises|odor|gill_attachment|gill_spacing|gill_size|gill_color|.
```

```
... ...
-----|---------|-----------|------------|----------|-------|-----|---------------|----------------|-----------|----------|.
... ...
e|f|y|n|t|n|f|c|b|p|t|b|s|s|p|w|p|w|o|p|k|y|d|e|{"p":3.1915386674427295E-5,"e":0.9999680846133256}
e|f|f|e|t|n|f|c|b|p|t|b|s|s|p|p|p|w|o|p|k|y|d|e|{"p":8.364168330966485E-5,"e":0.9999163583166915}
p|x|s|w|f|c|f|c|n|p|e|b|s|s|w|w|p|w|o|p|k|v|d|p|{"p":1.0,"e":0.0}
e|f|y|g|t|n|f|c|b|p|t|b|s|s|p|p|p|w|o|p|n|v|d|e|{"p":4.366093667648151E-4,"e":0.9995633906332355}
p|x|y|y|f|f|f|c|b|l|g|e|b|k|k|n|b|p|w|o|l|h|y|p|p|{"p":1.0,"e":0.0}
e|x|y|n|t|n|f|c|b|w|t|b|s|s|p|w|p|w|o|p|k|y|d|e|{"p":1.0517174774921838E-5,"e":0.9999894828252253}
e|x|y|w|t|a|f|c|b|w|e|c|s|s|w|w|p|w|o|p|n|s|m|e|{"p":0.0,"e":1.0}
e|x|s|w|f|n|f|w|b|k|t|e|s|s|w|w|p|w|o|e|k|a|g|e|{"p":0.0,"e":1.0}
e|x|s|w|f|n|f|w|b|p|t|e|s|s|w|w|p|w|o|e|n|s|g|e|{"p":3.613017366628178E-6,"e":0.9999963869826345}
e|f|f|e|t|n|f|c|b|u|t|b|s|s|w|g|p|w|o|p|n|v|d|e|{"p":0.0,"e":1.0}
----------------------------- Metrics: -----------------------------
Auc:0.9998 Accuracy:0.9889 Precision:0.9798     Recall:0.9974    F1:0.9885   LogLoss:0.0232
|Pred\Real|  p|  e|
|---------|---|---|
|        p|388|  8|
|        e|  1|415|
```

前面,我们训练朴素贝叶斯模型时用到了所有特征列。如果减少特征列的个数,会怎么样呢?比如,只关注卡方特征选择组件给出的前三个特征列:gill_color、spore_print_color 和 odor,会出现什么情况?

我们先了解一下这三个特征列。其中,菌褶颜色(gill_color)和气味(odor),一看名称就能理解,而且这两个特征是大家可以直接看到、闻到的。

孢子印(spore print)指的是蘑菇孢子散落而沉积的菌褶或菌管的印迹。孢子印及其颜色可作为蘑菇分类的依据之一。孢子印的制作方式如下:把菌褶或菌管上的子实层所产生的孢子接收在白纸或黑纸上。孢子印越厚,就越容易得到孢子颜色的准确判断。其制作过程需要数小时。

接下来的实验,我们选择菌褶颜色(gill_color)和气味(odor)这两个特征,得到的模型有助于我们在实际生活中对蘑菇是否有毒有一个初步判断。整个流程与前面对全部特征进行训练的代码基本相同,只是模型训练参数的设置不同,具体代码如下。训练特征列为"odor"、"gill_color",离散特征列为"odor"、"gill_color"。

```
NaiveBayesTrainBatchOp trainer =
    new NaiveBayesTrainBatchOp()
        .setFeatureCols("odor", "gill_color")
        .setCategoricalCols("odor", "gill_color")
        .setLabelCol(LABEL_COL_NAME);
```

使用 lazyCollectModelInfo 方法获取特征"odor"和"gill_color"的概率值,代码如下:

```
trainer.lazyCollectModelInfo(new Consumer <NaiveBayesModelInfo>() {
```

```java
@Override
public void accept(NaiveBayesModelInfo naiveBayesModelInfo) {
    StringBuilder sbd = new StringBuilder();
    for (String feature : new String[] {"odor", "gill_color"}) {
        HashMap <Object, HashMap <Object, Double>> map2 =
            naiveBayesModelInfo.getCategoryFeatureInfo().get(feature);
        sbd.append("\nfeature:").append(feature);
        for (Entry <Object, HashMap <Object, Double>> entry : map2.entrySet()) {
            sbd.append("\n").append(entry.getKey()).append(" : ")
                .append(entry.getValue().toString());
        }
    }
    System.out.println(sbd.toString());
}
});
```

预测结果如下：

```
< Prediction >
class|cap_shape|cap_surface|cap_color|bruises|odor|gill_attachment|gill_spacing|gill_size|gill_color|
... ...
-----|---------|-----------|---------|-------|----|---------------|------------|---------|----------|
... ...
e|f|y|n|t|n|f|c|b|p|t|b|s|s|p|w|p|w|o|p|k|y|d|e|{"p":0.027066242618865737,"e":0.9729337573811344}
e|f|f|e|t|n|f|c|b|p|t|b|s|s|p|p|p|w|o|p|k|y|d|e|{"p":0.027066242618865737,"e":0.9729337573811344}
p|x|s|w|f|c|f|c|n|p|e|b|s|s|w|w|p|w|o|p|k|v|d|p|{"p":1.0,"e":0.0}
e|f|y|g|t|n|f|c|b|p|t|b|s|s|p|p|p|w|o|p|n|v|d|e|{"p":0.027066242618865737,"e":0.9729337573811344}
p|x|y|y|f|f|f|c|b|g|e|b|k|k|n|b|p|w|o|l|h|y|p|p|{"p":1.0,"e":0.0}
e|x|y|n|t|n|f|c|b|w|t|b|s|s|p|w|p|w|o|p|k|y|d|e|{"p":0.009836338642934176,"e":0.990163661357066}
e|x|y|w|t|a|f|c|b|w|e|c|s|s|w|w|p|w|o|p|n|s|m|e|{"p":0.0,"e":1.0}
e|x|s|w|f|n|f|w|b|k|t|e|s|s|w|w|p|w|o|e|k|a|g|e|{"p":0.0062506869874194345,"e":0.9937493130125807}
e|x|s|w|f|n|f|w|b|p|t|e|s|s|w|w|p|w|o|e|n|s|g|e|{"p":0.027066242618865737,"e":0.9729337573811344}
e|f|f|e|t|n|f|c|b|u|t|b|s|s|w|g|p|w|o|p|n|v|d|e|{"p":0.0039845377693012935,"e":0.9960154622306986}
```

模型评估指标如下：

```
------------------------------ Metrics: ------------------------------
Auc:0.9937 Accuracy:0.9901 Precision:1    Recall:0.9794    F1:0.9896  LogLoss:0.0505
|Pred\Real|  p|  e|
|---------|---|---|
|        p|381|  0|
|        e|  8|423|
```

在此可发现整体指标与选择全部特征训练出的模型差异不大，准确率（Precision）为1.0（即判断为有毒的样本一定是有毒的）；召回率（Recall）为0.9794，略低于前面模型的0.9974。这两种模型都没做到将毒蘑菇全部分出来。

我们再看一下打印输出的朴素贝叶斯模型特征"odor"和"gill_color"的概率值，如下所示：

feature:odor

```
p : {p=1.0, c=1.0, s=1.0, f=1.0, y=1.0, m=1.0, n=0.033871478315922764}
e : {a=1.0, l=1.0, n=0.9661285216840773}
feature:gill_color
p : {p=0.42509225092250924, b=1.0, r=1.0, u=0.09610983981693363, g=0.668141592920354,
w=0.20888468809073724, h=0.7216338880484114, y=0.25675675675675674, k=0.14323607427055704,
n=0.10649627263045794}
e : {p=0.5749077490774908, e=1.0, u=0.9038901601830663, g=0.33185840707964603,
w=0.7911153119092628, h=0.2783661119515885, y=0.7432432432432432, k=0.8567639257294429,
n=0.8935037273695421, o=1.0}
```

对比前面使用全部特征得到的朴素贝叶斯模型，发现这两个特征"odor"和"gill_color"的概率值没有发生变化（了解一下朴素贝叶斯的原理，就很容易理解这一点了）。

9.5 蘑菇分类的决策树

本节会将决策树算法应用到蘑菇分类问题上。3 种常用的决策树算法及其对应的分裂指标如表 9-11 所示。

表 9-11 3 种常用的决策树算法及其对应的分裂指标

算法	分裂指标
ID3	信息增益（Information Gain）
C4.5	信息增益率（Information Gain Ratio）
CART	基尼指数（Gini Index）

Alink 对每种算法都单独提供了训练组件和预测组件：Id3TrainBatchOp、Id3PredictBatchOp、C45TrainBatchOp、C45PredictBatchOp、CartTrainBatchOp、CartPredictBatchOp。比如，前面示例中就使用了 C45TrainBatchOp。

Alink 还提供了一种使用方式，将 3 种算法整合为决策树组件（DecisionTreeTrainBatchOp、DecisionTreePredictBatchOp），通过设置树类型（TreeType）参数来选择使用哪种算法。

对于蘑菇数据集，我们会尝试使用 3 种决策树算法来看看具体分类效果。使用统一的决策树训练组件 DecisionTreeTrainBatchOp，各种算法通过调整树类型参数 TreeType 来进行切换。具体代码如下，决策树训练组件要设置特征列和标签列，还需要指出特征列中哪些列是离散特征列（对应参数 CategoricalCols）。在当前的分类问题中，所有特征都是离散特征，所以该参数填入所有的特征列 FEATURE_COL_NAMES。决策树预测组件需要指定预测结果列名，还可以通过设置预测详情列名，得到属于各分类值的概率，用于随后的二分类评估。

```
BatchOperator train_data = new AkSourceBatchOp().setFilePath(DATA_DIR + TRAIN_FILE);
```

```
BatchOperator test_data = new AkSourceBatchOp().setFilePath(DATA_DIR + TEST_FILE);

for (TreeType treeType : new TreeType[] {TreeType.GINI, TreeType.INFOGAIN, TreeType.INFOGAINRATIO}) {
    BatchOperator <?> model = train_data
        .link(
            new DecisionTreeTrainBatchOp()
                .setTreeType(treeType)
                .setFeatureCols(FEATURE_COL_NAMES)
                .setCategoricalCols(FEATURE_COL_NAMES)
                .setLabelCol(LABEL_COL_NAME)
                .lazyCollectModelInfo(new Consumer <DecisionTreeModelInfo>() {
                    @Override
                    public void accept(DecisionTreeModelInfo decisionTreeModelInfo) {
                        try {
                            decisionTreeModelInfo.saveTreeAsImage(
                                DATA_DIR + "tree_" + treeType.toString() + ".jpg", true);
                        } catch (IOException e) {
                            e.printStackTrace();
                        }
                    }
                })
        );

    DecisionTreePredictBatchOp predictor = new DecisionTreePredictBatchOp()
        .setPredictionCol(PREDICTION_COL_NAME)
        .setPredictionDetailCol(PRED_DETAIL_COL_NAME);

    predictor.linkFrom(model, test_data);

    predictor.link(
        new EvalBinaryClassBatchOp()
            .setPositiveLabelValueString("p")
            .setLabelCol(LABEL_COL_NAME)
            .setPredictionDetailCol(PRED_DETAIL_COL_NAME)
            .lazyPrintMetrics()
    );
}

BatchOperator.execute();
```

3 种决策树算法的评估指标对比如表 9-12 所示。

表 9-12　3 种决策树算法的评估指标对比

ID3	C4.5	CART
TreeType.INFOGAIN	TreeType.INFOGAINRATIO	TreeType.GINI
Auc:1 Accuracy:1 Precision:1 Recall:1 F1:1 LogLoss:0	Auc:1 Accuracy:1 Precision:1 Recall:1 F1:1 LogLoss:0	Auc:1 Accuracy:1 Precision:1 Recall:1 F1:1 LogLoss:0

续表

ID3	C4.5	CART
TreeType.INFOGAIN	TreeType.INFOGAINRATIO	TreeType.GINI
混淆矩阵 |Pred\Real| p | e | |---------|---|---| | p|389| 0| | e| 0|423|	混淆矩阵 |Pred\Real| p | e | |---------|---|---| | p|389| 0| | e| 0|423|	混淆矩阵 |Pred\Real| p | e | |---------|---|---| | p|389| 0| | e| 0|423|

从评估指标上看，这 3 种决策树算法都对测试集进行了完全正确的分类，比朴素贝叶斯算法更胜一筹。

我们打开树模型图片（见图 9-6 和见图 9-7），观察各算法模型的决策过程。ID3 算法（TreeType.INFOGAIN，图 9-6）和 C4.5 算法（TreeType.INFOGAINRATIO，图 9-7）的决策过程比较相似。先是根据特征 odor 进行划分，均只有一个节点的样本含有两个标签值，需要继续划分；其他各个节点都满足：节点内的样本标签值都相同，可以作为叶节点给出决策结果。继续进行划分，两种算法的选择还是一样的，都是根据特征 spore_print_color 进行划分。这两级之后，这两种算法展现出明显的差异。

我们再观察 CART 算法（TreeType.GINI），如图 9-8 所示。CART 算法每次都是 2 分划。在此可以看到，首先仍以特征 odor 做划分，分为两部分。左边的分枝有多个离散值，再次划分这些离散值的时候，可将两种标签值的数据完全分开。这个分枝就相当于前面 ID3 和 C4.5 第一层中那些样本标签值都相同的叶节点。右边的分枝是对特征 odor=n（即没有气味）的样本继续分类的结果。接下来使用特征 stalk_surface_below_ring 进行划分。

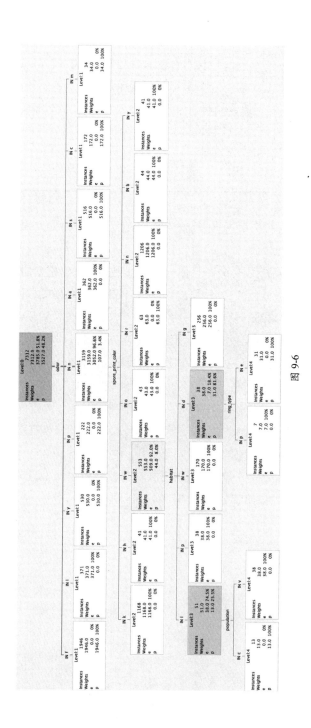

图 9-6

第 9 章 朴素贝叶斯模型与决策树模型

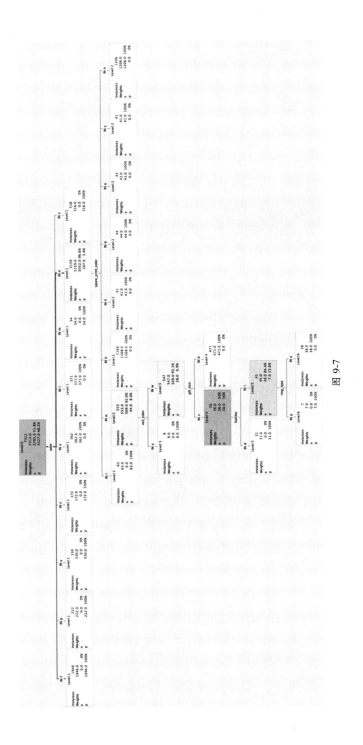

图 9-7

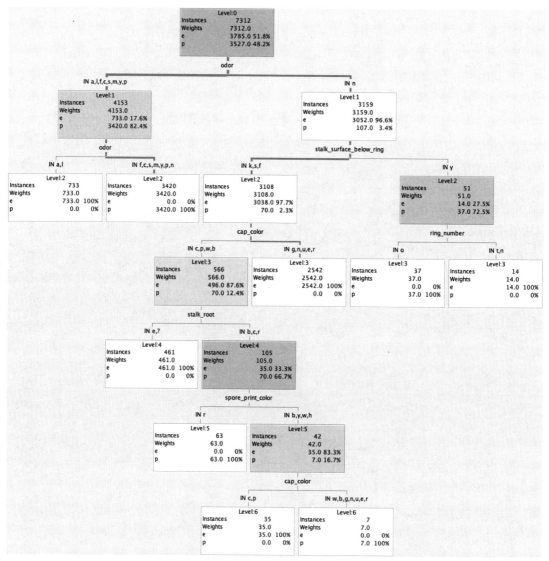

图 9-8

决策树的集成算法（比如，随机森林、GBDT）的分类效果会优于单棵决策树算法的分类效果。

10 特征的转化

前面介绍了如何使用数值类型的连续特征或枚举类型的离散特征建立二分类模型。大家或许已有了一个笼统的印象：所有分类模型都能处理数值类型的连续特征，部分分类模型能处理枚举类型的离散特征。比如，逻辑回归就是一个形式简单、效果不错的分类方法，但是其只能处理数值类型的数据。是否能通过某种变化或方法，让逻辑回归模型也可使用枚举类型的离散特征呢？

本章讲述特征哑元化的方法，实现特征由枚举类型向数值类型的转化，这样我们就可以使用逻辑回归等分类模型。特征哑元化的方法在给我们带来惊喜的同时，也使我们看到，特征的数量在急剧增长。在众多的特征中，哪些特征更重要呢，是否可以减少模型特征的个数呢？本章将通过实例分析，与大家一起探讨上述问题。

本章使用的数据集为德国信用数据集（German Credit Dataset），读者可从链接10-1下载。本数据集记录了1000名用户的20个属性，且从信用风险的角度，将客户分为好客户（Good=1，按时偿还贷款）与坏客户（Bad=2，违约）两类。

数据特征属性的字段说明，如表10-1所示。

表10-1 数据特征属性的字段说明

属性编号	数据类型	字段名称	含义	备注
1	枚举	status	现有的支票账户的状态	A11：…… <0 马克 A12：0≤ …… <200 马克 A13：…… ≥ 200 马克 A14：没有支票账户

属性编号	数据类型	字段名称	含义	备注
2	数值	duration	持续时间（单位：月）	
3	枚举	credit_history	信用记录	A30：没有贷款/所有贷款已及时还清 A31：在这家银行的所有贷款都已及时偿还 A32：时至今日仍有贷款，但该贷款目前处于及时还款之中 A33：过去有延迟还款的记录 A34：关键账户/存在其他贷款（不是在这家银行申请的贷款）
4	枚举	purpose	目的	A40：汽车（新） A41：汽车（使用过） A42：家具/设备 A43：广播/电视 A44：家用电器 A45：维修 A46：教育 A47：（假期 - 不存在吗？） A48：培训 A49：商务 A410：其他
5	数值	credit_amount	信贷量	
6	枚举	savings	储蓄账户/债券	A61：……<100 马克 A62：100≤……<500 马克 A63：500 ≤……<1000 马克 A64：..≥ 1000 马克 A65：不详/无主的储蓄账户
7	枚举	employment	职业	A71：失业 A72：……<1 年 A73：1≤ ……<4 年 A74：4 ≤ ……<7 年 A75：……≥7 年
8	数值	installment_rate	在可支配收入的百分比中，分期付款的比例	

续表

属性编号	数据类型	字段名称	含义	备注
9	枚举	marriage_sex	个人身份和性别	A91：男：离婚/分居 A92：女：离婚/分居/已婚 A93：男：未婚 A94：男：已婚/丧偶 A95：女：未婚
10	枚举	debtors	其他债务人/担保人	A101：无 A102：共同申请人 A103：担保人
11	数值	residence	目前住处	
12	枚举	property	财产	A121：房地产 A122：如果不是 A121 这种情况：具备建房互助协会的储蓄协定/人寿保险 A123：如果不是 A121/A122 这种情况：汽车或其他不在属性 6 中的储蓄账户/债券 A124：未知/无财产
13	数值	age	年龄	
14	枚举	other_plan	其他分期付款计划	A141：银行 A142：存储 A143：无
15	枚举	housing	住房	A151：租用 A152：自有 A153：免费
16	数值	number_credits	在这家银行的现有贷款数目	
17	枚举	job	工作职位	A171：无业/不熟练 - 非居民 A172：不熟练 - 居民 A173：熟练的员工/公务员 A174：经理/个体户/高级职员/官员
18	数值	maintenance_num	赡养人数	
19	枚举	telephone	电话	A191：无 A192：有，用客户的姓名进行注册
20	枚举	foreign_worker	外国工作者	A201：是 A202：无

该数据可以被看作以空格为字段分隔符的 CSV 格式数据。使用 CsvSourceBatchOp 读取的代码如下。随后打印统计信息，显示前 5 条数据，并将原始数据按 8∶2 的比例分为训练集和测试集。

```
CsvSourceBatchOp source = new CsvSourceBatchOp()
    .setFilePath(DATA_DIR + ORIGIN_FILE)
    .setSchemaStr(Utils.generateSchemaString(COL_NAMES, COL_TYPES))
    .setFieldDelimiter(" ");

source
    .lazyPrint(5, "< origin data >")
    .lazyPrintStatistics();

BatchOperator.execute();

Utils.splitTrainTestIfNotExist(source, DATA_DIR + TRAIN_FILE, DATA_DIR + TEST_FILE, 0.8);
```

注意：打印输出的原始数据和统计信息都为 Markdown 表格的形式。使用任意 Markdown 编辑器都可以将其转换为一般的表格形式进行展现。本书为了便于读者阅读，直接显示其转换后的内容。

前 5 条原始数据如图 10-1 所示。

status	duration	credit_hi story	purpose	credit_a mount	savings	employm ent	installme nt_rate	marriage _sex	debtors	residenc e	property	age	other_pl an	housing	number_ credits	job	maintena nce_num	telephon e	foreign_ worker	class
A13	6	A32	A42	2116	A61	A73	2	A93	A101	2	A121	41	A143	A152	1	A173	1	A192	A201	1
A12	9	A31	A40	1437	A62	A74	2	A93	A101	3	A124	29	A143	A152	1	A173	1	A191	A201	2
A14	42	A34	A42	4042	A63	A73	4	A93	A101	2	A121	36	A143	A152	2	A173	1	A192	A201	1
A14	9	A32	A46	3832	A65	A75	1	A93	A101	4	A121	64	A143	A152	1	A172	1	A191	A201	1
A11	24	A32	A43	3680	A61	A73	2	A92	A101	4	A123	28	A143	A152	1	A173	1	A191	A201	1

图 10-1

原始数据的统计信息如图 10-2 所示，共 1000 条数据，没有缺失值，不到一半的特征为数值特征。标签值列 class，均值为 1.3，可知标签值为 2 的样本占 30%，标签值为 1 的样本占 70%；特征 age，最小值为 19，最大值为 75，客户的平均年龄为 35.546；特征 credit_amount 的取值范围较大，其取值范围为 250～18 424。

colName	count	missing	sum	mean	variance	min	max
status	1000	0	NaN	NaN	NaN	NaN	NaN
duration	1000	0	20903	20.903	145.415	4	72
credit_history	1000	0	NaN	NaN	NaN	NaN	NaN
purpose	1000	0	NaN	NaN	NaN	NaN	NaN
credit_amount	1000	0	3271258	3271.258	7967843.4709	250	18424
savings	1000	0	NaN	NaN	NaN	NaN	NaN
employment	1000	0	NaN	NaN	NaN	NaN	NaN
installment_rate	1000	0	2973	2.973	1.2515	1	4
marriage_sex	1000	0	NaN	NaN	NaN	NaN	NaN
debtors	1000	0	NaN	NaN	NaN	NaN	NaN
residence	1000	0	2845	2.845	1.2182	1	4
property	1000	0	NaN	NaN	NaN	NaN	NaN
age	1000	0	35546	35.546	129.4013	19	75
other_plan	1000	0	NaN	NaN	NaN	NaN	NaN
housing	1000	0	NaN	NaN	NaN	NaN	NaN
number_credits	1000	0	1407	1.407	0.3337	1	4
job	1000	0	NaN	NaN	NaN	NaN	NaN
maintenance_num	1000	0	1155	1.155	0.1311	1	2
telephone	1000	0	NaN	NaN	NaN	NaN	NaN
foreign_worker	1000	0	NaN	NaN	NaN	NaN	NaN
class	1000	0	1300	1.3	0.2102	1	2

图 10-2

10.1 整体流程

将本章所要介绍的特征哑元化操作和特征的重要性等内容都整合到一个流程中，如下所示：

```
BatchOperator <?> train_data =
    new AkSourceBatchOp().setFilePath(DATA_DIR + TRAIN_FILE).select(CLAUSE_CREATE_FEATURES);

BatchOperator <?> test_data =
    new AkSourceBatchOp().setFilePath(DATA_DIR + TEST_FILE).select(CLAUSE_CREATE_FEATURES);

String[] new_features = ArrayUtils.removeElement(train_data.getColNames(), LABEL_COL_NAME);

train_data.lazyPrint(5, "< new features >");

LogisticRegressionTrainBatchOp trainer = new LogisticRegressionTrainBatchOp()
    .setFeatureCols(new_features)
    .setLabelCol(LABEL_COL_NAME);
```

```java
LogisticRegressionPredictBatchOp predictor = new LogisticRegressionPredictBatchOp()
    .setPredictionCol(PREDICTION_COL_NAME)
    .setPredictionDetailCol(PREDICTION_DETAIL_COL_NAME);

train_data.link(trainer);

predictor.linkFrom(trainer, test_data);

trainer
    .lazyPrintTrainInfo()
    .lazyCollectTrainInfo(new Consumer <LinearModelTrainInfo>() {
        @Override
        public void accept(LinearModelTrainInfo linearModelTrainInfo) {
            printImportance(
                linearModelTrainInfo.getColNames(),
                linearModelTrainInfo.getImportance()
            );
        }
    }
);

predictor.link(
    new EvalBinaryClassBatchOp()
        .setPositiveLabelValueString("2")
        .setLabelCol(LABEL_COL_NAME)
        .setPredictionDetailCol(PREDICTION_DETAIL_COL_NAME)
        .lazyPrintMetrics()
);

BatchOperator.execute();
```

这里需要说明如下几点：

（1）获取训练集和测试集的时候，就做了特征哑元化的操作，是通过执行 SQL 操作——select(CLAUSE_CREATE_FEATURES)实现的。

（2）不仅要对训练数据进行特征哑元化的操作，还应对所要预测的测试数据进行特征哑元化的操作。

（3）获取特征哑元化后新数据集的特征名称 new_features，用于后续训练预测。

（4）随后是设置训练、预测组件参数，连接整个流程。

（5）逻辑回归模型为线性模型，可以根据模型中各属性的系数及训练集中各特征的统计指标，得到每个属性的重要性。通过逻辑回归训练组件 trainer 的 lazyPrintTrainInfo 方法可以获得最重要的 3 个特征信息。若想要知道全部特征的信息，则需要使用 lazyCollectTrainInfo 方法，得到 LinearModelTrainInfo。

（6）最后使用二分类评估组件进行评估，并输出评估结果。

下面将详细介绍算法的细节。

10.1.1 特征哑元化

下面首先介绍一下特征哑元化。特征哑元化指的是将枚举类型的特征变量转换为多个二值特征变量。举个例子,如果特征属性 F 有 4 个属性值$\{a,b,c,d\}$,则可以通过定义 4 个新的特征变量 Fa、Fb、Fc、Fd,使得每个新变量只能取两个值,即 0 或 1,表示特征属性 F 是否取到了某个属性值。如果某条记录的特征 F 的取值为 b,则特征变量 Fa=0,Fb=1,Fc=0,Fd=0。

这种变换可以通过批式组件的 select(String clause)方法轻松实现。比如,本例中特征哑元化操作对应的 SQL 语句 CLAUSE_CREATE_FEATURES 定义如下:

```
static final String CLAUSE_CREATE_FEATURES
    = "(case status when 'A11' then 1 else 0 end) as status_A11,"
    + "(case status when 'A12' then 1 else 0 end) as status_A12,"
    + "(case status when 'A13' then 1 else 0 end) as status_A13,"
    + "(case status when 'A14' then 1 else 0 end) as status_A14,"
    + "duration,"
    + "(case credit_history when 'A30' then 1 else 0 end) as credit_history_A30,"
    + "(case credit_history when 'A31' then 1 else 0 end) as credit_history_A31,"
    + "(case credit_history when 'A32' then 1 else 0 end) as credit_history_A32,"
    + "(case credit_history when 'A33' then 1 else 0 end) as credit_history_A33,"
    + "(case credit_history when 'A34' then 1 else 0 end) as credit_history_A34,"
... ...
    + "age,"
    + "(case other_plan when 'A141' then 1 else 0 end) as other_plan_A141,"
    + "(case other_plan when 'A142' then 1 else 0 end) as other_plan_A142,"
    + "(case other_plan when 'A143' then 1 else 0 end) as other_plan_A143,"
    + "(case housing when 'A151' then 1 else 0 end) as housing_A151,"
    + "(case housing when 'A152' then 1 else 0 end) as housing_A152,"
    + "(case housing when 'A153' then 1 else 0 end) as housing_A153,"
    + "number_credits,"
    + "(case job when 'A171' then 1 else 0 end) as job_A171,"
    + "(case job when 'A172' then 1 else 0 end) as job_A172,"
    + "(case job when 'A173' then 1 else 0 end) as job_A173,"
    + "(case job when 'A174' then 1 else 0 end) as job_A174,"
    + "maintenance_num,"
    + "(case telephone when 'A192' then 1 else 0 end) as telephone,"
    + "(case foreign_worker when 'A201' then 1 else 0 end) as foreign_worker,"
    + "class ";
```

这里需要注意 3 个地方:

(1) SQL 语句中的 "case … when … then … else … end" 为条件选择语句。比如,语句 "case status when 'A11' then 1 else 0 end" 表示当 status 的值为'A11'时,选择 1,否则为 0。

(2) 对于具有多个属性值的 status,使用判断语句,生成 4 个新属性 status_A11、status_A12、status_A13 和 status_A14。

（3）对于只有 2 个属性值的 telephone 和 foreign_worker，则可以通过判断是否出现这 2 个属性值中的一个值，将属性转换为用 0-1 表示的二值属性。

我们再通过对比显示，看看该操作的影响。原始的数据如图 10-3 所示；执行哑元化操作后的数据如图 10-4 所示。

status	duration	credit_hi story	purpose	credit_a mount	savings	employm ent	installme nt_rate	marriage _sex	debtors	residenc e	property	age	other_pl an	housing	number_ credits	job	maintena nce_num	telephon e	foreign_ worker	class
A13	6	A32	A42	2116	A61	A73	2	A93	A101	2	A121	41	A143	A152	1	A173	1	A192	A201	1
A12	9	A31	A40	1437	A62	A74	2	A93	A101	3	A124	29	A143	A152	1	A173	1	A191	A201	2
A14	42	A34	A32	4042	A63	A74	4	A93	A101	4	A121	36	A143	A152	2	A173	1	A192	A201	1
A14	9	A32	A46	3832	A65	A75	1	A93	A101	4	A121	64	A143	A152	1	A172	1	A191	A201	1
A11	24	A32	A43	3660	A61	A73	2	A92	A101	4	A123	28	A143	A152	1	A173	1	A191	A201	1

图 10-3

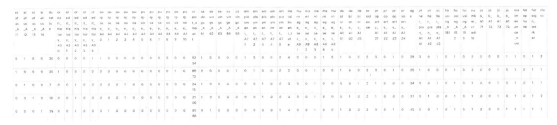

图 10-4

对比两图可知，执行哑元化操作后，特征的个数增加了很多。比如，原始数据特征 job，经过哑元化操作后，产生了 4 个特征：job_A171、job_A172、job_A173、job_A174。特征 maintenance_num 是数值字段，没有受到变换的影响；特征 telephone 和 foreign_worker 的字段名称没有发生变化，但内容发生了变化，它们都是只有 2 个属性值的特征，可通过判断是否出现这 2 个属性值中的一个值，将属性转换为用 0-1 表示的二值属性。

10.1.2 特征的重要性

模型中会涉及很多特征，但这些特征对模型的影响程度是不一样的。逻辑回归训练组件 trainer 使用 lazyPrintTrainInfo 打印的训练信息如下：

```
------------------------ train meta info ------------------------
{model name: Logistic Regression, num feature: 61}
------------------------ train importance info ------------------------
|           colName|importanceValue|          colName|weightValue|
|------------------|---------------|-----------------|-----------|
|         status_A14|     0.50808537|    foreign_worker| 1.33227696|
|           duration|     0.45647352|credit_history_A31| 0.90477436|
|   installment_rate|     0.38513890|        status_A11| 0.78638287|
```

```
|              ... ...|          ... ...|            ... ...|   ... ...|
|marriage_sex_A94|       0.00244852|         status_A14|-1.04164671|
|      purpose_A47|       0.00000000|         savings_A64|-1.30202577|
|marriage_sex_A95|       0.00000000|         purpose_A48|-1.31234705|
------------------------ train convergence info ------------------------
step:0  loss:0.67720572 gradNorm:0.40702294 learnRate:0.40000000
step:1  loss:0.57520002 gradNorm:0.38580895 learnRate:1.60000000
step:2  loss:0.46073005 gradNorm:0.24350819 learnRate:4.00000000
... ...
step:15 loss:0.44737146 gradNorm:0.00066504 learnRate:4.00000000
step:16 loss:0.44737055 gradNorm:0.00059793 learnRate:4.00000000
step:17 loss:0.44737029 gradNorm:0.00042434 learnRate:4.00000000
```

在中间部分中，左侧显示的是各特征的重要性（importance）排名；右侧显示的是各特征在模型中权重值（weight）的排名。线性模型特征重要性中的 weight 指标就是逻辑回归模型系数的绝对值；而 importance 指标，会在 weight 的基础上乘以该特征的标准差，这样可消除特征数值变化所带来的影响。

特征重要性排在前三位的是 status_A14、duration 和 installment_rate；排在后三位的是 marriage_sex_A95、purpose_A47 和 marriage_sex_A94。如果读者想知道更多特征的重要性数值，需要使用 lazyCollectTrainInfo 方法得到训练信息（LinearModelTrainInfo），从训练信息中获取各特征的重要性（importance）数值，进行排序并打印输出。

```java
public static void printImportance(String[] colNames, double[] importance) {
    ArrayList <Tuple2 <String, Double>> list = new ArrayList <>();
    for (int i = 0; i < colNames.length; i++) {
        list.add(Tuple2.of(colNames[i], importance[i]));
    }
    Collections.sort(list, new Comparator <Tuple2 <String, Double>>() {
        @Override
        public int compare(Tuple2 <String, Double> o1, Tuple2 <String, Double> o2) {
            return -(o1.f1).compareTo(o2.f1);
        }
    });
    StringBuilder sbd = new StringBuilder();
    for(int i=0; i<list.size();i++){
        sbd.append(i+1).append(" \t")
           .append(list.get(i).f0).append(" \t")
           .append(list.get(i).f1).append("\n");
    }
    System.out.print(sbd.toString());
}
```

得到的输出结果如下。输出结果有 61 条数据，这里只保留了前后 5 条数据。

```
1    status_A14            0.50808536663665
2    duration              0.45647351612984727
3    installment_rate      0.3851388963592921
```

```
4    status_A11              0.3533059193419253
5    savings_A64             0.2736303892679297
...
57   employment_A75          0.003980669438950036
58   purpose_A49             0.0029410949128441635
59   marriage_sex_A94        0.002448517402189116
60   purpose_A47             0.0
61   marriage_sex_A95        0.0
```

至此，我们可以知道哪些特征是重要的，但是特征 status_A14、status_A11 都只是标号，其对应的实际含义是什么？我们可以从原始数据的字段说明（见表 10-1）中，查找出经过哑元化操作后的特征属性字段的含义，如表 10-2 所示。

表 10-2 重要特征属性字段的含义

特征	含义
status_A14	现有支票账户的状态：没有支票账户
duration	持续时间（单位：月）
installment_rate	在可支配收入的百分比中，分期付款的比例
status_A11	现有支票账户的状态：…… <0 马克
savings_A64	储蓄账户/债券：…… ≥ 1000 马克

关注模型的效果，二分类评估的结果如下：

```
Auc:0.7904 Accuracy:0.78   Precision:0.5882    Recall:0.566    F1:0.5769 LogLoss:0.4799
|Pred\Real    2|  1|
|---------|---|---|
|         2| 30| 21|
|         1| 23|126|
```

10.2　减少模型特征的个数

我们在建模的时候，有时候不仅需要很高的精确度、召回率，还要考虑模型特征的可解释性、模型的大小等因素。本节的重点是如何减少特征的个数，选取少数重要的特征建模，且模型仍具有较高的精确度、召回率等指标。

对于逻辑回归模型，可以通过设置逻辑回归训练的 L1 正则项参数，使模型中一些特征的模型系数为 0。其代码与前面类似，只需要在逻辑回归训练组件 trainer 中设置参数 L1（该参数的默认值为 0，即对模型不起作用）。这里我们设置 L1=0.01，相关代码如下：

```
LogisticRegressionTrainBatchOp trainer = new LogisticRegressionTrainBatchOp()
```

```
.setFeatureCols(new_features)
.setLabelCol(LABEL_COL_NAME)
.setL1(0.01);
```

读者会问：为什么将正则项参数 L1 设置为 0.01？本书第 20 章中会介绍如何搜索最佳的 L1 值。

将本节的运行结果与前面不设置 L1 的结果进行对比，通过表 10-3 可知，与不设置 L1 相比，设置 L1 后，Auc、Recall 等指标都有不同程度的下降。但是，与此同时，重要性值为 0 的特征数量明显增加。不设置 L1 的时候只有 2 个重要性值为 0 的特征，现在有 23 个重要性值为 0 的特征。特征的重要性值为 0，就意味着该特征的逻辑回归系数为 0，这对分类结果没有影响。从数据中还可看到，在两次训练中，重要性排在前两位的两个特征没有变化，而排在第 3、4 位的两个特征交换了位置。感兴趣的读者，可以尝试继续调大训练参数 L1 的值，这样还会进一步减小非零模型特征的个数。

表 10-3 逻辑回归 L1 参数的不同设置

逻辑回归训练，不设置 L1		逻辑回归训练，设置 L1=0.01	
Auc:0.7904		Auc:0.7804	
Accuracy:0.78		Accuracy:0.77	
Precision:0.5882		Precision:0.5778	
Recall:0.566		Recall:0.4906	
F1:0.5769		F1:0.5306	
LogLoss:0.4799		LogLoss:0.4817	
混淆矩阵:		混淆矩阵:	
|Pred\Real| 2| 1|		|Pred\Real| 2| 1|	
|---------|---|---|		|---------|---|---|	
| 2| 30| 21|		| 2| 26| 19|	
| 1| 23|126|		| 1| 27|128|	
特征的重要性：		特征的重要性：	
1 status_A14	0.50808536663665	1 status_A14	0.5007428413144325
2 duration	0.45647351612984727	2 duration	0.42143922884968493
3 installment_rate	0.3851388963592921	3 status_A11	0.234015522923644477
4 status_A11	0.3533059193419253	4 installment_rate	0.20285203615332648
5 savings_A64	0.2736303892679297	5 purpose_A40	0.199010686608203715
……		……	
57 employment_A75	0.00398066943895036	37 employment_A75	0.0021207607321917523
58 purpose_A49	0.0029410949128441635	38 status_A13	$2.1658292055416037 \times 10^{-16}$

续表

逻辑回归训练，不设置 L1		逻辑回归训练，设置 L1=0.01	
59　marriage_sex_A94	0.002448517402189116	39　credit_history_A32	0.0
60　purpose_A47	0.0	40　credit_history_A33	0.0
61　marriage_sex_A95	0.0	……	
		60　job_A174	0.0
		61　maintenance_num	0.0

10.3　离散特征转化

前面通过 SQL SELECT 语句将离散特征转化为数值特征。本节将介绍另外两种算法，同样可以实现离散特征转化功能。

10.3.1　独热编码

独热（One-Hot）编码的处理过程如下：对于每一个特征，如果该特征有 m 个可能值，那么经过独热编码操作后，就会变成 m 个二元特征。并且，这些特征互斥，每次只有一个特征可被激活。在基本形式上又可以有一些变化，比如，因为 m 个二元特征的和为 1，则由前 $m-1$ 个值可以推断出第 m 个值，于是出现了 dropLast 选项；对于某些特征可能出现长尾的情况，允许设置离散个数阈值；输出的基本格式是稀疏向量，也可选择只输出索引值；对多个数据列同时进行 One-Hot 编码时，可以选择将各自生成的多个稀疏向量直接拼接成一个稀疏向量。

独热（One-Hot）编码组件的相关参数定义如表 10-4 所示。

表 10-4　独热（One-Hot）编码组件的参数

名称	描述
selectedCols	（必选）计算列所对应的列名列表
reservedCols	算法保留列。默认值为 null，即保留所有数据列
outputCols	输出结果列的列名数组。（可选），默认值为 null，即和输入数据列名数组一致，直接在原列上输出结果
handleInvalid	未知 Token 的处理策略，包括"keep"、"skip"和"error"。默认值为"keep"
encode	编码方式，包括"INDEX"、"VECTOR"、"ASSEMBLED_VECTOR"。默认采用"ASSEMBLED_VECTOR"方式

续表

名称	描述
dropLast	是否删除最后一个元素。默认值为 true
discreteThresholdsArray	离散个数阈值。每一列对应数组中的一个元素
discreteThresholds	离散个数阈值。低于该阈值的离散样本将不会单独成为一个组。默认值为 Integer.MIN_VALUE，即可理解为没有限制

下面使用独热（One-Hot）编码组件进行离散特征的转化，然后用逻辑回归模型进行训练。由于要进行独热（One-Hot）编码操作，因此需要对目标数据进行整体扫描，得到 One-Hot 模型，之后才能对数据进行编码。为了简化代码，我们会使用 Pipeline 的方式进行调用（参见 2.4 节）。

具体代码如下。OneHotEncoder 组件设置了 2 个参数：一个参数是所要编码的数据列名称（SelectedCols），这里填入了原数据集中所有离散特征的列名；另一个参数是编码输出方式（Encode），这里设置的是 Encode.VECTOR，每个特征都会被转化为一个稀疏向量。如果没有设置各输出列的名称，组件会将其输出列名称默认为输入列名称，即输入列的数据会被替换成编码后的结果。经过 OneHotEncoder 组件的处理，我们得到的数据集各特征列是原始数值，或者是编码后得到的稀疏向量。这时，还需要使用向量组装（VectorAssembler）组件，将这些数值和向量顺序拼接为一个向量，设置输出结果向量列名称为 VEC_COL_NAME。最后使用逻辑回归组件，设置特征向量列为 VEC_COL_NAME，再设置标签列、预测结果列、预测详细信息列。建立 Pipeline 后，通过对训练数据 train_data 使用 fit 方法，得到 PipelineModel，再对测试数据 test_data 使用 transform 方法进行预测，预测结果连接二分类评估组件。

```
BatchOperator <?> train_data = new AkSourceBatchOp().setFilePath(DATA_DIR + TRAIN_FILE);
BatchOperator <?> test_data = new AkSourceBatchOp().setFilePath(DATA_DIR + TEST_FILE);

Pipeline pipeline = new Pipeline()
   .add(
     new OneHotEncoder()
       .setSelectedCols(CATEGORY_FEATURE_COL_NAMES)
       .setEncode(Encode.VECTOR)
   )
   .add(
     new VectorAssembler()
       .setSelectedCols(FEATURE_COL_NAMES)
       .setOutputCol(VEC_COL_NAME)
   )
   .add(
     new LogisticRegression()
       .setVectorCol(VEC_COL_NAME)
       .setLabelCol(LABEL_COL_NAME)
       .setPredictionCol(PREDICTION_COL_NAME)
```

```
      .setPredictionDetailCol(PREDICTION_DETAIL_COL_NAME)
   );
pipeline
   .fit(train_data)
   .transform(test_data)
   .link(
      new EvalBinaryClassBatchOp()
         .setPositiveLabelValueString("2")
         .setLabelCol(LABEL_COL_NAME)
         .setPredictionDetailCol(PREDICTION_DETAIL_COL_NAME)
         .lazyPrintMetrics()
   );

BatchOperator.execute();
```

我们得到的运行结果如下。与前面使用 SQL SELECT 生成新特征所得模型的评估指标大致相同。

```
Auc:0.7904 Accuracy:0.78   Precision:0.5882   Recall:0.566   F1:0.5769   LogLoss:0.4801
|Pred\Real|  2|  1|
|---------|---|---|
|        2| 30| 21|
|        1| 23|126|
```

使用 One-Hot 编码的好处是，不用对每个离散特征分别编写逻辑代码进行处理，便于我们快速搭建模型。鉴于独热（One-Hot）编码方式仅针对离散特征，我们在使用该编码方式的同时还经常会配合使用 VectorAssembler，将离散特征和连续特征组合起来。

10.3.2 特征哈希

特征哈希（FeatureHasher）指的是，通过使用哈希技术生成特征向量，将离散特征或连续特征映射到向量索引。

- 连续特征：将列名称的哈希值映射到特征向量的索引。
- 离散特征：将字符串"column_name=value"的哈希值映射到向量索引，指标值为 1.0。

FeatureHasher 组件的参数如表 10-5 所示。

表 10-5　FeatureHasher 组件的参数

名称	描述
selectedCols	【必填】计算列对应的列名列表
outputCol	【必填】输出结果列的列名

续表

名称	描述
categoricalCols	【可选】离散特征列。默认选择 string 类型和 boolean 类型的数据作为离散特征
reservedCols	【可选】算法保留列。默认值为 null，保留所有列
numFeatures	【可选】生成向量的维度，默认值为 2^{18}，即 262 144

FeatureHasher 组件在使用上更加简单，不像 One-Hot 编码组件那样还需要一个向量拼接组件来配合；但 FeatureHasher 为了降低各 Hash 值间的冲突，往往选择较大的向量维度，这使得特征向量更加稀疏。

下面使用 FeatureHasher 组件进行特征变换，其整体流程与前面介绍的 One-Hot 编码组件类似，二者都是首先建立 Pipeline；然后使用与 Pipeline 相关的 fit 和 transform 方法。具体代码如下。这里主要关注 FeatureHasher 组件的参数设置。选择全部特征列作为输入列；设置参数 CategoricalCols，输入离散数据列的名称；输入列为一个稀疏向量，设置输出向量的列名。

```
BatchOperator <?> train_data = new AkSourceBatchOp().setFilePath(DATA_DIR + TRAIN_FILE);
BatchOperator <?> test_data = new AkSourceBatchOp().setFilePath(DATA_DIR + TEST_FILE);

Pipeline pipeline = new Pipeline()
  .add(
    new FeatureHasher()
      .setSelectedCols(FEATURE_COL_NAMES)
      .setCategoricalCols(CATEGORY_FEATURE_COL_NAMES)
      .setOutputCol(VEC_COL_NAME)
  )
  .add(
    new LogisticRegression()
      .setVectorCol(VEC_COL_NAME)
      .setLabelCol(LABEL_COL_NAME)
      .setPredictionCol(PREDICTION_COL_NAME)
      .setPredictionDetailCol(PREDICTION_DETAIL_COL_NAME)
  );

pipeline
  .fit(train_data)
  .transform(test_data)
  .link(
    new EvalBinaryClassBatchOp()
      .setPositiveLabelValueString("2")
      .setLabelCol(LABEL_COL_NAME)
      .setPredictionDetailCol(PREDICTION_DETAIL_COL_NAME)
      .lazyPrintMetrics()
  );

BatchOperator.execute();
```

计算得到的评估结果如下，与前面使用 One-Hot 方法所得的评估结果相似：

```
Auc:0.7904 Accuracy:0.78   Precision:0.5882    Recall:0.566    F1:0.5769 LogLoss:0.4799
|Pred\Real|  2|  1|
|---------|---|---|
|        2| 30| 21|
|        1| 23|126|
```

11 构造新特征

之前我们一直在实验如何使用已有的特征,包括特征哑元化、特征哈希,以及选择重要的特征。在本章中我们对"特征"的理解将更进一步,学习如何构造新特征。

本章的数据来源于 2014 年的阿里巴巴公司大数据竞赛第一赛季,下载地址为链接 11-1。在天猫平台中,每天都会有数千万用户通过在各品牌下搜索来发现自己喜欢的商品。品牌是连接消费者与商品的最重要纽带。本章的数据为用户 4 个月在天猫平台中的行为日志,包含 4 个字段。这些字段的具体说明详见表 11-1。

表 11-1 字段的具体说明

字 段	字段说明	提取说明
user_id	用户标记	抽样&字段加密
time	行为时间	精度到天级别&隐藏年份
action_type	用户对品牌的行为类型	包括点击、购买、加入购物车、收藏 4 种行为 (点击:0,购买:1,收藏:2,加入购物车:3)
brand_id	品牌 ID	抽样&字段加密

用户对任意商品的行为都会映射为一行数据。其中,所有商品的 ID 都已汇总为商品对应的品牌 ID。用户和品牌都分别做了一定程度的数据抽样,且数字 ID 均做了加密。所有行为的时间都精确到天级别(隐藏年份)。

本章将根据用户在天猫平台中的行为日志,建立用户的品牌偏好,并预测他们将来对该品牌旗下各商品的购买行为。

我们希望预测的品牌准确率越高越好,也希望覆盖的用户和品牌越多越好,所以用最常用的准确率与召回率作为排行榜的指标。

准确率（Precision）：

$$\text{Precision} = \frac{\sum_{i=1}^{N} \text{hitBrands}_i}{\sum_{i=1}^{N} \text{pBrands}_i}$$

其中，N 为预测的用户数；pBrands_i 的含义是，预测用户 i 会购买的商品品牌列表个数；hitBrands_i 的含义是，预测用户 i 可能购买的商品品牌列表与用户 i 真实购买的商品品牌交集的个数。

召回率（Recall）或称灵敏度（Sensitivity）：

$$\text{Recall} = \frac{\sum_{i=1}^{M} \text{hitBrands}_i}{\sum_{i=1}^{M} \text{bBrands}_i}$$

其中，M 为实际产生成交的用户数；bBrands_i 为用户 i 真实购买的商品品牌个数；hitBrands_i 为预测的品牌列表与用户 i 真实购买的商品品牌交集的个数。

最后我们用 F_1-Score 来拟合准确率与召回率，并且该大赛最终的比赛成绩排名以 F_1 得分为准。

$$F_1 = \frac{2 \times \text{Precision} \times \text{Recall}}{\text{Precision} + \text{Recall}}$$

11.1 数据探索

使用文本编辑器，打开下载的数据文件，如图 11-1 所示。该文件共有 4 列，每列均用 Tab 键作为分隔。由于 Tab 键具有对齐作用，因此这些数据看起来比较整齐。

```
tmall.csv
10944750    13451    0    2014-06-04 00:00:00
10944750    13451    2    2014-06-04 00:00:00
10944750    13451    2    2014-06-04 00:00:00
10944750    13451    0    2014-06-04 00:00:00
10944750    13451    0    2014-06-04 00:00:00
10944750    13451    0    2014-06-04 00:00:00
10944750    13451    0    2014-06-04 00:00:00
10944750    13451    0    2014-06-04 00:00:00
10944750    21110    0    2014-06-07 00:00:00
10944750    1131     0    2014-07-23 00:00:00
10944750    1131     0    2014-07-23 00:00:00
10944750    8689     0    2014-05-02 00:00:00
10944750    8689     2    2014-05-02 00:00:00
10944750    8689     2    2014-05-02 00:00:00
10944750    8689     0    2014-05-02 00:00:00
10944750    8689     0    2014-05-02 00:00:00
```

图 11-1

第 11 章 构造新特征

我们可以使用 CSV 格式的组件进行读取,因为该数据在本章的示例中会被多次用到。我们定义函数 getSource() 如下:

```
static CsvSourceBatchOp getSource() {
    return new CsvSourceBatchOp()
        .setFilePath(DATA_DIR + ORIGIN_FILE)
        .setSchemaStr("user_id long, brand_id long, type int, ts timestamp")
        .setFieldDelimiter("\t");
}
```

对原始数据进行打印、统计,以及计算相似系数,代码如下:

```
BatchOperator<?> source = getSource();

source.lazyPrint(10, "origin file");

source.lazyPrintStatistics("stat of origin file");

source.link(
    new CorrelationBatchOp()
        .setSelectedCols("user_id", "brand_id", "type")
        .lazyPrintCorrelation()
);
```

前 10 条数据打印显示如下:

```
Origin file
user_id |brand_id|type|ts
--------|--------|----|--
10944750|13451   |0   |2014-06-04 00:00:00.0
10944750|13451   |2   |2014-06-04 00:00:00.0
10944750|13451   |2   |2014-06-04 00:00:00.0
10944750|13451   |0   |2014-06-04 00:00:00.0
10944750|13451   |0   |2014-06-04 00:00:00.0
10944750|13451   |0   |2014-06-04 00:00:00.0
10944750|13451   |0   |2014-06-04 00:00:00.0
10944750|13451   |0   |2014-06-04 00:00:00.0
10944750|21110   |0   |2014-06-07 00:00:00.0
10944750|1131    |0   |2014-07-23 00:00:00.0
```

各列的基本统计指标如下:

```
stat of origin file
Summary:
| colName| count|missing|         sum|       mean|         variance|  min|     max|
|--------|------|-------|------------|-----------|-----------------|-----|--------|
| user_id|182880|      0|1083663976000|5925546.6754|12698777247108.18|19500|12417500|
|brand_id|182880|      0|   2595405857|  14191.8518|      71953590.661|   11|   29552|
|    type|182880|      0|        9851|      0.0539|           0.0692|    0|       3|
|      ts|182880|      0|         NaN|        NaN|              NaN|  NaN|     NaN|
```

在此可以看到，数据的总数为 182 880 条，没有缺失值；type 的变动范围为[0, 3]；ts 列没有什么统计指标。我们希望知道 ts 的开始时间和结束时间，后面会用 SQL 语句深入分析。

相关系数结果如下所示，显然，各字段不相关：

```
Correlation:
| colName|user_id|brand_id|   type|
|--------|-------|--------|-------|
| user_id|      1|  0.0054|  0.004|
|brand_id| 0.0054|       1| 0.0035|
|    type|  0.004|  0.0035|      1|
```

下面，我们使用 SQL 语句获取更多信息，代码如下。首先获取购买行为发生的时间范围，之后获取 type 字段各取值的分布情况。

```
source.select("min(ts) AS min_ts, max(ts) AS max_ts").lazyPrint(-1);
source.groupBy("type", "type, COUNT(*) AS cnt").lazyPrint(-1);
```

获得数据集中行为的开始时间、结束时间如下：

```
min_ts|max_ts
------|------
2014-04-15 00:00:00.0|2014-08-15 00:00:00.0
```

type 字段共有 4 个不同的值（点击：0；购买：1；收藏：2；加入购物车：3），它们各自出现的次数如下，显然，点击行为发生得最多。

```
type|cnt
----|---
0|174539
1|6984
2|1204
3|153
```

11.2 思路

预测用户在未来一个月内对该品牌旗下各商品的购买行为，即判断某个用户在下一个月是否发生购买某品牌商品的行为，也就是机器学习中典型的二分类预测问题。

解决的关键是如何得到关于用户和品牌的各种特征，并训练出二分类模型。

其整体流程主要是构造特征和标签，使用逻辑回归（LR）和随机森林（RF）模型进行训练、预测和评估。

11.2.1 用户和品牌的各种特征

某个用户是否购买某个品牌旗下商品的影响因素（特征）有哪些呢？

首先了解用户对品牌的关注度，比如，用户是否点击过该品牌的商品，是否有该品牌商品的购买行为，是否收藏了该品牌商品，以及是否将该品牌商品加入过购物车。而在这些因素中，用户的关注行为离现在越近，其即将购买该品牌商品的可能性就越大，所以我们要依次关注最近 3 天、最近 1 周、最近 1 个月、最近 2 个月、最近 3 个月以及有用户记录以来的情况，于是有了如下的一些特征：

- 最近 3 天的点击数、购买数、收藏数和加入购物车的次数。
- 最近 1 周的点击数、购买数、收藏数和加入购物车的次数。
- 最近 1 个月的点击数、购买数、收藏数和加入购物车的次数。
- 最近 2 个月的点击数、购买数、收藏数和加入购物车的次数。
- 最近 3 个月的点击数、购买数、收藏数和加入购物车的次数。
- 全部点击数、购买数、收藏数和加入购物车的次数。

有了按时间段细分的关注次数还不够，还需要知道该数值的变化率，以刻画该关注的持续程度。我们还可以构造如下特征：

- 最近 3 天点击数的变化率（最近 3 天的点击数 / 最近 4～6 天的点击数）、购买数的变化率、收藏数的变化率、加入购物车次数的变化率。
- 最近 1 周点击数的变化率（最近 1 周的点击数 / 上周的点击数）、购买数的变化率、收藏数的变化率、加入购物车次数的变化率。
- 最近 1 个月点击数的变化率（最近 1 个月的点击数 / 上个月的点击数）、购买数的变化率、收藏数的变化率、加入购物车次数的变化率。

如果用户对该品牌商品曾有过购买行为，我们希望了解，通过多少次点击产生了一次用户购买行为、通过多少次收藏转化为一次用户购买行为，即购买转化率（简称"转化率"）。构造特征如下：

- 最近 3 天的点击转化率、收藏转化率、加入购物车的转化率。
- 最近 1 周的点击转化率、收藏转化率、加入购物车的转化率。
- 最近 1 个月的点击转化率、收藏转化率、加入购物车的转化率。
- 整体的点击转化率、收藏转化率、加入购物车的转化率。

其次，我们将注意力放在用户上。需要构造特征，将用户的特点表现出来，重点是了解该用户对所关注的所有品牌的总体行为。用户最近对所有品牌的关注度，有如下特征：

- 最近3天的点击数、购买数、收藏数和加入购物车的次数。
- 最近1周的点击数、购买数、收藏数和加入购物车的次数。
- 最近1个月的点击数、购买数、收藏数和加入购物车的次数。
- 最近2个月的点击数、购买数、收藏数和加入购物车的次数。
- 最近3个月的点击数、购买数、收藏数和加入购物车的次数。
- 全部点击数、购买数、收藏数和加入购物车的次数。

每个用户都有自己的特点，有的人点击次数很多，却很少购买；有的人关注某商品很久，才会下单购买。这些特点可以用购买转化率来刻画：

- 最近3天的点击转化率、收藏转化率、加入购物车的转化率。
- 最近1周的点击转化率、收藏转化率、加入购物车的转化率。
- 最近1个月的点击转化率、收藏转化率、加入购物车的转化率。
- 整体的点击转化率、收藏转化率、加入购物车的转化率。

最后，单独看品牌这个因素的影响。有的热门品牌的关注度很高。我们更关心其近期的情况，包括如下特征：

- 最近3天的被点击数、被购买数、被收藏数和被加入购物车的次数。
- 最近1周的被点击数、被购买数、被收藏数和被加入购物车的次数。
- 最近1个月的被点击数、被购买数、被收藏数和被加入购物车的次数。
- 最近2个月的被点击数、被购买数、被收藏数和被加入购物车的次数。
- 最近3个月的被点击数、被购买数、被收藏数和被加入购物车的次数。
- 全部被点击数、被购买数、被收藏数和被加入购物车的次数。

有的品牌的受众较少。虽然其没有很高的关注度，但是购买量不少，可以用购买转化率来描述这些特征：

- 最近3天的点击转化率、收藏转化率、加入购物车的转化率。
- 最近1周的点击转化率、收藏转化率、加入购物车的转化率。
- 最近1个月的点击转化率、收藏转化率、加入购物车的转化率。
- 整体的点击转化率、收藏转化率、加入购物车的转化率。

综上，某个用户是否会购买某品牌商品的特征，由刻画该用户对该品牌商品关注的各种特征、描述该用户的特征及描述该品牌商品的特征共同构成。

11.2.2 二分类模型训练

本节的重点是如何构造训练集。训练集中的每一条记录均为某个用户针对某个品牌的数据，这些记录会包含前面提到的那些特征，但同时还要有一个标签项，表示在这个特征的前提下是否会发生购买行为。

我们知道的数据只是一些带发生时间的行为，比如，开始时间为 2014-4-15，结束时间为 2014-8-15，一共 4 个月。这怎么和特征、标签联系起来呢？

我们需要从一个新的角度来看数据，以 2014-07-16 00:00:00 为界将数据分为两段。由 2014-4-15 到 2014-7-15 这 3 个月的时间内发生的行为，我们可以得到用户与品牌的特征，比如，最近 3 天的点击数、最近 1 周点击数的购买转化率等；而某用户是否会在下个月发生购买某品牌商品的行为，我们可以通过查询其下个月的行为来获知。

这样我们就得到了训练集。对于二分类模型，我们可以有很多选择，比如逻辑回归、随机森林等。我们可以根据二分类问题的特点来选择这些模型。

11.3 计算训练集

由于整个构造特征、标签、模型训练和评估的流程比较长，首先以时间 2014-07-16 00:00:00 为界，将数据分为两段，该时间点以前的数据可用来构造特征，该时间点之后的数据可用来统计出在本段时间（2014-07-16 00:00:00 之后的一个月）内哪些用户、品牌有购买行为，用此统计数据来构造标签。

11.3.1 原始数据划分

要对数据按时间 2014-07-16 00:00:00 进行划分，相应的代码如下：

```
BatchOperator <?> source = getSource();

BatchOperator t1 = source.filter("ts < CAST('2014-07-16 00:00:00' AS TIMESTAMP)");
BatchOperator t2 = source.filter("ts >= CAST('2014-07-16 00:00:00' AS TIMESTAMP)");

t1.lazyPrint(3, "[ ts < '2014-07-16 00:00:00' ]")
    .lazyPrintStatistics();

t2.lazyPrint(3, "[ ts >= '2014-07-16 00:00:00' ]")
    .lazyPrintStatistics();

BatchOperator.execute();
```

运行结果如下，按时间 2014-07-16 00:00:00 划分的两个数据集，包含的数据条数分别为 131 720 和 51 160。

```
[ ts < '2014-07-16 00:00:00' ]
user_id|brand_id|type|ts
--------|--------|----|--
10944750|13451|0|2014-06-04 00:00:00.0
10944750|13451|2|2014-06-04 00:00:00.0
10944750|13451|2|2014-06-04 00:00:00.0
Summary:
| colName| count|missing|              sum|         mean|             variance|    min|      max|
|--------|------|-------|-----------------|-------------|---------------------|-------|---------|
| user_id|131720|      0|      776870757500|   5897895.2133|   12780375514149.984|  19500| 12417500|
|brand_id|131720|      0|         1854552448|    14079.5054|          72351905.181|     11|    29552|
|    type|131720|      0|               6819|        0.0518|               0.0649|      0|        3|
|      ts|131720|      0|                NaN|           NaN|                  NaN|    NaN|      NaN|

[ ts >= '2014-07-16 00:00:00' ]
user_id|brand_id|type|ts
--------|--------|----|--
10944750|1131|0|2014-07-23 00:00:00.0
10944750|1131|0|2014-07-23 00:00:00.0
10944750|24955|0|2014-07-26 00:00:00.0
Summary:
| colName|count|missing|              sum|         mean|             variance|    min|      max|
|--------|-----|-------|-----------------|-------------|---------------------|-------|---------|
| user_id|51160|      0|      306793218500|   5996740.002|   12481897307141.959|  19500| 12417500|
|brand_id|51160|      0|         740853409|    14481.1065|          70813290.0478|     15|    29552|
|    type|51160|      0|              3032|        0.0593|                 0.08|      0|        3|
|      ts|51160|      0|               NaN|           NaN|                  NaN|    NaN|      NaN|
```

11.3.2 计算特征

我们所要计算的特征大致分为三类：某时间段内的行为发生数、转化率和变化率。后两类特征可以由相应的行为发生数相除得出，所以特征计算可以分为两个阶段：

（1）计算出各时间段内点击、购买、收藏和加入购物车的次数。

（2）计算出相应的转化率和变化率。

在计算次数的时候，我们为了使计算过程清晰，并且在计算的时候有较高的效率，会根据每个用户对品牌的行为类型和发生时间，分为以下情况：是否为最近 3 天发生的点击行为，是否为最近 3 天发生的购买行为，……，是否为最近 1 个月发生的收藏行为……如果判断结果为"是"，该计数项就标识为 1；否则标识为 0。进行标识后，可以将用户和品牌的相同数据聚在同一组，将该组中所有记录对应的"是否为最近 3 天发生的点击行为"列的标识值相加，这样可得到"最近 3 天发生的点击次数"。

1. 数据预处理标识

预处理环节的作用是，为后面的特征生成做准备，比如标识了近 1 个月内、近 1 周内是否有点击、购买等行为。将划分出来构造特征的数据集，生成一些新的字段，按如下规则命名：

（1）以 is 开头，表明此列是用来标识行为的"是/否"的。

（2）之后是行为信息，包括点击（click）、购买（buy）、收藏（collect）和加入购物车（cart）等信息。

（3）最后为时间段信息，比如，1m 为最近 1 个月，3m 为最近 3 个月，m2nd 为倒数第 2 个月，3d 为最近 3 天，1w 为最近 1 周。若为整个时间段，则没有时间段信息。

（4）各部分之间用下画线"_"连接。

实现过程分为两步：第 1 步是，计算出每条样本的发生时间距离"2014-07-16 00:00:00"的天数，便于后面统计最近 3 天、最近 1 个月等特征；第 2 步是详细计算的过程。相关代码如下：

```
String clausePreProc = "user_id, brand_id, type, ts, past_days,"
  + "case when type=0 then 1 else 0 end AS is_click,"
  + "case when type=1 then 1 else 0 end AS is_buy,"
  + "case when type=2 then 1 else 0 end AS is_collect,"
  + "case when type=3 then 1 else 0 end AS is_cart,"
  + "case when type=0 and past_days<=30 then 1 else 0 end AS is_click_1m,"
  + "case when type=1 and past_days<=30 then 1 else 0 end AS is_buy_1m,"
  ... ...
  + "case when type=1 and past_days>14 and past_days<=21 then 1 else 0 end AS is_buy_w3th,"
  + "case when type=2 and past_days>14 and past_days<=21 then 1 else 0 end AS is_collect_w3th,"
  + "case when type=3 and past_days>14 and past_days<=21 then 1 else 0 end AS is_cart_w3th";

BatchOperator t1_preproc = t1
  .select("user_id, brand_id, type, ts, "
    + "TIMESTAMPDIFF(DAY, ts, TIMESTAMP '2014-07-16 00:00:00') AS past_days")
  .select(clausePreProc);
```

2. 用户-品牌联合特征

我们所关心的是在前 3 个月的行为数据中出现的用户-品牌对。对于每个用户-品牌对，可根据该用户在前 3 个月针对该品牌的所有行为数据，汇总统计、计算出如下的特征：

（1）计算出各时间段内点击、购买、收藏和加入购物车的次数。

（2）计算出相应的转化率和变化率。

我们先介绍第 1 部分的各种"次数"。在前面计算出来的标识数据表基础上，根据 user_id 和 brand_id 进行分组（group），使用 SQL 中的合计函数 SUM 计算次数。在得到的计数结果数据表中，产生的数据列名满足如下规则：

（1）以 cnt 开头，表明此列为计数信息。

（2）之后是行为信息，包括点击（click）、购买（buy）、收藏（collect）和加入购物车（cart）等信息。

（3）最后为时间段信息，比如，1m 为最近 1 个月，3m 为最近 3 个月，m2nd 为倒数第 2 个月，3d 为最近 3 天，1w 为最近 1 周。若为整个时间段，则没有时间段信息。

（4）各部分之间用下画线"_"连接。

详细的代码如下：

```
String clauseUserBrand = "user_id, brand_id, SUM(is_click) as cnt_click, SUM(is_buy) as cnt_buy, "
    + "SUM(is_collect) as cnt_collect, SUM(is_cart) as cnt_cart, "
    + "SUM(is_click_1m) as cnt_click_1m, SUM(is_buy_1m) as cnt_buy_1m, "
    + "SUM(is_collect_1m) as cnt_collect_1m, SUM(is_cart_1m) as cnt_cart_1m, "
    + "SUM(is_click_2m) as cnt_click_2m, SUM(is_buy_2m) as cnt_buy_2m, "
    + "SUM(is_collect_2m) as cnt_collect_2m, SUM(is_cart_2m) as cnt_cart_2m, "
    + "SUM(is_click_3m) as cnt_click_3m, SUM(is_buy_3m) as cnt_buy_3m, "
    + "SUM(is_collect_3m) as cnt_collect_3m, SUM(is_cart_3m) as cnt_cart_3m, "
    + "SUM(is_click_m2nd) as cnt_click_m2nd, SUM(is_buy_m2nd) as cnt_buy_m2nd, "
    + "SUM(is_collect_m2nd) as cnt_collect_m2nd, SUM(is_cart_m2nd) as cnt_cart_m2nd, "
    + "SUM(is_click_m3th) as cnt_click_m3th, SUM(is_buy_m3th) as cnt_buy_m3th, "
    + "SUM(is_collect_m3th) as cnt_collect_m3th, SUM(is_cart_m3th) as cnt_cart_m3th, "
    + "SUM(is_click_3d) as cnt_click_3d, SUM(is_buy_3d) as cnt_buy_3d, "
    + "SUM(is_collect_3d) as cnt_collect_3d, SUM(is_cart_3d) as cnt_cart_3d, "
    + "SUM(is_click_3d2nd) as cnt_click_3d2nd, SUM(is_buy_3d2nd) as cnt_buy_3d2nd, "
    + "SUM(is_collect_3d2nd) as cnt_collect_3d2nd, SUM(is_cart_3d2nd) as cnt_cart_3d2nd, "
    + "SUM(is_click_3d3th) as cnt_click_3d3th, SUM(is_buy_3d3th) as cnt_buy_3d3th, "
    + "SUM(is_collect_3d3th) as cnt_collect_3d3th, SUM(is_cart_3d3th) as cnt_cart_3d3th, "
    + "SUM(is_click_1w) as cnt_click_1w, SUM(is_buy_1w) as cnt_buy_1w, "
    + "SUM(is_collect_1w) as cnt_collect_1w, SUM(is_cart_1w) as cnt_cart_1w, "
    + "SUM(is_click_w2nd) as cnt_click_w2nd, SUM(is_buy_w2nd) as cnt_buy_w2nd, "
    + "SUM(is_collect_w2nd) as cnt_collect_w2nd, SUM(is_cart_w2nd) as cnt_cart_w2nd, "
    + "SUM(is_click_w3th) as cnt_click_w3th, SUM(is_buy_w3th) as cnt_buy_w3th, "
    + "SUM(is_collect_w3th) as cnt_collect_w3th, SUM(is_cart_w3th) as cnt_cart_w3th";

BatchOperator t1_userbrand = t1_preproc.groupBy("user_id, brand_id", clauseUserBrand);
```

有了计数信息后，下面可以进一步计算其转化率和变化率。我们将刚才计算的计数数据表作为输入，对其中的每一条记录进行变换。由于最终的特征中仍包括这些计算信息，因此在变换结果列中包括了原始的计数列，并将其对应的计算表达式写成该列的列名。新产生的转化率和变化率列，满足如下规则：

- 以 rt（即 rate 的缩写）开头，表明此列为比率信息。
- 对于变化率，中间为行为信息，包括点击（click）、购买（buy）、收藏（collect）和加入购物车（cart）等信息。
- 对于转化率，中间为行为信息加上"2buy"，包括点击转化率（click2buy）、购买转化率（buy2buy）、收藏转化率（collect2buy）和加入购物车的转化率（cart2buy）。

- 最后为时间段信息，比如 1m 为最近 1 个月，3m 为最近 3 个月，3d 为最近 3 天，1w 为最近 1 周。若为整个时间段，则没有时间段信息。
- 各部分之间用下画线"_"连接。

详细的代码如下：

```
String clauseUserBrand_Rate = "user_id,brand_id,"
  + "cnt_click,cnt_buy,cnt_collect,cnt_cart,"
  + "cnt_click_1m,cnt_buy_1m,cnt_collect_1m,cnt_cart_1m,"
  + "cnt_click_2m,cnt_buy_2m,cnt_collect_2m,cnt_cart_2m,"
  + "cnt_click_3m,cnt_buy_3m,cnt_collect_3m,cnt_cart_3m,"
... ...
  + "case when cnt_buy_1m=0 then 0.0 when cnt_buy_1m>=30.0*cnt_buy_m2nd then 30.0 else "
  + "cnt_buy_1m*1.0/cnt_buy_m2nd end AS rt_buy_1m,"
  + "case when cnt_collect_1m=0 then 0.0 when cnt_collect_1m>=30.0*cnt_collect_m2nd then 30.0 else "
  + "cnt_collect_1m*1.0/cnt_collect_m2nd end AS rt_collect_1m,"
  + "case when cnt_cart_1m=0 then 0.0 when cnt_cart_1m>=50.0*cnt_cart_m2nd then 50.0 else "
  + "cnt_cart_1m*1.0/cnt_cart_m2nd end AS rt_cart_1m";

t1_userbrand = t1_userbrand.select(clauseUserBrand_Rate);
```

3. 用户特征

本节的重点是计算出刻画用户的特征，用户范围为前 3 个月的行为数据中出现的全部用户。可根据每个用户在前 3 个月的所有行为数据，汇总统计、计算出如下的特征：

（1）计算出各时间段内点击、购买、收藏和加入购物车的次数。

（2）计算出相应的购买转化率。

与前面的计算方法相似，先计算第 1 部分的各种"次数"。在前面计算出来的标识数据表基础上，根据 user_id 进行分组（group），使用 SQL 中的合计函数 SUM 计算次数。产生的数据列名满足如下规则：

（1）以 user_cnt 开头，表明此列为用户计数信息。

（2）之后是行为信息，包括点击（click）、购买（buy）、收藏（collect）和加入购物车（cart）等信息。

（3）最后为时间段信息，比如，1m 为最近 1 个月，3m 为最近 3 个月，m2nd 为倒数第 2 个月，3d 为最近 3 天，1w 为最近 1 周。若为整个时间段，则没有时间段信息。

（4）各部分之间用下画线"_"连接。

详细的代码如下：

```
String clauseUser = "user_id, "
  + "SUM(is_click) as user_cnt_click, SUM(is_buy) as user_cnt_buy, "
  + "SUM(is_collect) as user_cnt_collect, SUM(is_cart) as user_cnt_cart, "
  + "SUM(is_click_1m) as user_cnt_click_1m, SUM(is_buy_1m) as user_cnt_buy_1m, "
```

```
    + "SUM(is_collect_1m) as user_cnt_collect_1m, SUM(is_cart_1m) as user_cnt_cart_1m, "
    + "SUM(is_click_2m) as user_cnt_click_2m, SUM(is_buy_2m) as user_cnt_buy_2m, "
    + "SUM(is_collect_2m) as user_cnt_collect_2m, SUM(is_cart_2m) as user_cnt_cart_2m, "
    + "SUM(is_click_3m) as user_cnt_click_3m, SUM(is_buy_3m) as user_cnt_buy_3m, "
    + "SUM(is_collect_3m) as user_cnt_collect_3m, SUM(is_cart_3m) as user_cnt_cart_3m, "
    + "SUM(is_click_m2nd) as user_cnt_click_m2nd, SUM(is_buy_m2nd) as user_cnt_buy_m2nd, "
    + "SUM(is_collect_m2nd) as user_cnt_collect_m2nd, SUM(is_cart_m2nd) as user_cnt_cart_m2nd, "
    + "SUM(is_click_m3th) as user_cnt_click_m3th, SUM(is_buy_m3th) as user_cnt_buy_m3th, "
    + "SUM(is_collect_m3th) as user_cnt_collect_m3th, SUM(is_cart_m3th) as user_cnt_cart_m3th, "
    + "SUM(is_click_3d) as user_cnt_click_3d, SUM(is_buy_3d) as user_cnt_buy_3d, "
    + "SUM(is_collect_3d) as user_cnt_collect_3d, SUM(is_cart_3d) as user_cnt_cart_3d, "
    + "SUM(is_click_3d2nd) as user_cnt_click_3d2nd, SUM(is_buy_3d2nd) as user_cnt_buy_3d2nd, "
    + "SUM(is_collect_3d2nd) as user_cnt_collect_3d2nd, SUM(is_cart_3d2nd) as user_cnt_cart_3d2nd, "
    + "SUM(is_click_3d3th) as user_cnt_click_3d3th, SUM(is_buy_3d3th) as user_cnt_buy_3d3th, "
    + "SUM(is_collect_3d3th) as user_cnt_collect_3d3th, SUM(is_cart_3d3th) as user_cnt_cart_3d3th, "
    + "SUM(is_click_1w) as user_cnt_click_1w, SUM(is_buy_1w) as user_cnt_buy_1w, "
    + "SUM(is_collect_1w) as user_cnt_collect_1w, SUM(is_cart_1w) as user_cnt_cart_1w, "
    + "SUM(is_click_w2nd) as user_cnt_click_w2nd, SUM(is_buy_w2nd) as user_cnt_buy_w2nd, "
    + "SUM(is_collect_w2nd) as user_cnt_collect_w2nd, SUM(is_cart_w2nd) as user_cnt_cart_w2nd, "
    + "SUM(is_click_w3th) as user_cnt_click_w3th, SUM(is_buy_w3th) as user_cnt_buy_w3th, "
    + "SUM(is_collect_w3th) as user_cnt_collect_w3th, SUM(is_cart_w3th) as user_cnt_cart_w3th";

BatchOperator t1_user = t1_preproc.groupBy("user_id", clauseUser);
```

有了计数信息后，再进一步计算其转化率。我们将刚才计算的计数数据表作为输入，对其中的每一条记录进行变换。由于最终的特征中仍包括这些计算信息，因此在变换结果列中包括了原始的计数列，并将其对应的计算表达式写成该列的列名。新产生的购买转化率列满足如下规则：

- 以 user_rt 开头，表明此列为用户的比率信息。
- 中间为行为信息加上 "2buy"，包括点击转化率（click2buy）、购买转化率（buy2buy）、收藏转化率（collect2buy）和加入购物车的转化率（cart2buy）等信息。
- 最后为时间段信息，比如 1m 为最近 1 个月，3m 为最近 3 个月，3d 为最近 3 天，1w 为最近 1 周。若为整个时间段，则没有时间段信息。
- 各部分之间用下画线 "_" 连接。

详细的代码如下：

```
String clauseUser_Rate = "user_id AS user_id4join,"
    + "user_cnt_click,user_cnt_buy,user_cnt_collect,user_cnt_cart,"
    + "user_cnt_click_1m,user_cnt_buy_1m,user_cnt_collect_1m,user_cnt_cart_1m,"
    + "user_cnt_click_2m,user_cnt_buy_2m,user_cnt_collect_2m,user_cnt_cart_2m,"
    + "user_cnt_click_3m,user_cnt_buy_3m,user_cnt_collect_3m,user_cnt_cart_3m,"
    + "user_cnt_click_m2nd,user_cnt_buy_m2nd,user_cnt_collect_m2nd,user_cnt_cart_m2nd,"
    + "user_cnt_click_m3th,user_cnt_buy_m3th,user_cnt_collect_m3th,user_cnt_cart_m3th,"
    + "user_cnt_click_3d,user_cnt_buy_3d,user_cnt_collect_3d,user_cnt_cart_3d,"
    + "user_cnt_click_3d2nd,user_cnt_buy_3d2nd,user_cnt_collect_3d2nd,user_cnt_cart_3d2nd,"
    + "user_cnt_click_3d3th,user_cnt_buy_3d3th,user_cnt_collect_3d3th,user_cnt_cart_3d3th,"
    + "user_cnt_click_1w,user_cnt_buy_1w,user_cnt_collect_1w,user_cnt_cart_1w,"
    + "user_cnt_click_w2nd,user_cnt_buy_w2nd,user_cnt_collect_w2nd,user_cnt_cart_w2nd,"
    + "user_cnt_click_w3th,user_cnt_buy_w3th,user_cnt_collect_w3th,user_cnt_cart_w3th,"
```

```
+ "case when user_cnt_buy>user_cnt_click then 1.0 when user_cnt_buy=0 then 0.0 else "
+ "user_cnt_buy*1.0/user_cnt_click end AS user_rt_click2buy,"
+ "case when user_cnt_buy>user_cnt_collect then 1.0 when user_cnt_buy=0 then 0.0 else "
+ "user_cnt_buy*1.0/user_cnt_collect end AS user_rt_collect2buy,"
+ "case when user_cnt_buy>user_cnt_cart then 1.0 when user_cnt_buy=0 then 0.0 else "
+ "user_cnt_buy*1.0/user_cnt_cart end AS user_rt_cart2buy,"
+ "case when user_cnt_buy_3d>user_cnt_click_3d then 1.0 when user_cnt_buy_3d=0 then 0.0 else "
+ "user_cnt_buy_3d*1.0/user_cnt_click_3d end AS user_rt_click2buy_3d,"
+ "case when user_cnt_buy_3d>user_cnt_collect_3d then 1.0 when user_cnt_buy_3d=0 then 0.0 else "
+ "user_cnt_buy_3d*1.0/user_cnt_collect_3d end AS user_rt_collect2buy_3d,"
+ "case when user_cnt_buy_3d>user_cnt_cart_3d then 1.0 when user_cnt_buy_3d=0 then 0.0 else "
+ "user_cnt_buy_3d*1.0/user_cnt_cart_3d end AS user_rt_cart2buy_3d,"
+ "case when user_cnt_buy_1w>user_cnt_click_1w then 1.0 when user_cnt_buy_1w=0 then 0.0 else "
+ "user_cnt_buy_1w*1.0/user_cnt_click_1w end AS user_rt_click2buy_1w,"
+ "case when user_cnt_buy_1w>user_cnt_collect_1w then 1.0 when user_cnt_buy_1w=0 then 0.0 else "
+ "user_cnt_buy_1w*1.0/user_cnt_collect_1w end AS user_rt_collect2buy_1w,"
+ "case when user_cnt_buy_1w>user_cnt_cart_1w then 1.0 when user_cnt_buy_1w=0 then 0.0 else "
+ "user_cnt_buy_1w*1.0/user_cnt_cart_1w end AS user_rt_cart2buy_1w,"
+ "case when user_cnt_buy_1m>user_cnt_click_1m then 1.0 when user_cnt_buy_1m=0 then 0.0 else "
+ "user_cnt_buy_1m*1.0/user_cnt_click_1m end AS user_rt_click2buy_1m,"
+ "case when user_cnt_buy_1m>user_cnt_collect_1m then 1.0 when user_cnt_buy_1m=0 then 0.0 else "
+ "user_cnt_buy_1m*1.0/user_cnt_collect_1m end AS user_rt_collect2buy_1m,"
+ "case when user_cnt_buy_1m>user_cnt_cart_1m then 1.0 when user_cnt_buy_1m=0 then 0.0 else "
+ "user_cnt_buy_1m*1.0/user_cnt_cart_1m end AS user_rt_cart2buy_1m";
t1_user = t1_user.select(clauseUser_Rate);
```

4. 品牌特征

本节重点计算描述品牌的特征，品牌选择范围为前 3 个月的行为数据中出现的全部品牌。对于每个品牌，可根据前 3 个月中所有用户对该品牌商品的行为数据，汇总统计、计算出如下的特征：

（1）计算出各时间段内该品牌的商品被点击、被购买、被收藏和被加入购物车的次数。

（2）计算出相应的购买转化率。

与计算用户特征的方法相似，先计算第 1 部分的各种"次数"。在前面计算出来的标识数据表基础上，根据 brand_id 进行分组（group），使用 SQL 中的合计函数 SUM 计算次数。产生的数据列名满足如下规则：

（1）以 brand_cnt 开头，表明此列为用户计数信息。

（2）之后是行为信息，包括点击（click）、购买（buy）、收藏（collect）和加入购物车（cart）等信息。

（3）最后为时间段信息，比如，1m 为最近 1 个月，3m 为最近 3 个月，m2nd 为倒数第 2 个月，3d 为最近 3 天，1w 为最近 1 周。若为整个时间段，则没有时间段信息。

（4）各部分之间用下画线"_"连接。

详细的代码如下：

```
String clauseBrand = "brand_id, "
    + "SUM(is_click) as brand_cnt_click, SUM(is_buy) as brand_cnt_buy, "
    + "SUM(is_collect) as brand_cnt_collect, SUM(is_cart) as brand_cnt_cart, "
    + "SUM(is_click_1m) as brand_cnt_click_1m, SUM(is_buy_1m) as brand_cnt_buy_1m, "
    + "SUM(is_collect_1m) as brand_cnt_collect_1m, SUM(is_cart_1m) as brand_cnt_cart_1m, "
    + "SUM(is_click_2m) as brand_cnt_click_2m, SUM(is_buy_2m) as brand_cnt_buy_2m, "
    + "SUM(is_collect_2m) as brand_cnt_collect_2m, SUM(is_cart_2m) as brand_cnt_cart_2m, "
    + "SUM(is_click_3m) as brand_cnt_click_3m, SUM(is_buy_3m) as brand_cnt_buy_3m, "
    + "SUM(is_collect_3m) as brand_cnt_collect_3m, SUM(is_cart_3m) as brand_cnt_cart_3m, "
    + "SUM(is_click_m2nd) as brand_cnt_click_m2nd, SUM(is_buy_m2nd) as brand_cnt_buy_m2nd, "
    + "SUM(is_collect_m2nd) as brand_cnt_collect_m2nd, SUM(is_cart_m2nd) as brand_cnt_cart_m2nd, "
    + "SUM(is_click_m3th) as brand_cnt_click_m3th, SUM(is_buy_m3th) as brand_cnt_buy_m3th, "
    + "SUM(is_collect_m3th) as brand_cnt_collect_m3th, SUM(is_cart_m3th) as brand_cnt_cart_m3th, "
    + "SUM(is_click_3d) as brand_cnt_click_3d, SUM(is_buy_3d) as brand_cnt_buy_3d, "
    + "SUM(is_collect_3d) as brand_cnt_collect_3d, SUM(is_cart_3d) as brand_cnt_cart_3d, "
    + "SUM(is_click_3d2nd) as brand_cnt_click_3d2nd, SUM(is_buy_3d2nd) as brand_cnt_buy_3d2nd, "
    + "SUM(is_collect_3d2nd) as brand_cnt_collect_3d2nd, SUM(is_cart_3d2nd) as brand_cnt_cart_3d2nd, "
    + "SUM(is_click_3d3th) as brand_cnt_click_3d3th, SUM(is_buy_3d3th) as brand_cnt_buy_3d3th, "
    + "SUM(is_collect_3d3th) as brand_cnt_collect_3d3th, SUM(is_cart_3d3th) as brand_cnt_cart_3d3th, "
    + "SUM(is_click_1w) as brand_cnt_click_1w, SUM(is_buy_1w) as brand_cnt_buy_1w, "
    + "SUM(is_collect_1w) as brand_cnt_collect_1w, SUM(is_cart_1w) as brand_cnt_cart_1w, "
    + "SUM(is_click_w2nd) as brand_cnt_click_w2nd, SUM(is_buy_w2nd) as brand_cnt_buy_w2nd, "
    + "SUM(is_collect_w2nd) as brand_cnt_collect_w2nd, SUM(is_cart_w2nd) as brand_cnt_cart_w2nd, "
    + "SUM(is_click_w3th) as brand_cnt_click_w3th, SUM(is_buy_w3th) as brand_cnt_buy_w3th, "
    + "SUM(is_collect_w3th) as brand_cnt_collect_w3th, SUM(is_cart_w3th) as brand_cnt_cart_w3th";

BatchOperator t1_brand = t1_preproc.groupBy("brand_id", clauseBrand);
```

有了计数信息后，再进一步计算其转化率。我们将刚才计算的计数数据表作为输入，对其中的每一条记录进行变换。由于最终的特征中仍包括这些计算信息，因此在变换结果列中包括了原始的计数列，并将其对应的计算表达式写成该列的列名。新产生的购买转化率列满足如下规则：

- 以 brand_rt 开头，表明此列为品牌的比率信息。
- 中间为行为信息加上 "2buy"，包括点击转化率（click2buy）、购买转化率（buy2buy）、收藏转化率（collect2buy）和加入购物车的转化率（cart2buy）等信息。
- 最后为时间段信息，比如，1m 为最近 1 个月，3m 为最近 3 个月，3d 为最近 3 天，1w 为最近 1 周。若为整个时间段，则没有时间段信息。
- 各部分之间用下画线 "_" 连接。

详细的代码如下：

```
String clauseBrand_Rate = "brand_id AS brand_id4join,"
    + "brand_cnt_click,brand_cnt_buy,brand_cnt_collect,brand_cnt_cart,"
    + "brand_cnt_click_1m,brand_cnt_buy_1m,brand_cnt_collect_1m,brand_cnt_cart_1m,"
    + "brand_cnt_click_2m,brand_cnt_buy_2m,brand_cnt_collect_2m,brand_cnt_cart_2m,"
    + "brand_cnt_click_3m,brand_cnt_buy_3m,brand_cnt_collect_3m,brand_cnt_cart_3m,"
    + "brand_cnt_click_m2nd,brand_cnt_buy_m2nd,brand_cnt_collect_m2nd,brand_cnt_cart_m2nd,"
    + "brand_cnt_click_m3th,brand_cnt_buy_m3th,brand_cnt_collect_m3th,brand_cnt_cart_m3th,"
    + "brand_cnt_click_3d,brand_cnt_buy_3d,brand_cnt_collect_3d,brand_cnt_cart_3d,"
    + "brand_cnt_click_3d2nd,brand_cnt_buy_3d2nd,brand_cnt_collect_3d2nd,brand_cnt_cart_3d2nd,"
```

```
        + "brand_cnt_click_3d3th,brand_cnt_buy_3d3th,brand_cnt_collect_3d3th,brand_cnt_cart_3d3th,"
        + "brand_cnt_click_1w,brand_cnt_buy_1w,brand_cnt_collect_1w,brand_cnt_cart_1w,"
        + "brand_cnt_click_w2nd,brand_cnt_buy_w2nd,brand_cnt_collect_w2nd,brand_cnt_cart_w2nd,"
        + "brand_cnt_click_w3th,brand_cnt_buy_w3th,brand_cnt_collect_w3th,brand_cnt_cart_w3th,"
        + "case when brand_cnt_buy>brand_cnt_click then 1.0 when brand_cnt_buy=0 then 0.0 else "
        + "brand_cnt_buy*1.0/brand_cnt_click end AS brand_rt_click2buy,"
        + "case when brand_cnt_buy>brand_cnt_collect then 1.0 when brand_cnt_buy=0 then 0.0 else "
        + "brand_cnt_buy*1.0/brand_cnt_collect end AS brand_rt_collect2buy,"
        + "case when brand_cnt_buy>brand_cnt_cart then 1.0 when brand_cnt_buy=0 then 0.0 else "
        + "brand_cnt_buy*1.0/brand_cnt_cart end AS brand_rt_cart2buy,"
        + "case when brand_cnt_buy_3d>brand_cnt_click_3d then 1.0 when brand_cnt_buy_3d=0 then 0.0 else "
        + "brand_cnt_buy_3d*1.0/brand_cnt_click_3d end AS brand_rt_click2buy_3d,"
        + "case when brand_cnt_buy_3d>brand_cnt_collect_3d then 1.0 when brand_cnt_buy_3d=0 then 0.0 else "
        + "brand_cnt_buy_3d*1.0/brand_cnt_collect_3d end AS brand_rt_collect2buy_3d,"
        + "case when brand_cnt_buy_3d>brand_cnt_cart_3d then 1.0 when brand_cnt_buy_3d=0 then 0.0 else "
        + "brand_cnt_buy_3d*1.0/brand_cnt_cart_3d end AS brand_rt_cart2buy_3d,"
        + "case when brand_cnt_buy_1w>brand_cnt_click_1w then 1.0 when brand_cnt_buy_1w=0 then 0.0 else "
        + "brand_cnt_buy_1w*1.0/brand_cnt_click_1w end AS brand_rt_click2buy_1w,"
        + "case when brand_cnt_buy_1w>brand_cnt_collect_1w then 1.0 when brand_cnt_buy_1w=0 then 0.0 else "
        + "brand_cnt_buy_1w*1.0/brand_cnt_collect_1w end AS brand_rt_collect2buy_1w,"
        + "case when brand_cnt_buy_1w>brand_cnt_cart_1w then 1.0 when brand_cnt_buy_1w=0 then 0.0 else "
        + "brand_cnt_buy_1w*1.0/brand_cnt_cart_1w end AS brand_cart2buy_1w,"
        + "case when brand_cnt_buy_1m>brand_cnt_click_1m then 1.0 when brand_cnt_buy_1m=0 then 0.0 else "
        + "brand_cnt_buy_1m*1.0/brand_cnt_click_1m end AS brand_rt_click2buy_1m,"
        + "case when brand_cnt_buy_1m>brand_cnt_collect_1m then 1.0 when brand_cnt_buy_1m=0 then 0.0 else "
        + "brand_cnt_buy_1m*1.0/brand_cnt_collect_1m end AS brand_rt_collect2buy_1m,"
        + "case when brand_cnt_buy_1m>brand_cnt_cart_1m then 1.0 when brand_cnt_buy_1m=0 then 0.0 else "
        + "brand_cnt_buy_1m*1.0/brand_cnt_cart_1m end AS brand_rt_cart2buy_1m";

t1_brand = t1_brand.select(clauseBrand_Rate);
```

5. 整合训练数据的特征

前面分别计算出了用户-品牌联合特征、用户特征和品牌特征。在判断一个用户是否会购买一个品牌的时候，用户对该品牌的关注度是重要的因素，它可以通过用户-品牌联合特征进行刻画。另外，该用户的购买特点和该品牌被购买的情况，也是重要的因素。所以，我们需要将所有的特征整合起来进行分析。

整合的方法如下：以 user_id 和 brand_id 作为联合主键，包括用户-品牌联合特征，再使用 SQL 的 JOIN（连接）操作，加入 user_id 对应的用户特征，并加入 brand_id 对应的品牌特征。

详细的代码如下：

```
BatchOperator t1_join = new JoinBatchOp()
    .setSelectClause("*")
    .setJoinPredicate("user_id=user_id4join")
    .linkFrom(t1_userbrand, t1_user);

t1_join = new JoinBatchOp()
    .setSelectClause("*")
    .setJoinPredicate("brand_id=brand_id4join")
    .linkFrom(t1_join, t1_brand);
```

11.3.3 计算标签

首先计算出最后一个月中发生了购买行为的用户-品牌对。这可以使用 SQL DISTINCT 语句轻松得到。

如果前面计算出来的特征数据表中某一条记录的 user_id 和 brand_id 在由原始数据划分出的 t2 中出现，就说明该 user_id 用户在最后一个月中购买了该 brand_id 品牌的商品，可以将 label 列赋值为 1；反之该用户就没有购买该品牌的商品。使用 LEFT OUTER JOIN（左连接）方法，将有购买行为的那些记录的 label 列赋值为 1，其他项为缺失值状态。相应的代码如下：

```
BatchOperator t2_label = t2
    .filter("type=1")
    .select("user_id AS user_id4label, brand_id AS brand_id4label, 1 as label")
    .distinct();

BatchOperator feature_label = new LeftOuterJoinBatchOp()
    .setSelectClause("*")
    .setJoinPredicate("user_id = user_id4label AND brand_id = brand_id4label")
    .linkFrom(t1_join, t2_label);
```

然后，我们再使用缺失值填充组件，选择 label 字段，缺失值用 0 填充，具体代码如下：

```
Imputer imputer = new Imputer()
    .setStrategy("value")
    .setFillValue("0")
    .setSelectedCols("label");

feature_label = imputer.fit(feature_label).transform(feature_label);
```

我们再看一下当前数据集的 Schema，代码如下：

```
System.out.println(feature_label.getSchema());
```

结果如下。由于列数较多，因此这里选择了一些有代表性的列名称和类型。

```
root
 |-- user_id: BIGINT
 |-- brand_id: BIGINT
 |-- cnt_click: INT
 |-- cnt_buy: INT
 |-- cnt_collect: INT
 |-- cnt_cart: INT
 |-- cnt_click_1m: INT
 |-- cnt_buy_1m: INT
```

```
... ...
 |-- brand_id4join: BIGINT
 |-- brand_cnt_click: INT
 |-- brand_cnt_buy: INT
... ...
 |-- brand_rt_click2buy_1m: LEGACY(BigDecimal)
 |-- brand_rt_collect2buy_1m: LEGACY(BigDecimal)
 |-- brand_rt_cart2buy_1m: LEGACY(BigDecimal)
 |-- user_id4label: BIGINT
 |-- brand_id4label: BIGINT
 |-- label: INT
```

在此，我们看到两个问题：

- 有些数据列对于后面的分类问题是没有用的，比如 brand_id4join、user_id4label 等。
- 这里有 LEGACY(BigDecimal) 类型，该类型需要转化为基本的 double 类型，方便后面分类器使用。

为解决这两个问题，可以构造 SQL 语句，选取所有需要的特征列，将其类型转化为 double 类型，再加上标签列。具体代码如下。最后将生成的数据集保存起来，供后面的模型训练使用。将整个数据集导出到文件 feature_label.ak 中。

```
String[] featureColNames =
   ArrayUtils.removeElements(
      feature_label.getColNames(),
      new String[] {
         "user_id", "brand_id",
         "user_id4join", "brand_id4join",
         "user_id4label", "brand_id4label",
         LABEL_COL_NAME
      }
   );

StringBuilder sbd = new StringBuilder();
for (String name : featureColNames) {
   sbd.append("CAST(").append(name).append(" AS DOUBLE) AS ").append(name).append(", ");
}
sbd.append(LABEL_COL_NAME);

feature_label
   .select(sbd.toString())
   .link(
      new AkSinkBatchOp()
         .setFilePath(DATA_DIR + FEATURE_LABEL_FILE)
         .setOverwriteSink(true)
   );
```

11.4 正负样本配比

前面已经构造出很多特征,本节将尝试多种二分类模型,并选出最适合的模型。这里需要提醒用户注意的一点是,训练样本中正负样本的比例如果相差太悬殊,会对训练出来的模型产生影响。我们会在模型训练前先关注一下,训练数据的正负样本比例,并进行调整。

前面构造了特征和标签,现在我们先深入了解一下数据:

```
AkSourceBatchOp all_data = new AkSourceBatchOp().setFilePath(DATA_DIR + FEATURE_LABEL_FILE);
all_data
    .lazyPrintStatistics()
    .groupBy("label", "label, COUNT(*) AS cnt")
    .print();
```

打印统计信息如下,样本总条数为 42 531,没有缺失值;计数类特征的值域变化较大,比率类特征取值为[0, 1]。

colName	count	missing	sum	mean	variance	min	max
cnt_click	42531	0	125865	2.9594	76.9641	0	542
cnt_buy	42531	0	4971	0.1169	0.3815	0	40
cnt_collect	42531	0	804	0.0189	0.0324	0	13
cnt_cart	42531	0	80	0.0019	0.0025	0	4
cnt_click_1m	42531	0	48025	1.1292	22.1792	0	302
cnt_buy_1m	42531	0	1880	0.0442	0.1202	0	17
cnt_collect_1m	42531	0	307	0.0072	0.0115	0	6
cnt_cart_1m	42531	0	43	0.001	0.0013	0	4
... ...							
brand_rt_click2buy_1w	42531	0	600.978	0.0141	0.0036	0	1
brand_rt_collect2buy_1w	42531	0	6259	0.1472	0.1252	0	1
brand_cart2buy_1w	42531	0	6288	0.1478	0.126	0	1
brand_rt_click2buy_1m	42531	0	1243.6492	0.0292	0.0052	0	1
brand_rt_collect2buy_1m	42531	0	15682.119	0.3687	0.2307	0	1
brand_rt_cart2buy_1m	42531	0	15860	0.3729	0.2339	0	1
label	42531	0	259	0.0061	0.0061	0	1

对标签列进行分组聚合计数,得到如下的结果,正负样本的数量有 2 个数量级的差距。

```
label|cnt
-----|---
0|42272
1|259
```

接下来，我们对数据进行分类建模。首先对数据集进行拆分，得到训练集和测试集，代码如下：

```
Utils.splitTrainTestIfNotExist(all_data, DATA_DIR + TRAIN_FILE, DATA_DIR + TEST_FILE, 0.8);
```

随后，使用基本的二分类算法（逻辑回归），试验分类效果：

```
AkSourceBatchOp train_data = new AkSourceBatchOp().setFilePath(DATA_DIR + TRAIN_FILE);
AkSourceBatchOp test_data = new AkSourceBatchOp().setFilePath(DATA_DIR + TEST_FILE);

String[] featureColNames = ArrayUtils.removeElement(train_data.getColNames(), LABEL_COL_NAME);

new LogisticRegression()
   .setFeatureCols(featureColNames)
   .setLabelCol(LABEL_COL_NAME)
   .setPredictionCol(PREDICTION_COL_NAME)
   .setPredictionDetailCol(PRED_DETAIL_COL_NAME)
   .fit(train_data)
   .transform(test_data)
   .link(
      new EvalBinaryClassBatchOp()
          .setLabelCol(LABEL_COL_NAME)
          .setPredictionDetailCol(PRED_DETAIL_COL_NAME)
          .lazyPrintMetrics("LogisticRegression")
   );
BatchOperator.execute();
```

得到二分类评估指标如下。其中，精确度（Accuracy）指标很高，为 0.9945；但是召回率（Recall）指标很低，只有 0.0435，46 个正样本只有 2 个被正确分类。

```
-------------------------- Metrics: --------------------------
Auc:0.7629  Accuracy:0.9945  Precision:0.4   Recall:0.0435   F1:0.0784  LogLoss:0.0342
|Pred\Real|   1|    0|
|---------|---|----|
|        1|  2|   3|
|        0| 44|8457|
```

这里召回率（Recall）指标很低的主要原因在于，训练数据中正负样本的比例相差悬殊。一个解决方法是保持正样本的同时，通过采样来降低负样本的数量。可以使用分层采样组件 StratifiedSampleBatchOp，对正负样本分别设置采样率，正样本全部保留，负样本采样 5%，参数 setStrataRatios 设置为"0:0.05,1:1.0"。将该结果保存到文件 DATA_DIR + TRAIN_SAMPLE_FILE 中，方便后续多次试验时调用。相关代码如下：

```
train_data
  .link(
     new StratifiedSampleBatchOp()
        .setStrataRatios("0:0.05,1:1.0")
        .setStrataCol(LABEL_COL_NAME)
```

```
    )
    .link(
        new AkSinkBatchOp()
            .setFilePath(DATA_DIR + TRAIN_SAMPLE_FILE)
    );
BatchOperator.execute();
```

我们再对调整了正负样本比例的数据进行逻辑回归训练,并在测试集上进行二分类评估。具体代码如下:

```
AkSourceBatchOp train_sample =
    new AkSourceBatchOp().setFilePath(DATA_DIR + TRAIN_SAMPLE_FILE);

new LogisticRegression()
    .setFeatureCols(featureColNames)
    .setLabelCol(LABEL_COL_NAME)
    .setPredictionCol(PREDICTION_COL_NAME)
    .setPredictionDetailCol(PRED_DETAIL_COL_NAME)
    .fit(train_sample)
    .transform(test_data)
    .link(
        new EvalBinaryClassBatchOp()
            .setLabelCol(LABEL_COL_NAME)
            .setPredictionDetailCol(PRED_DETAIL_COL_NAME)
            .lazyPrintMetrics("LogisticRegression with Stratified Sample")
    );
BatchOperator.execute();
```

运行结果如下。其中,召回率(Recall)提升为 0.2391,46 个正样本中有 11 个预测对了。

```
-------------------------- Metrics: --------------------------
Auc:0.6724  Accuracy:0.9638  Precision:0.0387  Recall:0.2391  F1:0.0667  LogLoss:0.19
|Pred\Real|   1|   0|
|---------|---|----|
|        1|  11| 273|
|        0|  35|8187|
```

11.5 决策树

本节将继续对分层采样后的训练数据进行建模。下面我们看看 3 种决策树算法的效果。

使用决策树组件 DecisionTreeClassifier,遍历 3 种决策树类型 TreeType.GINI、TreeType.INFOGAIN、TreeType.INFOGAINRATIO,具体代码如下:

```
for (TreeType treeType : new TreeType[] {TreeType.GINI,
    TreeType.INFOGAIN, TreeType.INFOGAINRATIO}) {
```

```
new DecisionTreeClassifier()
    .setTreeType(treeType)
    .setFeatureCols(featureColNames)
    .setLabelCol(LABEL_COL_NAME)
    .setPredictionCol(PREDICTION_COL_NAME)
    .setPredictionDetailCol(PRED_DETAIL_COL_NAME)
    .fit(train_sample)
    .transform(test_data)
    .link(
        new EvalBinaryClassBatchOp()
            .setPositiveLabelValueString("1")
            .setLabelCol(LABEL_COL_NAME)
            .setPredictionDetailCol(PRED_DETAIL_COL_NAME)
            .lazyPrintMetrics(treeType.toString())
    );
}
```

将运行结果放在一起对比，如表 11-2 所示。

表 11-2 3 种决策树算法的运行结果对比

ID3	C4.5	CART
TreeType.INFOGAIN	TreeType.INFOGAINRATIO	TreeType.GINI
Auc:0.6532	Auc:0.6465	Auc:0.6632
Accuracy:0.8895	Accuracy:0.8655	Accuracy:0.8924
Precision:0.0194	Precision:0.0167	Precision:0.0199
Recall:0.3913	Recall:0.413	Recall:0.3913
F1:0.0369	F1:0.0321	F1:0.0379
LogLoss:2.9092	LogLoss:2.0405	LogLoss:2.3753
混淆矩阵： |Pred\Real| 1| 0| |---------|---|----| | 1| 18| 912| | 0| 28|7548|	混淆矩阵： |Pred\Real| 1| 0| |---------|---|----| | 1| 19|1117| | 0| 27|7343|	混淆矩阵： |Pred\Real| 1| 0| |---------|---|----| | 1| 18| 887| | 0| 28|7573|

这 3 种决策树算法在各项指标上的差异不大。和前面的逻辑回归算法相比，这 3 种决策树算法在召回率上有一些提升，但由于其 Precision 指标的下降，致使 F1 指标偏低。

11.6 集成学习

本节介绍一个新的思路：集成学习（Ensemble Learning），以便在决策树模型的基础上，构建更为强大的分类器：随机森林模型和 GBDT 模型。

人们常说"三个臭皮匠赛过诸葛亮"，这种思想在机器学习领域的体现，就是集成学习。

集成学习（Ensemble Learning），用多个弱分类器构成一个强分类器，目的是增强集成模型的预测性能。弱分类器可以由决策树、神经网络、贝叶斯、K 最近邻等构成。

下面着重讨论两种广泛使用的集成技术：Bagging（装袋法）和 Boosting（提升法）。

11.6.1 Bootstrap aggregating

Bootstrap aggregating，经常使用缩写形式"Bagging"（由 Bootstrap AGGregatING 得来）。由于这种缩写的关系，其也被称为"装袋法"。

"Bootstrap"这个名字来自谚语"pull up by your own bootstraps"，意思是"improve your situation by your own efforts"，即依靠自身努力得到提高。因此，Bootstrap 被称为自助法，或自举法。Bootstrap 是非参数统计中一种重要的估计统计量方差，进而进行区间估计的统计方法，基本步骤如下：

（1）采用重复抽样技术，从原始样本中抽取出 N 个样本。

（2）根据抽取出的样本计算给定的统计量 T。

（3）重复上述步骤 M 次（一般大于 1000 次），得到 M 个统计量 T。

（4）计算上述 M 个统计量 T 的样本方差，得到统计量的方差。

Bootstrap aggregating（Bagging）将统计学中常用的 Bootstrap 方法应用到分类模型上，成为 Bagging 分类器（Bagging Classifier）。Bagging 分类器的基本思想是对训练集进行有放回的抽样，得到 M 个子训练集，分别对每个子训练集使用若干分类器进行训练，得到 M 个模型；预测时，对 M 个模型预测的结果进行投票，获得最终的分类结果。具体步骤如下：

（1）采用重复抽样技术，从训练样本中抽取 N 个样本。

（2）对这 N 个样本建立 M 个弱分类器（CART、SVM 等）。

（3）重复以上两步 M 次，训练这 M 个弱分类器（CART、SVM 等）。

（4）预测时，M 个分类器各自得到分类结果。通过投票方式，确定最终的分类结果。

Bagging 分类器的训练如图 11-2 所示，各个子分类器的训练是可以并发进行的。

在预测阶段，首先是各子分类器分别进行预测，然后对

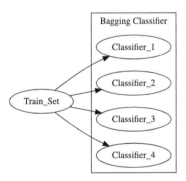

图 11-2　Bagging Classifier 训练示意图

各预测结果通过投票（Vote）的方式进行汇总，作为 Bagging 分类器（Bagging Classifier）的预测结果，如图 11-3 所示。

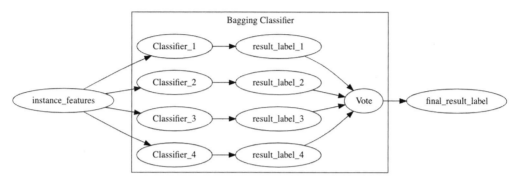

图 11-3　Bagging Classifier 预测示意图

11.6.2　Boosting

前面介绍的 Bagging（装袋法）采用并行的方法组合多个弱分类器，而 Boosting（提升法）则采样串行的方式组合多个弱分类器。

采用 Boosting 方法时，先使用一个弱分类器建立一个模型，以便对训练数据有一个初步的预测；然后针对预测结果，侧重那些被错误预测的样本，通过调整其权重或者关注其残差等方式，重新生成新的训练集；之后再使用一个弱分类器对新生成的训练集进行训练，得到新的模型及新的预测结果；随后重新生成新的训练集……一直重复下去，直到得到指定数量的弱分类器模型。我们可以想象成，Boosting 方法得到了一个弱分类器序列；第一个弱分类器在训练原始数据；其后的每个弱分类器都是针对之前弱分类器没有训练好的情形，进行加强训练。

如图 11-4 的左子图所示，Train_Set 为原始训练数据，训练第一个弱分类器 Classifier_1 并得到一个弱分类器模型；然后对 Train_Set 进行预测得到 Prediction_1，结合训练数据 Train_Set 与预测结果 Prediction_1，通过某种关注训练错误的方式 Update_Data_1，得到新的训练集 Train_Set_2；随后，再由 Train_Set_2 得到 Classifier_2；之后预测出 Prediction_2，使用方式 Update_Data_2 得到新的数据集 Train_Set_3；直到得到最后一个弱分类器 Classifier_3。弱分类器序列{ Classifier_1, Classifier_2, Classifier_3}就是最终集成的分类器。

使用 Boosting 思想的具体算法主要有 AdaBoost 算法和 Gradient Boosting 算法。

采用 AdaBoost 算法，可通过对训练失败的样本赋以较大权重的方式，得到新的训练集。如图 11-4 的中间子图所示，对原始数据进行训练并得到 Classifier_1 后，结合训练数据 Train_Set 与预测结果 Prediction_1，通过调整训练样本权重（Update_Weight_1）的方式，得到新的训练集 Train_Set_2。此外，还计算出了弱分类器 Classifier_1 的权重参数 Classifier_Weight_1（权重参数会在介绍 AdaBoost 模型如何执行预测操作时说明）；然后基于 Train_Set_2，得到弱分类器 Classifier_2，再结合训练数据 Train_Set_2 与预测结果 Prediction_2，通过调整训练样本权重（Update_Weight_2）的方式，得到新的训练集 Train_Set_3 及模型权重参数 Classifier_Weight_3……直到得到最后一个弱分类器 Classifier_3 的权重参数 Classifier_Weight_3。

图 11-4

训练中得到的各弱分类器及其权重构成了 AdaBoost 模型：{ Classifier_1, Classifier_Weight_1; Classifier_2, Classifier_Weight_2; Classifier_3，Classifier_Weight_3}。

下面来看看使用 AdaBoost 模型的预测过程。如图 11-5 所示，待预测样本的特征值 instance_features 分别被各弱分类器{ Classifier_1, Classifier_2, Classifier_3}预测，得到各自的预测结果{ result_label_1, result_label_2, result_label_3}，然后考虑各模型的权重系数 {Classifier_Weight_1, Classifier_Weight_2，Classifier_Weight_3}，通过投票的方式得到 AdaBoost 模型的预测结果 final_result_label。

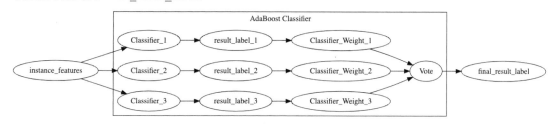

图 11-5　AdaBoost 预测示意图

注意：想了解 AdaBoost 算法更多细节的读者，可以参考以下内容：

Freund, Y. & R.E. Schapire (1996), *Experiments with a New Boosting Algorithm, in Proceedings of the Thirteenth International Conference on Machine Learning, Morgan Kaufmann*（参见链接 11-2）。

Gradient Boosting 算法通过损失函数（loss function）来度量预测值与真实值的不一致程度。损失函数越小，则模型预测得越准确。如果随着更多弱分类器的加入，损失函数能持续下降，就说明模型在持续改进。最好的方式是，使损失函数在其梯度（Gradient）方向下降。Gradient Boosting 算法将二分类问题看作值为+1和−1的回归问题，并最终根据回归值是否大于 0 来将回归值转换输出成分类标签值。

Gradient Boosting 使用的是弱回归器。如图 11-4 的右子图所示，Train_Set 为原始训练数据，训练第一个弱回归器 Regressor_1 并得到模型；然后对 Train_Set 进行回归预测并得到 Prediction_1，结合训练数据 Train_Set 与回归预测结果 Prediction_1，通过计算残差 Residual_Error_1，得到新的训练集 Train_Set_2；随后，由 Train_Set_2 得到 Regressor_2；之后预测出 Prediction_2，计算残差 Residual_Error_2，得到新的数据集 Train_Set_3……直到得到最后一个弱回归器 Regressor_3。弱回归器模型序列{ Regressor_1, Regressor_2, Regressor_3}就是 Gradient Boosting 模型。

使用 Gradient Boosting 模型进行预测的流程如图 11-6 所示。待预测样本的特征值

instance_features 分别被各弱回归器 { Regressor_1, Regressor_2, Regressor_3 } 预测，得到各自的预测结果 { result_value_1, result_value_2, result_value_3 }，求和汇总出 Gradient Boosting 模型的预测结果 final_result，根据结果值是否大于 0，便可将结果值转换为二分类标签值。

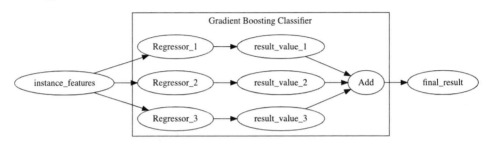

图 11-6　Gradient Boosting 预测示意图

注意：想了解 Gradient Boosting 算法更多细节的读者，可以参考以下内容：

Friedman, J. H, *Greedy Function Approximation: A Gradient Boosting Machine*（February 1999），参见链接 11-3。

11.6.3　随机森林与GBDT

知道了集成方法，就很容易理解随机森林算法和 GBDT 算法了。

随机森林（random forests）算法是由 Leo Breiman 和 Adele Cutler 提出的，并被注册成了商标。随机森林使用了 Bagging 方法，由多个分类决策树（弱分类器）构成，每个分类决策树的地位是平等的，并通过投票的方式确定最终的预测结果。关于该算法的详细内容可以参考以下内容：

- Ho, Tin Kam (1995). *Random Decision Forests.* Proceedings of the 3rd International Conference on Document Analysis and Recognition, Montreal, QC, 14–16 August 1995. pp. 278–282
- Breiman, Leo (2001). "Random Forests". *Machine Learning*，45 (1): 5–32

GBDT 为 Gradient Boosting Decision Tree 的缩写，该算法又被称为 MART(Multiple Additive Regression Tree)算法，在其名称中就已经包含了所使用的集成方法（Gradient Boosting）。GBDT 是由多棵回归树（Regression Tree）组成的，所有树的结论累加起来，通过是否大于 0，将回归值转换为二分类标签值。

随机森林与 GBDT 的简单对比如表 11-3 所示。

表 11-3 随机森林算法和 GBDT 算法的对比

算法	集成方法	弱分类/回归器
随机森林	Bagging	分类树
GBDT	Gradient Boosting	回归树

11.7 使用随机森林算法

随机森林是由多棵分类决策树构成的，可以通过调节决策树的棵数、每棵决策树的深度等参数来取得较好的分类效果。

如下代码所示，使用随机森林组件 RandomForestClassifier，对分层采样的训练数据进行训练，并使用测试集进行评估。这里设置树的棵数为 20，每棵树的深度为 4，连续特征会被分成 512 个 bin。

```
new RandomForestClassifier()
    .setNumTrees(20)
    .setMaxDepth(4)
    .setMaxBins(512)
    .setFeatureCols(featureColNames)
    .setLabelCol(LABEL_COL_NAME)
    .setPredictionCol(PREDICTION_COL_NAME)
    .setPredictionDetailCol(PRED_DETAIL_COL_NAME)
    .fit(train_sample)
    .transform(test_data)
    .link(
        new EvalBinaryClassBatchOp()
            .setLabelCol(LABEL_COL_NAME)
            .setPredictionDetailCol(PRED_DETAIL_COL_NAME)
            .lazyPrintMetrics("RandomForest with Stratified Sample")
    );
```

使用随机森林算法的评估结果如下。与逻辑回归算法相比，召回率指标下降，但其他指标都有所提升，F1 指标超过逻辑回归算法中的 F1 数值。

```
-------------------------------- Metrics: --------------------------------
Auc:0.8428 Accuracy:0.9763 Precision:0.0465    Recall:0.1739  F1:0.0734  LogLoss:0.1261
|Pred\Real|  1|   0|
|---------|---|----|
|        1|  8| 164|
|        0| 38|8296|
```

11.8　使用GBDT算法

前面使用随机森林算法获得了较好的效果，这次尝试使用 GBDT 算法。代码如下，使用组件 GbdtClassifier，设置树的棵数为 100，每棵树的深度为 5，连续特征会被分成 256 个 bin。

```
new GbdtClassifier()
   .setNumTrees(100)
   .setMaxDepth(5)
   .setMaxBins(256)
   .setFeatureCols(featureColNames)
   .setLabelCol(LABEL_COL_NAME)
   .setPredictionCol(PREDICTION_COL_NAME)
   .setPredictionDetailCol(PRED_DETAIL_COL_NAME)
   .fit(train_sample)
   .transform(test_data)
   .link(
      new EvalBinaryClassBatchOp()
         .setLabelCol(LABEL_COL_NAME)
         .setPredictionDetailCol(PRED_DETAIL_COL_NAME)
         .lazyPrintMetrics("GBDT with Stratified Sample")
   );
```

使用 GBDT 算法的评估结果如下。与使用随机森林算法相比，使用 GBDT 算法的 F1 指标与其接近，但召回率有较大提升。

```
-------------------------------- Metrics: --------------------------------
Auc:0.8052 Accuracy:0.9559 Precision:0.0418     Recall:0.3261   F1:0.0741  LogLoss:0.1386
|Pred\Real|   1|   0|
|---------|----|----|
|        1|  15| 344|
|        0|  31|8116|
```

随机森林算法和 GBDT 算法的分类效果较好，但在使用中二者也有不方便的地方，这两种算法在调参方面花费的时间较多。随机森林算法和 GBDT 算法都有多个参数，这些参数对模型效果的影响较大，需要根据具体的问题选择适合的参数值。在第 20 章中会介绍参数搜索方面的内容，以帮助读者掌握这类复杂模型参数的选择方法。

12 从二分类到多分类

本章仍然演示分类的例子,但类别的个数要多于两个,即我们常说的多分类问题。前面介绍的朴素贝叶斯、决策树、随机森林模型都能用在多分类的场景中,本章及第 13 章会在实例分析的过程中陆续介绍一些新的多分类模型。

多分类模型的评估指标包含一些二分类评估指标(例如精确度、Kappa、LogLoss),还有一些基于多个二分类指标的整合指标。评估指标涉及的内容比较通用,与具体分类算法无关,将在 12.1 节中进行介绍。

12.1 多分类模型评估方法

在我们使用测试数据样本集评判模型的分类能力时,每个测试样本都有一个已知的实际分类值,而模型通过对每个样本的特征进行预测,每个测试样本又可以得到其预测分类值。混淆矩阵(Confusion Matrix)就是采用统计上交叉表的做法,将实际分类值与预测分类值分别作为矩阵的 2 个维度,统计各种组合情况下样本的个数作为矩阵值。

假设有 K 个分类,则分别统计测试数据样本中被预测为第 i 个分类,而实际上属于第 j 个分类的样本数 $n_{i,j}$。这样我们就得到了如表 12-1 所示的混淆矩阵。

表 12-1 K 分类的混淆矩阵

预测结果	实际值			
	分类 1	分类 2	⋯	分类 K
分类 1	$n_{1,1}$	$n_{1,2}$	⋯	$n_{1,K}$

续表

预测结果	实际值			
	分类 1	分类 2	...	分类 K
分类 2	$n_{2,1}$	$n_{2,2}$	...	$n_{2,K}$
⋮	⋮	⋮	⋮	⋮
分类 K	$n_{K,1}$	$n_{K,2}$	...	$n_{K,K}$

显然，上述混淆矩阵有如下的性质。

- 第 i 行代表预测结果是分类 i，该行的总数，即 $\sum_{j=1}^{K} n_{i,j}$，为预测为分类 i 的样本数目。
- 第 j 列代表实际属于分类 j，该列的总数，即 $\sum_{i=1}^{K} n_{i,j}$，为实际属于分类 j 的样本数目。
- 测试样本的总数为

$$N = \sum_{i=1}^{K} \sum_{j=1}^{K} n_{i,j}$$

下面介绍另一种常用的混淆矩阵的展现方式——使用比例来描述实际属于分类 j 的样本，被预测为分类 i 的比例，即矩阵中第 i 行 j 列的值为

$$\frac{n_{i,j}}{\sum_{i=1}^{K} n_{i,j}}$$

由比值构成的矩阵如表 12-2 所示。

表 12-2　由比值构成的混淆矩阵

预测结果	实际值			
	分类 1	分类 2	...	分类 K
分类 1	$\dfrac{n_{1,1}}{\sum_{i=1}^{K} n_{i,1}}$	$\dfrac{n_{1,2}}{\sum_{i=1}^{K} n_{i,2}}$	...	$\dfrac{n_{1,K}}{\sum_{i=1}^{K} n_{i,K}}$
分类 2	$\dfrac{n_{2,1}}{\sum_{i=1}^{K} n_{i,1}}$	$\dfrac{n_{2,2}}{\sum_{i=1}^{K} n_{i,2}}$	...	$\dfrac{n_{2,K}}{\sum_{i=1}^{K} n_{i,K}}$
⋮	⋮	⋮	⋮	⋮
分类 K	$\dfrac{n_{K,2}}{\sum_{i=1}^{K} n_{i,1}}$	$\dfrac{n_{K,2}}{\sum_{i=1}^{K} n_{i,2}}$	...	$\dfrac{n_{K,K}}{\sum_{i=1}^{K} n_{i,K}}$

由于矩阵对角线上的各项实际上表示的是属于分类 k，预测结果也为分类 k 的情形，因此这些值越大，模型的精确度越高。

12.1.1 综合指标

本节将介绍三个多分类指标：精确度、Kappa 系数、LogLoss。

后面会陆续介绍使用 Micro、Macro、Weighted 方法，以及基于各标签值的二分类指标给出的整体多分类指标。

1. 精确度

精确度（Accuracy，ACC）的定义为，预测正确的样本在整个样本中的比例。结合混淆矩阵的定义，其公式为

$$\text{ACC} = \frac{\sum_{k=1}^{K} n_{k,k}}{\sum_{i=1}^{K} \sum_{j=1}^{K} n_{i,j}}$$

2. Kappa系数

1960 年，Cohen 等人提出用 Kappa 值（用希腊字母 κ 表示）作为评判诊断试验一致性程度的指标。建议的参考标准如下。

- κ 的值域为 $[-1, +1]$。
- $0.75 \leqslant \kappa$，说明一致性较好。
- $0.4 \leqslant \kappa < 0.75$，说明一致性一般。
- $\kappa < 0.4$，说明一致性较差。

多分类情况下，Kappa 系数的定义可以由混淆矩阵来描述：把样本总数（N）乘以混淆矩阵对角线（$n_{k,k}$）的和，再减去各分类中属于该分类的样本数（$\sum_{i=1}^{K} n_{i,k}$）与被预测为该分类的样本数（$\sum_{j=1}^{K} n_{k,j}$）之积之后，再除以样本总数（N）的平方，减去各分类中属于该分类的样本数（$\sum_{i=1}^{K} n_{i,k}$）与被预测为该分类的样本数（$\sum_{j=1}^{K} n_{k,j}$）之积。对所有类别求和的结果为

$$\kappa = \frac{N \sum_{k=1}^{K} n_{k,k} - \sum_{k=1}^{K} \left\{ \left(\sum_{i=1}^{K} n_{i,k} \right) \cdot \left(\sum_{j=1}^{K} n_{k,j} \right) \right\}}{N^2 - \sum_{k=1}^{K} \left\{ \left(\sum_{i=1}^{K} n_{i,k} \right) \cdot \left(\sum_{j=1}^{K} n_{k,j} \right) \right\}}$$

在此定义表达式上进行恒等变换，我们可以得到新的描述方式。将分子、分母同时除以 N^2，可以得到

$$\kappa = \frac{\frac{\sum_{k=1}^{K} n_{k,k}}{N} - \sum_{k=1}^{K} \left\{ \frac{\left(\sum_{i=1}^{K} n_{i,k} \right)}{N} \cdot \frac{\left(\sum_{j=1}^{K} n_{k,j} \right)}{N} \right\}}{1 - \sum_{k=1}^{K} \left\{ \frac{\left(\sum_{i=1}^{K} n_{i,k} \right)}{N} \cdot \frac{\left(\sum_{j=1}^{K} n_{k,j} \right)}{N} \right\}}$$

变换后，我们可以看到分子、分母中有相同的部分，设

$$P_\mathrm{a} = \frac{\sum_{k=1}^{K} n_{k,k}}{N} \quad 和 \quad P_\mathrm{e} = \sum_{k=1}^{K} \left\{ \frac{\left(\sum_{i=1}^{K} n_{i,k}\right)}{N} \cdot \frac{\left(\sum_{j=1}^{K} n_{k,j}\right)}{N} \right\}$$

则

$$\kappa = \frac{P_\mathrm{a} - P_\mathrm{e}}{1 - P_\mathrm{e}}$$

其中，P_a 为实际分类情况与预测分类结果一致的样本数同测试数据样本总数的比值，即分类的精确度，被称为实际一致率；P_e 被称为期望一致率。

3. LogLoss

多分类线性模型 Softmax 的损失函数可以看作二分类逻辑回归模型的对数损失函数的扩展，具体的比较可以参见 12.5 节关于 Softmax 算法的介绍。二分类的 LogLoss 评估指标可以看作来源于逻辑回归，多分类的 LogLoss 评估指标可以看作来源于 Softmax。

Softmax 的经验损失函数为

$$L(W) = -\frac{1}{n} \sum_{i=1}^{n} \sum_{k=0}^{K-1} \left[I(y^{(i)} = k) \cdot \log\left(\phi_k(\boldsymbol{x}^{(i)})\right) \right]$$

其中，n 为样本总数，K 为分类数，$\boldsymbol{x}^{(i)}$ 为第 i 个样本的特征向量，$y^{(i)}$ 为第 i 个样本的标签值，标签值的取值范围为 $\{0, 1, 2, \cdots, K-1\}$，各标签值对应的预测概率为 $\{\phi_0(\boldsymbol{x}), \phi_1(\boldsymbol{x}), \cdots, \phi_{K-1}(\boldsymbol{x})\}$。

我们引入如下定义来简化表达式。

$y_{i,k}$ 表示第 i 个样本的真实分类（即标签值）是否为第 k 个类，"是"对应值为 1，"否"对应值为 0。$P_{i,k}$ 是模型预测出来的第 i 个样本为第 k 个类的概率。

则 LogLoss 指标为

$$\text{LogLoss} = -\frac{1}{n} \sum_{i=1}^{n} \sum_{k=0}^{K-1} \left[y_{i,k} \cdot \log(P_{i,k}) \right]$$

12.1.2 关于每个标签值的二分类指标

对于一个多分类问题（分类个数为 K），可以使用 One-vs-Rest 方法——将一个标签（Label）看作 1，其余标签都看作 0。这样得到 K 个二分类问题，每个二分类问题都能得到基础计数。

- TruePositive(TP)：Predicted=True, Actual=True
- TrueNegative(TN)：Predicted=False, Actual=False
- FalsePositive(FP)：Predicted=True, Actual=False
- FalseNegative(FN)：Predicted=False, Actual=True

通过基础计数 TP、FP、TN、FN，就可以计算得到后续指标。

（1）Sensitivity or TruePositiveRate(TPR)：$TPR = \frac{TP}{TP+FN}$

（2）Specificity(SPC) or TrueNegativeRate(TNR)：$SPC = \frac{TN}{FP+TN}$

（3）Precision or PositivePredictiveValue(PPV)：$PPV = \frac{TP}{TP+FP}$

（4）NegativePredictiveValue(NPV)：$NPV = \frac{TN}{TN+FN}$

（5）Fall-out or FalsePositiveRate(FPR)：$FPR = \frac{FP}{FP+TN} = 1 - SPC$

（6）FalseDiscoveryRate(FDR)：$FDR = \frac{FP}{FP+TP} = 1 - PPV$

（7）MissRate or FalseNegativeRate(FNR)：$FNR = \frac{FN}{FN+TN}$

（8）Accuracy(ACC)：$ACC = \frac{TP+TN}{TP+FN+TN+FP}$

（9）F1-Score：$F1 = \frac{2TP}{2TP+FP+FN}$

（10）Cohen's Kappa：$\kappa = \frac{P_a - P_e}{1 - P_e}$

其中，$P_a = \frac{TP+TN}{TP+FN+FP+TN}$，$P_e = \frac{(TP+FN)\times(TP+FP)+(FP+TN)\times(FN+TN)}{(TP+FN+FP+TN)\times(TP+FN+FP+TN)}$。

12.1.3 Micro、Macro、Weighted计算的指标

12.1.2节提到，一个多分类（分类个数为 K）问题使用One-vs-Rest方法能得到 K 个二分类问题，每个二分类问题都可以计算出相应的指标。那么如何将 K 个二分类指标整合为一个反映原始多分类的指标呢？

常用的有 Micro、Macro、Weighted 三种算法，下面将分别介绍。为了表示方便，我们引入 B，代表由二分类的基础计数 TP、FP、TN、FN可以计算的指标，譬如召回率、F1 等。

（1）Micro 算法：核心想法是将各个二分类的基础计数TP_i、FP_i、TN_i、FN_i进行求和，作为整个多分类的 TP、FP、TN、FN，然后就可以利用二分类中的公式计算召回率等各种指标。

$$B_{\text{Micro}} = B(\sum_{i=1}^{K} TP_i, \sum_{i=1}^{K} FP_i, \sum_{i=1}^{K} TN_i, \sum_{i=1}^{K} FN_i)$$

（2）Macro 算法：先计算各个二分类的指标，然后将其算术平均值作为多分类的指标。

$$B_{\text{Macro}} = \frac{1}{K} \sum_{i=1}^{K} B(\text{TP}_i, \text{FP}_i, \text{TN}_i, \text{FN}_i)$$

（3）Weighted 算法：可以看作 Macro 算法的扩展，考虑到各类在总体的占比，将加权平均值作为多分类的指标。

$$B_{\text{weighted}} = \sum_{i=1}^{K} \frac{N_i}{N} B(\text{TP}_i, \text{FP}_i, \text{TN}_i, \text{FN}_i)$$

其中，N_i 表示第 i 个标签值的样本条数，N 表示全体评估数据的条数。

前面从不同角度讲了两个指标——Accuracy 和 Micro F1，其实这两个指标的值是相等的。下面是关于此结论的证明。

将混淆矩阵与二分类的基础计数 TP_i、FP_i、TN_i、FN_i 联系起来。混淆矩阵元素 $n_{i,j}$ 被预测为第 i 个分类，而实际上属于第 j 个分类的样本数。基础计数 TP_i、FP_i、TN_i、FN_i 可以用 $n_{i,j}$ 表示。

- TP_k：将第 k 个分类值看作 1（正例），其余为 0（负例）时，预测为第 k 个分类，而实际上属于第 k 个分类的样本数，即

$$\text{TP}_k = n_{k,k}$$

- FP_k：将第 k 个分类值看作 1（正例），其余为 0（负例）时，预测为第 k 个分类，而实际上不属于第 k 个分类的样本数，即

$$\text{FP}_k = \sum_{i \neq k} n_{k,i}$$

- FN_k：为将第 k 个分类值看作 1（正例），其余为 0（负例）时，没有被预测为第 k 个分类，而实际上是第 k 个分类的样本数，即

$$\text{FN}_k = \sum_{i \neq k} n_{i,k}$$

由 Micro 的定义，有

$$\text{Micro F1} = \frac{2 \sum_k \text{TP}_k}{2 \sum_k \text{TP}_k + \sum_k \text{FP}_k + \sum_k \text{FN}_k}$$

结合基础计数 TP_i、FP_i、FN_i 与 $n_{i,j}$ 的关系，有

$$\sum_k \text{TP}_k = \sum_k n_{k,k}$$

$$\sum_k \text{FP}_k = \sum_k \sum_{i \ne k} n_{k,i}$$

$$\sum_k \text{FN}_k = \sum_k \sum_{i \ne k} n_{i,k}$$

显然，$\sum_k \text{TP}_k$ 为混淆矩阵中对角线 $\{n_{1,1}, \cdots, n_{K,K}\}$ 元素的和，$\sum_k \text{FP}_k$ 和 $\sum_k \text{FN}_k$ 都是混淆矩阵中非对角线 $\{n_{1,1}, \cdots, n_{K,K}\}$ 元素的和。于是

$$2\sum_k \text{TP}_k + \sum_k \text{FP}_k + \sum_k \text{FN}_k$$

$$= 2\sum_k n_{k,k} + \sum_k \sum_{i \ne k} n_{k,i} + \sum_k \sum_{i \ne k} n_{i,k}$$

$$= \left(\sum_k n_{k,k} + \sum_k \sum_{i \ne k} n_{k,i}\right) + \left(\sum_k n_{k,k} + \sum_k \sum_{i \ne k} n_{i,k}\right)$$

$$= \sum_k \sum_i n_{k,i} + \sum_k \sum_i n_{i,k} = 2\sum_k \sum_i n_{k,i}$$

即

$$\text{Micro F1} = \frac{2\sum_k n_{k,k}}{2\sum_k \sum_i n_{k,i}} = \frac{\sum_k n_{k,k}}{\sum_k \sum_i n_{k,i}} = \text{Accuracy}$$

12.2 数据探索

本章使用的数据集是鸢尾花（Iris）数据集。该数据集包含鸢尾属下的 3 个亚属（山鸢尾、变色鸢尾和维吉尼亚鸢尾），每个亚属各包含 50 个样本，该数据集共有 150 条数据。数据集有 4 个特征，分别是花萼（Sepal）和花瓣（Petal）的长度和宽度；包含一列分类信息，取值为亚属的名称（Iris setosa、Iris virginica 和 Iris versicolor）。该数据集是由 Edgar Anderson 在加拿大加斯帕半岛测量收集的。Fisher, R.A.在 1936 年以此数据集为例演示了判别分析方法。

为了给读者一些直观的印象，图 12-1 中给出了这 3 种鸢尾花的照片。如果读者想看到更清

晰的图片并了解更多内容，可以访问链接 12-1。

山鸢尾（Iris setosa）　　　变色鸢尾（Iris virginica）　　　维吉尼亚鸢尾（Iris versicolor）

图 12-1

下载数据文件（参见链接 12-2），使用文本编辑器打开，如图 12-2 所示。

图 12-2

该数据文件是典型的 CSV 格式，可以使用 CsvSourceBatchOp 读取，并输出 5 条数据，计算、显示基本统计量和相关系数。相关代码如下。

```
static final String SCHEMA_STRING
    = "sepal_length double, sepal_width double, petal_length double, petal_width double, category string";

... ...

CsvSourceBatchOp source =
    new CsvSourceBatchOp()
        .setFilePath(DATA_DIR + ORIGIN_FILE)
        .setSchemaStr(SCHEMA_STRING);

source
    .lazyPrint(5, "origin file")
    .lazyPrintStatistics("stat of origin file")
    .link(
        new CorrelationBatchOp()
            .setSelectedCols(FEATURE_COL_NAMES)
            .lazyPrintCorrelation()
    );
```

显示 5 条数据如下。

```
origin file
sepal_length|sepal_width|petal_length|petal_width|category
------------|-----------|------------|-----------|--------
5.1000|3.5000|1.4000|0.2000|Iris-setosa
4.9000|3.0000|1.4000|0.2000|Iris-setosa
4.7000|3.2000|1.3000|0.2000|Iris-setosa
4.6000|3.1000|1.5000|0.2000|Iris-setosa
5.0000|3.6000|1.4000|0.2000|Iris-setosa
```

输出统计结果如下，共 150 条数据，没有缺失值。

```
|     colName|count|missing|   sum|  mean|variance|min|max|
|------------|-----|-------|------|------|--------|---|---|
|sepal_length|  150|      0| 876.5|5.8433|  0.6857|4.3|7.9|
| sepal_width|  150|      0| 458.1| 3.054|   0.188|  2|4.4|
|petal_length|  150|      0| 563.8|3.7587|  3.1132|  1|6.9|
| petal_width|  150|      0| 179.8|1.1987|  0.5824|0.1|2.5|
|    category|  150|      0|   NaN|   NaN|     NaN|NaN|NaN|
```

相关系数矩阵如下，petal_length 和 petal_width 的相关系数最高，为 0.9628。

```
Correlation:
|     colName|sepal_length|sepal_width|petal_length|petal_width|
|------------|------------|-----------|------------|-----------|
|sepal_length|           1|    -0.1094|      0.8718|      0.818|
| sepal_width|     -0.1094|          1|     -0.4205|    -0.3565|
|petal_length|      0.8718|    -0.4205|           1|     0.9628|
| petal_width|       0.818|    -0.3565|      0.9628|          1|
```

在使用 groupBy 时，要看一下标签值的分布情况。

```
source.groupBy(LABEL_COL_NAME, LABEL_COL_NAME + ", COUNT(*) AS cnt").lazyPrint(-1);
```

运行结果如下，各个标签值的分布平均都是 50 条样本。

```
category|cnt
--------|---
Iris-setosa|50
Iris-versicolor|50
Iris-virginica|50
```

在开始实验前，将原始数据按 9:1 的比例分为训练集和测试集，具体代码如下。

```
Utils.splitTrainTestIfNotExist(source, DATA_DIR + TRAIN_FILE, DATA_DIR + TEST_FILE, 0.9);
```

12.3 使用朴素贝叶斯进行多分类

下面使用我们熟悉的朴素贝叶斯算法展开本书的第一次多分类试验。

使用朴素贝叶斯训练组件 NaiveBayesTrainBatchOp，与处理二分类问题的设置相似，还是设置特征列和标签列，只不过这里的标签列的标签值多于 2 个；设置朴素贝叶斯训练组件 NaiveBayesPredictBatchOp，设置预测结果列名称和预测详细信息列名称。将各组件连接起来，并设置训练组件，输出模型信息，预测组件输出 1 条数据，观察预测结果的具体内容，相关代码如下。

```
NaiveBayesTrainBatchOp trainer =
    new NaiveBayesTrainBatchOp()
        .setFeatureCols(FEATURE_COL_NAMES)
        .setLabelCol(LABEL_COL_NAME);

NaiveBayesPredictBatchOp predictor =
    new NaiveBayesPredictBatchOp()
        .setPredictionCol(PREDICTION_COL_NAME)
        .setPredictionDetailCol(PRED_DETAIL_COL_NAME);

train_data.link(trainer);

predictor.linkFrom(trainer, test_data);

trainer.lazyPrintModelInfo();

predictor.lazyPrint(1, "< Prediction >");
```

输出的朴素贝叶斯模型信息如下，依次是特征列和标签列的基本信息、各个标签值所占比例、离散特征（categorical feature）和连续特征（gaussian feature）。当前数据集只有连续特征，输出了各特征和标签值所对应的均值和标准差，由这些信息可以计算高斯分布的概率。

```
============================ model meta info ============================
{label number: 3, feature size: 4, feature col names:
["sepal_length","sepal_width","petal_length","petal_width"], labels:
["Iris-setosa","Iris-versicolor","Iris-virginica"]}
====================== label proportion information ======================

label info:[Iris-versicolor, Iris-virginica, Iris-setosa]
proportion:[0.35, 0.3167, 0.3333]
========================= category information =========================

categorical features: []
gaussian features: [sepal_length, sepal_width, petal_length, petal_width]
=============== categorical features proportion information ===============
```

```
There is no category feature.
============== continuous features mean sigma information ==============
Mean of features of each label:
|                 |sepal_length|sepal_width|petal_length|petal_width|
|-----------------|------------|-----------|------------|-----------|
|     Iris-setosa |      4.9925|     3.3875|       1.455|     0.2425|
| Iris-versicolor |      5.9571|     2.7952|      4.2857|     1.3286|
|  Iris-virginica |      6.5211|     2.9184|      5.4947|     2.0079|

Std of features of each label:
|                 |sepal_length|sepal_width|petal_length|petal_width|
|-----------------|------------|-----------|------------|-----------|
|     Iris-setosa |      0.1242|     0.1421|      0.0255|     0.0124|
| Iris-versicolor |      0.1996|     0.0857|      0.1917|     0.0359|
|  Iris-virginica |      0.4048|      0.101|      0.2973|     0.0755|
```

预测结果输出如下。

```
sepal_length|sepal_width|petal_length|petal_width|category|pred|pred_info
------------|-----------|------------|-----------|--------|----|---------
4.4000|2.9000|1.4000|0.2000|Iris-setosa|Iris-setosa|{"Iris-virginica":6.932034333184479E-25,"Iris-versicolor":3.811131021451496E-20,"Iris-setosa":1.0}
```

右数第二列为预测标签值，可以直接和右数第三列的原始标签值进行比对；右数第一列为预测详细信息列，给出了每个标签值对应的概率，概率最大的那个就是预测结果标签值。

与二分类评估相比，多分类评估没有参数 PositiveLabelValueString（作为正例的标签值）；预测详情列是二分类评估的必选参数，但在多分类评估中是可选的，而且输入该信息只能对少数几个指标的计算有帮助，譬如 LogLoss 指标。所以在使用多分类评估组件时，我们一般只输入如下两个参数：标签列、分类预测结果列。

下面对朴素贝叶斯模型的预测结果进行多分类评估，使用多分类评估组件 EvalMultiClassBatchOp，设置上标签列 LabelCol 和分类预测结果列 PredictionCol，这里也设置了可选参数预测详细信息列 PredictionDetailCol，并选择使用 Lazy 的方式输出评估结果。具体代码如下。

```
predictor
    .link(
      new EvalMultiClassBatchOp()
        .setLabelCol(LABEL_COL_NAME)
        .setPredictionCol(PREDICTION_COL_NAME)
        .setPredictionDetailCol(PRED_DETAIL_COL_NAME)
        .lazyPrintMetrics("NaiveBayes")
    );
```

运行结果如下，输出了 Accuracy 和 Kappa，还有多分类评估时常关注的 Macro F1 和 Micro F1。

```
------------------------------- Metrics: -------------------------------
Accuracy:0.9333 Macro F1:0.9267 Micro F1:0.9333 Kappa:0.8973   LogLoss:0.125
|---------------|-------------|---------------|-----------|
|     Pred\Real |Iris-virginica|Iris-versicolor|Iris-setosa|
|---------------|-------------|---------------|-----------|
|  Iris-virginica|          12|              2|          0|
|Iris-versicolor|            0|              6|          0|
|     Iris-setosa|           0|              0|         10|
```

因为这两个 F1 是通过将所有标签值分别作为正例标签计算再汇总得到的，所以不需要参数指定正例标签值。LogLoss 指标的计算需要依赖 PredictionDetailCol，如果没有输入该参数列，则该指标为 null，就不会被输出。

从上面多分类评估混淆矩阵可以看出，有 2 个标签值为 Iris-versicolor 的样本被错误地预测为标签值 Iris-virginica，其他样本都被正确预测。后面我们还会介绍其他多分类器，看看能否有更好的效果。

12.4 二分类器组合

表 12-3 列出了常用的分类器所支持的分类问题。所有分类器都支持二分类情况，但是只有部分算法同时支持多分类情况。为了描述方便，下面将只支持二分类情况的分类器称为二分类器。本节将介绍如何将二分类器进行组合，解决多分类问题。

表 12-3 常用分类器支持二分类、多分类的情况

分类器名称	二分类	多分类
朴素贝叶斯	√	√
逻辑回归	√	×
线性 SVM	√	×
Softmax	√	√
多层感知器	√	√
FM	√	×
决策树	√	√
随机森林	√	√
GBDT	√	×
KNN	√	√

设问题的分类个数为 K，也就是标签列有不同的标签值，常用的方式有如下两种。

（1）一对多（One-vs-Rest 或 One-vs-All）：使用二分类器的数量为K，与分类个数相同，但要求每个二分类器不仅给出分类结果，还要给出概率，用来确定最终的多分类结果。

（2）一对一（One-vs-One 或 pairwise）：使用二分类器数量较多，为$\frac{K(K-1)}{2}$，但每个二分类器的训练样本数较少，最终由各个二分类器投票确定多分类结果。

举个例子，假设有 3 个类（也就是 3 个 Label 值）要划分，分别是 A、B、C。采用一对多方式，将其分为 3 个二分类子问题，然后将 3 个子问题的分类概率综合得到最终的结果。

- 二分类子问题 1：抽取 A 所对应的样本，对应的标签值改为 1；抽取 B、C 所对应的样本，对应的标签值改为 0。
- 二分类子问题 2：抽取 B 所对应的样本，对应的标签值改为 1；抽取 C、A 所对应的样本，对应的标签值改为 0。
- 二分类子问题 3：抽取 C 所对应的样本，对应的标签值改为 1；抽取 A、B 所对应的样本，对应的标签值改为 0。

分别进行训练后，得到 3 个二分类子模型，分别利用这 3 个子模型进行预测，得到各自对于正例标签值 1 的概率$P_A(x)$、$P_B(x)$、$P_C(x)$，其中概率最大的便是最终的分类结果。

如果采用一对一方式，会将各个类别的数据进行两两组合，对于标签值 A、B、C，会组合成 3 个二分类子问题。

- 二分类子问题 1：将标签值为 A 和 B 的样本放在一起，是关于 A 和 B 的二分类问题。
- 二分类子问题 2：将标签值为 A 和 C 的样本放在一起，是关于 A 和 C 的二分类问题。
- 二分类子问题 3：将标签值为 B 和 C 的样本放在一起，是关于 B 和 C 的二分类问题。

分别进行训练后，得到 3 个二分类子模型，每个模型都会给出一个预测的标签值，然后采用投票的方式得到最终的分类结果。

Alink 提供了 One-vs-Rest 方法的实现，具体代码如下。重点在于 setClassifier 函数，可以在这里设置我们熟悉的二分类器，示例中使用的是逻辑回归算法，按照平常使用逻辑回归组件的方式设置其相应参数。对于 OneVsRest 组件，需要指定其多分类的个数（这个参数是必需的），还要设置组件最终输出的预测结果列，也可选择设置输出预测详细信息列。

```
new OneVsRest()
  .setClassifier(
    new LogisticRegression()
      .setFeatureCols(FEATURE_COL_NAMES)
      .setLabelCol(LABEL_COL_NAME)
      .setPredictionCol(PREDICTION_COL_NAME)
  )
  .setNumClass(3)
```

```
    .fit(train_data)
    .transform(test_data)
    .link(
        new EvalMultiClassBatchOp()
            .setLabelCol(LABEL_COL_NAME)
            .setPredictionCol(PREDICTION_COL_NAME)
            .lazyPrintMetrics("OneVsRest_LogisticRegression")
    );
```

评估结果如下，相比于 12.3 节中朴素贝叶斯的评估结果，这次没有设置预测详细信息列参数，所以没有输出 LogLoss 指标。在预测指标的数值上，此次结果较朴素贝叶斯有全面提升，只有 1 个标签值为 Iris-versicolor 的样本被错误地预测为标签值 Iris-virginica。

```
------------------------------ Metrics: ------------------------------
Accuracy:0.9667 Macro F1:0.9644 Micro F1:0.9667 Kappa:0.949
|     Pred\Real|Iris-virginica|Iris-versicolor|Iris-setosa|
|--------------|--------------|---------------|-----------|
| Iris-virginica|           12|              1|          0|
|Iris-versicolor|            0|              7|          0|
|    Iris-setosa|            0|              0|         10|
```

如果我们将逻辑回归二分类器换为分类能力更强的 GBDT 二分类器，分类效果是否会更好呢？试验代码如下。

```
new OneVsRest()
    .setClassifier(
        new GbdtClassifier()
            .setFeatureCols(FEATURE_COL_NAMES)
            .setLabelCol(LABEL_COL_NAME)
            .setPredictionCol(PREDICTION_COL_NAME)
    )
    .setNumClass(3)
    
    .fit(train_data)
    .transform(test_data)
    .link(
        new EvalMultiClassBatchOp()
            .setLabelCol(LABEL_COL_NAME)
            .setPredictionCol(PREDICTION_COL_NAME)
            .lazyPrintMetrics("OneVsRest_GBDT")
    );
```

与使用逻辑回归二分类器的代码相比，只是在 setClassifier 函数内部有变化，设置为 GBDT 二分类器 GbdtClassifier，并可设置其特有的参数。这里使用了参数的默认值，只设置了必需的特征列和标签列名称。

运行结果如下，各项指标与使用逻辑回归二分类器相同。

```
------------------------------ Metrics: ------------------------------
Accuracy:0.9667 Macro F1:0.9644 Micro F1:0.9667 Kappa:0.949
|          Pred\Real|Iris-virginica|Iris-versicolor|Iris-setosa|
|-------------------|--------------|---------------|-----------|
|     Iris-virginica|            12|              1|          0|
|    Iris-versicolor|             0|              7|          0|
|        Iris-setosa|             0|              0|         10|
```

尝试修改了几次 GBDT 参数,但是没能进一步优化分类效果。这时我们尝试使用一个二分类器——LinearSVM。输入 setClassifier 函数,设置为线性 SVM 分类器 LinearSvm,使用参数的默认值,并设置必需的特征列和标签列名称。相关代码如下。

```
new OneVsRest()
  .setClassifier(
    new LinearSvm()
      .setFeatureCols(FEATURE_COL_NAMES)
      .setLabelCol(LABEL_COL_NAME)
      .setPredictionCol(PREDICTION_COL_NAME)
  )
  .setNumClass(3)

  .fit(train_data)
  .transform(test_data)
  .link(
    new EvalMultiClassBatchOp()
      .setLabelCol(LABEL_COL_NAME)
      .setPredictionCol(PREDICTION_COL_NAME)
      .lazyPrintMetrics("OneVsRest_LinearSvm")
  );
```

运行结果如下,全部实现正确分类!

```
------------------------------ Metrics: ------------------------------
Accuracy:1 Macro F1:1 Micro F1:1 Kappa:1
|          Pred\Real|Iris-virginica|Iris-versicolor|Iris-setosa|
|-------------------|--------------|---------------|-----------|
|     Iris-virginica|            12|              0|          0|
|    Iris-versicolor|             0|              8|          0|
|        Iris-setosa|             0|              0|         10|
```

我们看到了二分类器组合可以解决多分类问题,同时也能感受到,二分类器分类能力的强弱不是绝对的,需要根据具体的问题灵活选择合适的二分类器。

12.5 Softmax算法

Softmax 将多分类问题用一个损失函数描述,当分类数为 2 时,其损失函数与逻辑回归的损失函数相同。我们可以将 Softmax 算法看作逻辑回归算法在多分类问题上的扩展。

设分类个数为 K，标签值的取值范围为 $\{0, 1, 2, \cdots, K-1\}$。

设示性函数（Indicative Function）为

$$I(y=k) = \begin{cases} 1, & \text{当 } y = k \text{ 为真} \\ 0, & \text{当 } y = k \text{ 为假} \end{cases}$$

用一个 $(K-1) \times (m+1)$ 维的向量来表示模型权重系数 W，该向量被等分为 $K-1$ 段，分别为

$$w^{(1)}, w^{(2)}, \cdots, w^{(K-1)}$$

则

$$\eta(w^{(k)}, x) = w_0^{(k)} + w_1^{(k)} x_1 + \cdots + w_m^{(k)} x_m$$

定义当 x 分类为标签值 $1 \leq k \leq K-1$ 时的概率为

$$\phi_k(x) = \frac{e^{\eta(w^{(k)}, x)}}{1 + \sum_{i=1}^{K-1} e^{\eta(w^{(i)}, x)}}$$

并定义

$$\phi_0(x) = \frac{1}{1 + \sum_{i=1}^{K-1} e^{\eta(w^{(i)}, x)}}$$

则 $\{\phi_0(x), \phi_1(x), \cdots, \phi_{K-1}(x)\}$ 满足如下等式

$$\sum_{k=0}^{K-1} \phi_k(x) = \frac{1}{1 + \sum_{i=1}^{K-1} e^{\eta(w^{(i)}, x)}} + \frac{\sum_{k=1}^{K-1} e^{\eta(w^{(k)}, x)}}{1 + \sum_{i=1}^{K-1} e^{\eta(w^{(i)}, x)}} = \frac{1 + \sum_{k=1}^{K-1} e^{\eta(w^{(k)}, x)}}{1 + \sum_{i=1}^{K-1} e^{\eta(w^{(i)}, x)}} = 1$$

定义损失函数为

$$L(W, x^{(i)}, y^{(i)}) = -\log \left(\prod_{k=0}^{K-1} \left(\phi_k(x^{(i)}) \right)^{I(y^{(i)}=k)} \right)$$

注意：

当 $K = 2$ 时，有

$$L(W, x^{(i)}, y^{(i)}) = -\log \left(\prod_{k=0}^{1} \left(\phi_k(x^{(i)}) \right)^{I(y^{(i)}=k)} \right)$$

当 $y^{(i)} = 1$ 时，有
$$L(W, x^{(i)}, y^{(i)}) = -\log\left(\frac{e^{\eta(w^{(1)}, x)}}{1 + e^{\eta(w^{(1)}, x)}}\right) = \log\left(1 + e^{-\eta(w^{(1)}, x)}\right)$$

当 $y^{(i)} = 0$ 时，有
$$L(W, x^{(i)}, y^{(i)}) = -\log\left(\frac{1}{1 + e^{\eta(w^{(1)}, x)}}\right) = \log\left(1 + e^{\eta(w^{(1)}, x)}\right)$$

将此结果与二分类逻辑回归算法的损失函数（标签值为 0、1 的情形）对比，会发现结果一致，即在分类数 $K = 2$ 时，多分类器的损失函数与逻辑回归的损失函数相同。

Softmax 的经验损失函数为
$$L(W) = \frac{1}{n}\sum_{i=1}^{n} L(W, x^{(i)}, y^{(i)})$$
$$= \frac{1}{n}\sum_{i=1}^{n}\left[-\log\left(\prod_{k=0}^{K-1}\left(\phi_k(x^{(i)})\right)^{I(y^{(i)}=k)}\right)\right]$$
$$= -\frac{1}{n}\sum_{i=1}^{n}\sum_{k=0}^{K-1}\left[I(y^{(i)} = k) \cdot \log\left(\phi_k(x^{(i)})\right)\right]$$

则其结构风险函数为
$$J(W) = L(W) + \lambda \cdot \Omega(W)$$

Softmax 算法会输出标签值 $\{0, 1, 2, \cdots, K-1\}$ 对应的概率，即
$$\{\phi_0(x), \phi_1(x), \cdots, \phi_{K-1}(x)\}$$

并选出概率最大的标签值作为预测结果。在实际的多分类问题中，标签值往往是一些有实际意义的标签，Alink 的 Softmax 组件会在训练前、预测后将实际标签值与 $\{0, 1, 2, \cdots, K-1\}$ 进行自动转换，用户在使用时不需要关注标签值的形式问题。

我们使用 Softmax 组件解决多分类问题，设置 Softmax 组件的参数，包括特征列、标签列和预测结果列名称，设置使用 Lazy 方式输出训练过程信息和模型信息，具体代码如下。

```
new Softmax()
  .setFeatureCols(FEATURE_COL_NAMES)
  .setLabelCol(LABEL_COL_NAME)
  .setPredictionCol(PREDICTION_COL_NAME)
  .enableLazyPrintTrainInfo()
  .enableLazyPrintModelInfo()
  .fit(train_data)
```

```
    .transform(test_data)
    .link(
      new EvalMultiClassBatchOp()
        .setLabelCol(LABEL_COL_NAME)
        .setPredictionCol(PREDICTION_COL_NAME)
        .lazyPrintMetrics("Softmax")
    );
```

训练信息如下。

```
-------------------------- train meta info --------------------------
{model name: softmax, num feature: 4}
-------------------------- train convergence info --------------------------
step:0  loss:1.03883853  gradNorm:0.78880934  learnRate:0.40000000
step:1  loss:0.79029177  gradNorm:0.72704714  learnRate:1.60000000
step:2  loss:0.46450488  gradNorm:0.45492455  learnRate:4.00000000
...  ....
step:33  loss:0.04431295  gradNorm:0.00016576  learnRate:4.00000000
step:34  loss:0.04431267  gradNorm:0.00007317  learnRate:4.00000000
step:35  loss:0.04431267  gradNorm:0.00000423  learnRate:4.00000000
```

该组件默认迭代次数为 100，可以看到在 step:35 时就已达到了收敛条件。

对于 K 分类问题，Softmax 会有 $K-1$ 组线性参数，模型信息如下。

```
-------------------------- model meta info --------------------------
{hasInterception: true, model name: softmax, num feature: 4}
-------------------------- model label values --------------------------
[Iris-setosa, Iris-versicolor, Iris-virginica]
-------------------------- model weight info --------------------------
|intercept|sepal_length|sepal_width|petal_length| petal_width|
|---------|------------|-----------|------------|------------|
| 115.6539|-22.13402422|32.05813365|-14.44968333|-31.20994729|
|  43.1648|  1.57043487| 6.21071882| -8.54629750|-17.20480213|
```

对于本章的三分类问题，有 2 组线性参数，详见模型系数信息（model weight info）显示的内容。

模型评估结果如下，也做到了完全分类！

```
-------------------------- Metrics: --------------------------
Accuracy:1 Macro F1:1 Micro F1:1 Kappa:1
|      Pred\Real|Iris-virginica|Iris-versicolor|Iris-setosa|
|---------------|--------------|---------------|-----------|
| Iris-virginica|            12|              0|          0|
|Iris-versicolor|             0|              8|          0|
|    Iris-setosa|             0|              0|         10|
```

Softmax 有很好的分类效果，而且使用简单，基本不需要调整参数。在处理多分类问题时，我们可以将其作为首选的算法。

12.6 多层感知器分类器

多层感知器分类器（Multi-Layer Perceptron Classifier，MLPC）是一种基于前向神经网络的分类器。使用常用的设置等方式，简化了使用前向神经网络进行分类操作的过程，用户只需设置各隐藏层的节点个数。

MLPC 由 $L+1$ 层节点组成，每层节点都可以对应一个向量。第一层为输入层，对应输入特征向量，节点个数即为特征数，也是特征向量的维度；最后一层为输出层，对应各个分类标签的概率向量，节点个数为分类标签的个数，也是概率向量的维度；中间的 $L-1$ 层为隐藏层。

每层完全连接到网络中的下一层，第 i 层通过节点权重矩阵 $\boldsymbol{W}_i$ 和偏差向量 $\boldsymbol{b}_i$ 的线性组合，并应用激活函数或 Softmax 函数，得到第 $i+1$ 层节点。对于具有 $L+1$ 层的 MLPC，可以采用矩阵形式编写，如下所示。

$$f(x) = s_L(\boldsymbol{W}_L(\cdots s_2(\boldsymbol{W}_2 s_1(\boldsymbol{W}_1 \boldsymbol{x} + \boldsymbol{b}_1) + \boldsymbol{b}_2) \cdots) + \boldsymbol{b}_L)$$

其中，对于 $k = 1, 2, \cdots, L-1$，即前面 $L-1$ 层，激活（Sigmoid）函数选择 Logistic 函数，即

$$s_k(\boldsymbol{z}_i) = \frac{1}{1+\mathrm{e}^{-z_i}} = \frac{\mathrm{e}^{z_i}}{1+\mathrm{e}^{z_i}}$$

Logistic 函数将 $(-\infty, +\infty)$ 的值映射到区间 $(0,1)$，其函数曲线如图 12-3 所示。

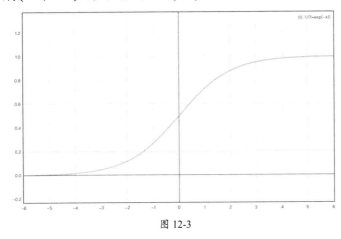

图 12-3

第 L 层使用 Softmax 函数，得到最终的输出层节点，各标签值的概率为

$$s_k(\boldsymbol{z}_i) = \frac{\mathrm{e}^{z_i}}{\sum_{i=1}^{K} \mathrm{e}^{z_i}}$$

其中，K 为不同标签值的个数，即分类总数，也是输出层的节点数。

下面通过示例了解多层感知器分类器的使用，具体代码如下。

```
new MultilayerPerceptronClassifier()
    .setLayers(new int[] {4, 20, 3})
    .setFeatureCols(FEATURE_COL_NAMES)
    .setLabelCol(LABEL_COL_NAME)
    .setPredictionCol(PREDICTION_COL_NAME)
    .fit(train_data)
    .transform(test_data)
    .link(
        new EvalMultiClassBatchOp()
            .setLabelCol(LABEL_COL_NAME)
            .setPredictionCol(PREDICTION_COL_NAME)
            .lazyPrintMetrics("MultilayerPerceptronClassifier [4, 20, 3]")
    );
```

使用多层感知器分类器组件 MultilayerPerceptronClassifier，设置特征列和标签列、预测结果列名称，该组件特有的参数是 Layers（神经网络各层神经元的数量）。第一层（输入层）神经元的数量要与特征数相同，Iris 数据集有 4 个特征；最后一层（输出层）神经元的数量等于分类的类别数，Iris 数据集的类别数（标签值个数）为 3；中间为若干个隐层，这里设置了一个隐层，有 20 个神经元。

模型评估结果如下，也做到了完全分类。

```
-------------------------- Metrics: --------------------------
Accuracy:1 Macro F1:1 Micro F1:1 Kappa:1
|      Pred\Real|Iris-virginica|Iris-versicolor|Iris-setosa|
|---------------|--------------|---------------|-----------|
| Iris-virginica|            12|              0|          0|
|Iris-versicolor|             0|              8|          0|
|    Iris-setosa|             0|              0|         10|
```

多层感知器分类器的最后一层与倒数第二层之间使用的激活函数是 Softmax 函数。如果没有隐层，那么相当于直接对输入层的原始特征进行线性组合，并使用 Softmax 激活函数，等价于 12.5 节的 Softmax 算法。多层感知器分类器可以被看作 Softmax 分类器的扩展。

我们在前面多层感知器分类器试验的基础上，调整 Layers 参数，使其只包含输入层和输出层，即 Layers={4, 3}。相关代码如下。

```
new MultilayerPerceptronClassifier()
    .setLayers(new int[] {4, 3})
    .setFeatureCols(FEATURE_COL_NAMES)
    .setLabelCol(LABEL_COL_NAME)
    .setPredictionCol(PREDICTION_COL_NAME)
    .fit(train_data)
    .transform(test_data)
```

```
.link(
    new EvalMultiClassBatchOp()
        .setLabelCol(LABEL_COL_NAME)
        .setPredictionCol(PREDICTION_COL_NAME)
        .lazyPrintMetrics("MultilayerPerceptronClassifier [4, 3]")
);
```

运行结果如下，同样得到了 Softmax 分类器的效果。

```
-------------------------------- Metrics: --------------------------------
Accuracy:1 Macro F1:1 Micro F1:1 Kappa:1
|     Pred\Real|Iris-virginica|Iris-versicolor|Iris-setosa|
|--------------|--------------|---------------|-----------|
| Iris-virginica|           12|              0|          0|
|Iris-versicolor|            0|              8|          0|
|    Iris-setosa|            0|              0|         10|
```

13 常用多分类算法

MNIST 是常用的手写数字识别数据集，包含一个 60000 样本的训练集和 10000 样本的测试集。如图 13-1 所示，这些手写数字都位于图像中心，图像尺寸都被规范到 28 像素×28 像素的固定大小。像素值范围为 0~255，0 表示背景（白色），255 表示前景（黑色）。数据集下载地址为链接 13-1。

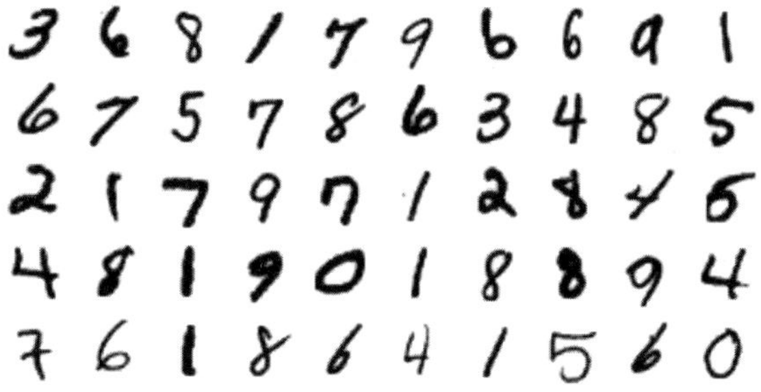

图 13-1

注意：本章只是为了演示通用的多分类算法，将图像的每个元素都看作一个特征。使用图像方面的深度学习算法可以获得更好的分类效果，但不在本章的讨论范围内，感兴趣的读者可以参阅相关材料。

13.1 数据准备

MNIST 的原始数据是一种自定义的格式，需要按照网站上的格式说明，单独写程序读取。

首先将 4 个数据文件下载到本地文件夹，训练数据的图像数据部分与标签部分是独立的两个文件，预测数据也分为两个文件。

（1）train-images-idx3-ubyte.gz：training set images (9912422 bytes)

（2）train-labels-idx1-ubyte.gz：training set labels (28881 bytes)

（3）t10k-images-idx3-ubyte.gz：test set images (1648877 bytes)

（4）t10k-labels-idx1-ubyte.gz：test set labels (4542 bytes)

13.1.1　读取MNIST数据文件

在附录代码中，有一段读取 MNIST 数据的代码，是根据指定的图像数据文件与标签文件生成的形式为"特征向量+标签"的数据表，定义为 Alink Source 形式。代码的主体部分如下，构造函数需要输入图像数据文件路径、标签文件路径和参数 isSparse。

```java
public static class MnistGzFileSourceBatchOp extends BaseSourceBatchOp
<MnistGzFileSourceBatchOp> {

    private final String imageGzFile;
    private final String labelGzFile;
    private final boolean isSparse;

    public MnistGzFileSourceBatchOp(String imageGzFile, String labelGzFile, boolean isSparse) {
        super(null, null);
        this.imageGzFile = imageGzFile;
        this.labelGzFile = labelGzFile;
        this.isSparse = isSparse;
    }

    @Override
    protected Table initializeDataSource() {
        try {
            ArrayList <Row> rows = new ArrayList <>();
            String[] images = getImages();
            Integer[] labels = getLabels();
            int n = images.length;
            if (labels.length != n) {
                throw new RuntimeException("The size of images IS NOT EQUAL WITH the size of labels.");
            }
            for (int i = 0; i < n; i++) {
                rows.add(Row.of(images[i], labels[i]));
            }
            return new MemSourceBatchOp(rows, new String[] {"vec", "label"}).getOutputTable();
        } catch (Exception ex) {
            ex.printStackTrace();
            throw new RuntimeException(ex.getMessage());
        }
    }
}
```

```
    ... ...
}
```

initializeDataSource 方法包含了主要流程，使用 getImages 方法从文件读取图像数据向量序列化后的 String 类型数据；使用 getLabels 方法从文件获取标签信息，然后组装为数据表的形式，返回产生的数据表。

其中，读取数据文件的核心函数如下。

```
private Integer[] getLabels() throws IOException {
    BufferedInputStream bis = new BufferedInputStream(
        new GZIPInputStream(new FileInputStream(this.labelGzFile)));
    ... ...
    return labels;
}
private String[] getImages() throws IOException {
    BufferedInputStream bis = new BufferedInputStream(
        new GZIPInputStream(new FileInputStream(this.imageGzFile)));
    ... ...
    String[] images = new String[record_number];

    if (isSparse) {
        ... ...
        for (int i = 0; i < record_number; i++) {
            ... ...
            images[i] = new SparseVector(nPixels, pixels).toString();
        }
    } else {
        ... ...
        for (int i = 0; i < record_number; i++) {
            ... ...
            images[i] = new DenseVector(image).toString();
        }
    }

    bis.close();
    return images;
}
```

通过先获取原始文件流 FileInputStream，然后使用 GZIPInputStream 进行解压，再使用 BufferedInputStream 提高读数据的性能，之后便是各自按格式进行解析、输出。注意：变量 isSparse 决定了生成的向量格式为稀疏或稠密，函数返回向量序列化后的 String 类型。

13.1.2 稠密向量与稀疏向量

基于 MnistGzFileSourceBatchOp 可以轻松得到稠密、稀疏向量的训练和预测数据。为了便于后续操作使用，我们直接将其存为 AK 格式，具体代码如下。

```
new MnistGzFileSourceBatchOp
    (
        DATA_DIR + "train-images-idx3-ubyte.gz",
        DATA_DIR + "train-labels-idx1-ubyte.gz",
        true
    )
    .link(
        new AkSinkBatchOp().setFilePath(DATA_DIR + SPARSE_TRAIN_FILE)
    );
new MnistGzFileSourceBatchOp
    (
        DATA_DIR + "t10k-images-idx3-ubyte.gz",
        DATA_DIR + "t10k-labels-idx1-ubyte.gz",
        true
    )
    .link(
        new AkSinkBatchOp().setFilePath(DATA_DIR + SPARSE_TEST_FILE)
    );
new MnistGzFileSourceBatchOp
    (
        DATA_DIR + "train-images-idx3-ubyte.gz",
        DATA_DIR + "train-labels-idx1-ubyte.gz",
        false
    )
    .link(
        new AkSinkBatchOp().setFilePath(DATA_DIR + DENSE_TRAIN_FILE)
    );
new MnistGzFileSourceBatchOp
    (
        DATA_DIR + "t10k-images-idx3-ubyte.gz",
        DATA_DIR + "t10k-labels-idx1-ubyte.gz",
        false
    )
    .link(
        new AkSinkBatchOp().setFilePath(DATA_DIR + DENSE_TEST_FILE)
    );
BatchOperator.execute();
```

对于稠密训练数据，输出一条数据，看看其具体内容，并对其向量列进行统计，具体代码如下。

```
new AkSourceBatchOp()
    .setFilePath(DATA_DIR + DENSE_TRAIN_FILE)
    .lazyPrint(1, "MNIST data")
    .link(
        new VectorSummarizerBatchOp()
            .setSelectedCol(VECTOR_COL_NAME)
            .lazyPrintVectorSummary()
    );
```

数据内容如下。

```
vec|label
---|-----
0.0 0.0 0.0 0.0 ··· 230.0 132.0 133.0 132.0 132.0 ··· 0.0 0.0 0.0 0.0|3
```

右边一列为标签值，这条数据的标签值为 3；左边为稠密向量序列化为字符串的形式，以空格分隔各个数字，由于向量内容较多，这里只保留少量内容进行示意。

向量列统计结果如下。

```
------------------------------- Summary -------------------------------
id|count|sum|mean|variance|standardDeviation|min|max|normL1|normL2
--|-----|---|----|--------|-----------------|---|---|------|------
0|60000|0.0000|0.0000|0.0000|0.0000|0.0000|0.0000|0.0000|0.0000
1|60000|0.0000|0.0000|0.0000|0.0000|0.0000|0.0000|0.0000|0.0000
2|60000|0.0000|0.0000|0.0000|0.0000|0.0000|0.0000|0.0000|0.0000
... ...
774|60000|12026.0000|0.2004|36.5115|6.0425|0.0000|254.0000|12026.0000|1480.8991
775|60000|5332.0000|0.0889|15.6514|3.9562|0.0000|254.0000|5332.0000|969.3008
776|60000|2738.0000|0.0456|8.0647|2.8398|0.0000|253.0000|2738.0000|695.7011
777|60000|1157.0000|0.0193|2.8452|1.6868|0.0000|253.0000|1157.0000|413.1961
778|60000|907.0000|0.0151|2.8166|1.6783|0.0000|254.0000|907.0000|411.1070
779|60000|120.0000|0.0020|0.1201|0.3466|0.0000|62.0000|120.0000|84.8999
780|60000|0.0000|0.0000|0.0000|0.0000|0.0000|0.0000|0.0000|0.0000
781|60000|0.0000|0.0000|0.0000|0.0000|0.0000|0.0000|0.0000|0.0000
782|60000|0.0000|0.0000|0.0000|0.0000|0.0000|0.0000|0.0000|0.0000
783|60000|0.0000|0.0000|0.0000|0.0000|0.0000|0.0000|0.0000|0.0000
```

其中给出了各个维度的统计量，最左边为向量的索引号，后面是该索引位置的各数据的统计指标。

再来看稀疏训练数据，输出一条数据，看看其具体内容，并对其向量列进行统计，具体代码如下。

```
new AkSourceBatchOp()
    .setFilePath(DATA_DIR + SPARSE_TRAIN_FILE)
    .lazyPrint(1, "MNIST data")
    .link(
        new VectorSummarizerBatchOp()
            .setSelectedCol(VECTOR_COL_NAME)
            .lazyPrintVectorSummary()
    );
```

数据内容如下。

```
vec|label
---|-----
$784$158:124.0 159:253.0 160:255.0 161:63.0 ··· 683:220.0|1
```

右边一列为标签值，这条数据的标签值为 1；左边为稀疏向量序列化为字符串的形式，左边以 "$" 起始的为稀疏向量维度，向量索引和数值间以 ":" 间隔，以空格分隔各个索引数值对。由于向量内容较多，这里只保留首尾内容进行展示。

稀疏向量列统计结果如下。

```
------------------------------- Summary -------------------------------
id|count|sum|mean|variance|standardDeviation|min|max|normL1|normL2
--|-----|---|----|--------|-----------------|---|---|-----|------
0|60000|0.0000|0.0000|0.0000|0.0000|0.0000|0.0000|0.0000|0.0000
1|60000|0.0000|0.0000|0.0000|0.0000|0.0000|0.0000|0.0000|0.0000
2|60000|0.0000|0.0000|0.0000|0.0000|0.0000|0.0000|0.0000|0.0000
3|60000|0.0000|0.0000|0.0000|0.0000|0.0000|0.0000|0.0000|0.0000
4|60000|0.0000|0.0000|0.0000|0.0000|0.0000|0.0000|0.0000|0.0000
5|60000|0.0000|0.0000|0.0000|0.0000|0.0000|0.0000|0.0000|0.0000
6|60000|0.0000|0.0000|0.0000|0.0000|0.0000|0.0000|0.0000|0.0000
7|60000|0.0000|0.0000|0.0000|0.0000|0.0000|0.0000|0.0000|0.0000
8|60000|0.0000|0.0000|0.0000|0.0000|0.0000|0.0000|0.0000|0.0000
9|60000|0.0000|0.0000|0.0000|0.0000|0.0000|0.0000|0.0000|0.0000
10|60000|0.0000|0.0000|0.0000|0.0000|0.0000|0.0000|0.0000|0.0000
11|60000|0.0000|0.0000|0.0000|0.0000|0.0000|0.0000|0.0000|0.0000
12|60000|126.0000|0.0021|0.2259|0.4753|0.0000|116.0000|126.0000|116.4302
13|60000|470.0000|0.0078|1.8528|1.3612|0.0000|254.0000|470.0000|333.4247
... ...
776|60000|2738.0000|0.0456|8.0647|2.8398|0.0000|253.0000|2738.0000|695.7011
777|60000|1157.0000|0.0193|2.8452|1.6868|0.0000|253.0000|1157.0000|413.1961
778|60000|907.0000|0.0151|2.8166|1.6783|0.0000|254.0000|907.0000|411.1070
779|60000|120.0000|0.0020|0.1201|0.3466|0.0000|62.0000|120.0000|84.8999
```

其中给出了各个维度的统计量，最左边为向量的索引号，后面是该索引位置的各数据的统计指标。

对比稀疏向量统计结果与稠密向量统计结果，我们会发现一点差异：稀疏向量统计输出的是数据中实际出现的最大索引值，稠密向量统计输出的是全部索引值。

13.1.3 标签值的统计信息

对稀疏训练数据执行一般的全表统计，向量列会被忽略，代码如下。

```
new AkSourceBatchOp()
  .setFilePath(DATA_DIR + SPARSE_TRAIN_FILE)
  .lazyPrintStatistics()
  .groupBy(LABEL_COL_NAME, LABEL_COL_NAME + ", COUNT(*) AS cnt")
  .orderBy("cnt", 100)
  .lazyPrint(-1);
```

统计结果如下。

colName	count	missing	sum	mean	variance	min	max
vec	60000	0	NaN	NaN	NaN	NaN	NaN
label	60000	0	267236	4.4539	8.3479	0	9

共有 60000 条训练数据，没有缺失值，主要统计了标签值的信息，从 0 到 9，均值为 4.4539。关于每个标签值对应的样本数，还需要看 groupBy 的结果。

```
label|cnt
-----|---
5|5421
4|5842
8|5851
6|5918
0|5923
9|5949
2|5958
3|6131
7|6265
1|6742
```

显然，数字"5"的样本数最少，数字"1"的样本数最多，各数字的样本数在平均样本数 6000 左右浮动。

13.2　Softmax算法

对于向量特征的多分类，我们先尝试使用 Softmax 算法。选择稀疏向量格式的训练和预测数据集，在本章后续的试验中也会使用同样的数据集，就不再重复描述。

使用 Softmax 组件，设置向量列名称和标签列名称，以及预测列的名称；选择使用 Lazy 方法输出显示训练过程的信息及模型信息，以便了解更多信息。具体代码如下。

```
AkSourceBatchOp train_data = new AkSourceBatchOp().setFilePath(DATA_DIR + SPARSE_TRAIN_FILE);
AkSourceBatchOp test_data = new AkSourceBatchOp().setFilePath(DATA_DIR + SPARSE_TEST_FILE);

new Softmax()
    .setVectorCol(VECTOR_COL_NAME)
    .setLabelCol(LABEL_COL_NAME)
    .setPredictionCol(PREDICTION_COL_NAME)
    .enableLazyPrintTrainInfo()
    .enableLazyPrintModelInfo()
    .fit(train_data)
```

```
    .transform(test_data)
    .link(
        new EvalMultiClassBatchOp()
            .setLabelCol(LABEL_COL_NAME)
            .setPredictionCol(PREDICTION_COL_NAME)
            .lazyPrintMetrics("Softmax")
    );
```

训练信息如下。

```
------------------------- train meta info ------------------------
{model name: softmax, num feature: 784}
------------------------- train convergence info ------------------
step:0  loss:2.20390243 gradNorm:1.00628876 learnRate:0.40000000
step:1  loss:1.19250130 gradNorm:0.95834731 learnRate:1.60000000
step:2  loss:0.90806251 gradNorm:0.94712724 learnRate:1.60000000
... ...
step:97 loss:0.24306077 gradNorm:0.00665607 learnRate:4.00000000
step:98 loss:0.24283406 gradNorm:0.00800750 learnRate:4.00000000
step:99 loss:0.24251477 gradNorm:0.00579344 learnRate:4.00000000
```

训练是达到最大迭代轮数（默认值为 100）而退出的，loss 的变化已经很小了。

模型信息显示如下。

```
------------------------- model meta info -------------------------
{hasInterception: true, model name: softmax, num feature: 784, vector colName: vec}
------------------------- model label values ----------------------
[0, 1, 2, ..., 7, 8, 9]
------------------------- model weight info -----------------------
|intercept|     1|        2|        3|        4|        5|        6|        7|        8|... ...|
|---------|----------|----------|----------|----------|----------|----------|----------|----------|-------|
| -1.6538|0.00000000|0.00000000|0.00000000|0.00000000|0.00000000|0.00000000|0.00000000|0.00000000|... ...|
|  0.8954|0.00000000|0.00000000|0.00000000|0.00000000|0.00000000|0.00000000|0.00000000|0.00000000|... ...|
| -0.0636|0.00000000|0.00000000|0.00000000|0.00000000|0.00000000|0.00000000|0.00000000|0.00000000|... ...|
| -0.825 |0.00000000|0.00000000|0.00000000|0.00000000|0.00000000|0.00000000|0.00000000|0.00000000|... ...|
|    ... |    ...   |    ...   |    ...   |    ...   |    ...   |    ...   |    ...   |    ...   |
```

在模型系数权重信息中显示了部分系数，截距项系数都有数值，但向量前 8 项的系数都为 0.0，是不是算错了呢？我们看一下稀疏向量的统计结果，发现前 12 项的最大值、最小值都为 0.0，即这些项对分类结果都没有影响。

最后，我们看一下分类评估结果。

```
------------------------- Metrics: -------------------------
Accuracy:0.9252 Macro F1:0.9242 Micro F1:0.9252 Kappa:0.9169
|Pred\Real|  9|   8|   7|...|   2|   1|   0|
|---------|---|----|----|---|----|----|----|
|        9|920|  11|  31|...|   5|   0|   0|
|        8| 10| 860|   3|...|  35|  11|   1|
|        7| 22|   6| 949|...|   9|   2|   4|
```

```
|   ...|...|...|...|...|  ...|...|
|     2|  0|  9| 22|...| 929|  4|  1|
|     1|  7| 10|  8|...|   8|1111|  0|
|     0|  8|  8|  1|...|   3|  0|958|
```

精确度为 0.9252，各数字的绝大部分样本被正确预测。

13.3　二分类器组合

下面我们使用二分类器组合 One-vs-Rest 方法，并选择逻辑回归作为二分类器。因为 One-vs-Rest 会将 10 个逻辑回归二分类器连成一个任务，在内存资源消耗上要远高于 Softmax，而且当前训练集的数据量也较多，所以我们通过减小并发度来减少资源的需求。建议在分类数较少的情况下使用 One-vs-Rest 方法。

> **注意**：如何设置并行度（Parallelism）？
> 设置当前环境执行的并行度，当前任务所涉及的所有批式组件或流式组件都会按此并行度进行并行执行。
> 向集群提交分布式任务时，并行度是必填参数，整个任务分布运行在多台机器上。在本地环境执行时，如果不指定并行度，那么它会根据机器硬件配置的不同而被默认指定，并行度的默认值为本地环境的 CPU 核心/线程数。
> 可以使用如下方法设置批式任务、流式任务的并行度参数。
> *BatchOperator.setParallelism(3);*
> *StreamOperator.setParallelism(3);*

实验代码如下，需要注意的是，第一行设置了批式组件的并发度为 1。

```
BatchOperator.setParallelism(1);

new OneVsRest()
   .setClassifier(
      new LogisticRegression()
         .setVectorCol(VECTOR_COL_NAME)
         .setLabelCol(LABEL_COL_NAME)
         .setPredictionCol(PREDICTION_COL_NAME)
   )
   .setNumClass(10)

   .fit(train_data)
   .transform(test_data)
   .link(
      new EvalMultiClassBatchOp()
         .setLabelCol(LABEL_COL_NAME)
```

```
    .setPredictionCol(PREDICTION_COL_NAME)
    .lazyPrintMetrics("OneVsRest - LogisticRegression")
);
```

运行结果如下，精确度为 0.919。

```
------------------------------- Metrics: -------------------------------
Accuracy:0.919  Macro F1:0.9176 Micro F1:0.919  Kappa:0.91
|Pred\Real|   9|   8|   7|...|   2|   1|   0|
|---------|---|---|---|---|---|----|---|
|        9|896| 13| 27|...|   4|   0|   2|
|        8| 10|842|  4|...|  39|  12|   1|
|        7| 25| 10|947|...|   9|   1|   4|
|      ...|...|...|...|...|...| ...|...|
|        2|  1|  6| 22|...|921|   3|   1|
|        1|  8| 11|  8|...|  5|1112|   0|
|        0|  8| 11|  3|...|  8|   0|962|
```

再将逻辑回归换成线性 SVM，运行结果如下。

```
------------------------------- Metrics: -------------------------------
Accuracy:0.9214 Macro F1:0.9202 Micro F1:0.9214 Kappa:0.9126
|Pred\Real|   9|   8|   7|...|   2|   1|   0|
|---------|---|---|---|---|---|----|---|
|        9|899| 11| 25|...|   3|   0|   1|
|        8| 12|853|  3|...|  40|   9|   1|
|        7| 23| 11|952|...|  11|   1|   3|
|      ...|...|...|...|...|...| ...|...|
|        2|  1|  5| 21|...|921|   3|   1|
|        1|  8| 10|  8|...|  7|1113|   0|
|        0|  7| 11|  2|...|  8|   0|958|
```

精确度比使用逻辑回归二分类器的精确度高 0.0024，但比 Softmax 低 0.0038。

13.4 多层感知器分类器

前面介绍过，多层感知器分类器（MLPC）在没有隐层时相当于 Softmax 分类器，下面我们就先从没有隐层的网络开始。具体代码如下，输入向量的维度是 784，预测的类别数是 10，设置 Layers={784, 10}。

```
new MultilayerPerceptronClassifier()
    .setLayers(new int[] {784, 10})
    .setVectorCol(VECTOR_COL_NAME)
    .setLabelCol(LABEL_COL_NAME)
    .setPredictionCol(PREDICTION_COL_NAME)
```

```
      .fit(train_data)
      .transform(test_data)
      .link(
        new EvalMultiClassBatchOp()
          .setLabelCol(LABEL_COL_NAME)
          .setPredictionCol(PREDICTION_COL_NAME)
          .lazyPrintMetrics("MultilayerPerceptronClassifier {784, 10}")
      );
```

运行结果如下。

```
-------------------------------- Metrics: --------------------------------
Accuracy:0.9241  Macro F1:0.923  Micro F1:0.9241  Kappa:0.9156
|Pred\Real|   9|   8|   7|...|   2|   1|   0|
|---------|---|---|---|---|---|---|---|
|        9|920| 12| 28|...|  4|  0|  0|
|        8|  9|858|  4|...| 39| 11|  1|
|        7| 20|  7|951|...|  7|  2|  4|
|      ...|...|...|...|...|...|...|...|
|        2|  1|  8| 23|...|926|  3|  0|
|        1|  7|  9|  7|...|  9|1112|  0|
|        0| 10|  9|  1|...|  4|  0|958|
```

前面使用 Softmax 算法的精确度是 0.9252，与本次试验的精确度 0.9241 相比，可以说是相近的。感兴趣的读者可以尝试提高迭代次数，即使用 setMaxIter 方法，默认迭代 100 次，提高到 150 次后，得到的精确度是 0.9259。

下面尝试设计多层网络，优化分类效果。设置向量列名称、标签列名称，以及预测列的名称，神经网络的输入层为 784 个特征，输出层对应 10 个标签值，中间设置了 2 个隐层，神经元个数分别为 256 和 128。具体代码如下。

```
new MultilayerPerceptronClassifier()
  .setLayers(new int[] {784, 256, 128, 10})
  .setVectorCol(VECTOR_COL_NAME)
  .setLabelCol(LABEL_COL_NAME)
  .setPredictionCol(PREDICTION_COL_NAME)
  .fit(train_data)
  .transform(test_data)
  .link(
    new EvalMultiClassBatchOp()
      .setLabelCol(LABEL_COL_NAME)
      .setPredictionCol(PREDICTION_COL_NAME)
      .lazyPrintMetrics("MultilayerPerceptronClassifier {784, 256, 128, 10}")
  );
```

由于该训练的时间较长，大约 9 分钟，建议做如下设置。

```
AlinkGlobalConfiguration.setPrintProcessInfo(true);
```

这样可以看到训练过程中的一些输出内容，了解训练过程的进度。

运行结束后，评估结果如下。

```
Accuracy:0.971   Macro F1:0.9708 Micro F1:0.971   Kappa:0.9678
|Pred\Real|   9|   8|   7|...|   2|   1|   0|
|---------|---|---|----|---|----|----|---|
|        9|960|   1|   7|...|   0|   1|   2|
|        8|   3|944|   1|...|   4|   5|   2|
|        7|  14|   5|1006|...|   6|   0|   3|
|      ...|...|...|...|...|...|...|...|
|        2|   4|   5|   5|...|1007|   3|   0|
|        1|   3|   0|   3|...|   2|1121|   0|
|        0|   3|   2|   0|...|   4|   0|962|
```

精确度为 0.971，各项指标明显高于 Softmax 算法。

我们也可以尝试更多的迭代次数，评估指标还可以再提高，如表 13-1 所示。

表 13-1　不同迭代次数对应的评估指标对比

迭代次数	Accuracy	Macro F1	Micro F1	Kappa	运行时间
100	0.971	0.9708	0.971	0.9678	9 分
150	0.9774	0.9772	0.9774	0.9749	15 分 37 秒
200	0.9781	0.9779	0.9781	0.9757	21 分 40 秒

13.5　决策树与随机森林

决策树也可以进行多分类，但通常单棵决策树的分类效果不是最优的，实际中常用多棵决策树构成的随机森林算法。

决策树和随机森林算法组件不支持输入向量特征，需要使用 Alink 组件 VectorToColumns 进行一次转换。需要设置输入的向量 VectorCol，设置输出的参数 SchemaStr，选择保留标签列（如果不设置保留列，那么默认会保留所有列），如下代码所示。

```
new VectorToColumns()
    .setVectorCol(VECTOR_COL_NAME)
    .setSchemaStr(sbd.toString())
    .setReservedCols(LABEL_COL_NAME)
    .transform(train_sparse)
    .link(
        new AkSinkBatchOp().setFilePath(DATA_DIR + TABLE_TRAIN_FILE)
    );
```

其中参数 SchemaStr 的信息是这样生成的，使用 "c_" 作为列名前缀，后面接向量索引值，如下所示。

```
StringBuilder sbd = new StringBuilder();
sbd.append("c_0 double");
for (int i = 1; i < 784; i++) {
    sbd.append(", c_").append(i).append(" double");
}
```

转换后的数据包含 1 个标签列（label）、784 个特征列（c_0 到 c_783），输出数据后，每行无法完整显示，所以截图显示了左边的部分列，如图 13-2 所示。

```
label|c_0|c_1|c_2|c_3|c_4|c_5|c_6|c_7|c_8|c_9|c_10|c_11|c_12|c_13|c_14|c_15|c_16|c_17|c_18|c_19|c_20
-----|---|---|---|---|---|---|---|---|---|---|----|----|----|----|----|----|----|----|----|----|----
1|0.0000|0.0000|0.0000|0.0000|0.0000|0.0000|0.0000|0.0000|0.0000|0.0000|0.0000|0.0000|0.0000|0.0000|0.0000|0.0000|0.0000|0.0000|0.0000|0.0000|0.0000
7|0.0000|0.0000|0.0000|0.0000|0.0000|0.0000|0.0000|0.0000|0.0000|0.0000|0.0000|0.0000|0.0000|0.0000|0.0000|0.0000|0.0000|0.0000|0.0000|0.0000|0.0000
2|0.0000|0.0000|0.0000|0.0000|0.0000|0.0000|0.0000|0.0000|0.0000|0.0000|0.0000|0.0000|0.0000|0.0000|0.0000|0.0000|0.0000|0.0000|0.0000|0.0000|0.0000
1|0.0000|0.0000|0.0000|0.0000|0.0000|0.0000|0.0000|0.0000|0.0000|0.0000|0.0000|0.0000|0.0000|0.0000|0.0000|0.0000|0.0000|0.0000|0.0000|0.0000|0.0000
1|0.0000|0.0000|0.0000|0.0000|0.0000|0.0000|0.0000|0.0000|0.0000|0.0000|0.0000|0.0000|0.0000|0.0000|0.0000|0.0000|0.0000|0.0000|0.0000|0.0000|0.0000
```

图 13-2

最左边为标签列，随后是各特征列。

接下来，我们试验各决策树的多分类效果，由于数据量较大，训练时间偏长，我们定义一个停表对象 Stopwatch sw = **new** Stopwatch()，用来记录各次试验的时间。具体代码如下所示。

```
for (TreeType treeType : new TreeType[] {TreeType.GINI, TreeType.INFOGAIN, TreeType.INFOGAINRATIO}) {
    sw.reset();
    sw.start();
    new DecisionTreeClassifier()
        .setTreeType(treeType)
        .setFeatureCols(featureColNames)
        .setLabelCol(LABEL_COL_NAME)
        .setPredictionCol(PREDICTION_COL_NAME)
        .enableLazyPrintModelInfo()
        .fit(train_data)
        .transform(test_data)
        .link(
            new EvalMultiClassBatchOp()
                .setLabelCol(LABEL_COL_NAME)
                .setPredictionCol(PREDICTION_COL_NAME)
                .lazyPrintMetrics("DecisionTreeClassifier " + treeType.toString())
        );
    BatchOperator.execute();
    sw.stop();
    System.out.println(sw.getElapsedTimeSpan());
}
```

计算结果汇总如表 13-2 所示。

表 13-2 各决策树的评估指标对比

决策树	Accuracy	Macro F1	Micro F1	Kappa	时间
GINI	0.876	0.8746	0.876	0.8622	1 分 59 秒
INFOGAIN	0.881	0.8793	0.881	0.8677	1 分 32 秒
INFOGAINRATIO	0.863	0.861	0.863	0.8477	5 分 42 秒

可以看到，对于当前数据集，INFOGAIN 类型决策树（即 ID3 算法）的各项指标最高，而且用时最短。单棵决策树的多分类指标与前面介绍的几种算法有很大差距，但是利用多棵树构成随机森林可以获得更好的分类效果。

下面的试验选择了 INFOGAIN 类型决策树，使用 setNumTreesOfInfoGain 方法，调整随机森林中 INFOGAIN 类型决策树的棵数。森林由 2 棵树开始，不断翻倍，直到 128 棵树。设置了当前批式任务的并发度为 4，对于随机森林算法的实现，每个并发的 worker 会执行一棵决策树的训练；在计算任务的前后设置了停表计时；设置了参数 SubsamplingRatio=0.6，每棵决策树会对整个训练数据采样 60% 作为其训练数据。具体代码如下。

```
BatchOperator.setParallelism(4);

... ...

for(int numTrees: new int[]{2, 4, 8, 16, 32, 64, 128}){
   sw.reset();
   sw.start();
   new RandomForestClassifier()
      .setSubsamplingRatio(0.6)
      .setNumTreesOfInfoGain(numTrees)
      .setFeatureCols(featureColNames)
      .setLabelCol(LABEL_COL_NAME)
      .setPredictionCol(PREDICTION_COL_NAME)
      .enableLazyPrintModelInfo()
      .fit(train_data)
      .transform(test_data)
      .link(
         new EvalMultiClassBatchOp()
            .setLabelCol(LABEL_COL_NAME)
            .setPredictionCol(PREDICTION_COL_NAME)
            .lazyPrintMetrics("RandomForestClassifier : InfoGain - " + numTrees)
      );
   BatchOperator.execute();
   sw.stop();
   System.out.println(sw.getElapsedTimeSpan());
}
```

汇总计算结果如表 13-3 所示，可以看到一些特点。

- 从精确度（Accuracy）等指标上看，随着决策树棵数的增加，指标在持续增长，但增速也逐渐放缓。
- 2 棵决策树和 4 棵决策树的训练时间分别为 55 秒和 59 秒，这是因为每个并发的 worker 负责训练一棵树。我们设置任务并发度为 4，当有 2 棵决策树时，只有 2 个 worker 在运行，另外 2 个在闲置；而当有 4 棵决策树时，全部 4 个 worker 都在运行。任务所用的时间基本上是单棵决策树的训练时间。
- 当有 4 棵决策树及以上时，所有的并发 worker 都在满负荷运行，任务运行的总体时间与决策树棵数成正比。

注意：细心的读者会发现精确度与 Micro F1 在数值上相等，如表 13-3 所示，再看前面章节中的多分类评估的结果，也同样是相等的。其实，这两个指标是可以在理论上证明相等的，具体内容详见 12.1.3 节。

表 13-3　不同决策树棵数对应的评估指标对比

决策树棵数	Accuracy	Macro F1	Micro F1	Kappa	时间
2	0.8797	0.8774	0.8797	0.8663	55 秒
4	0.931	0.9299	0.931	0.9233	59 秒
8	0.9449	0.944	0.9449	0.9388	1 分 42 秒
16	0.9585	0.958	0.9585	0.9539	2 分 50 秒
32	0.9632	0.9628	0.9632	0.9591	5 分 39 秒
64	0.964	0.9636	0.964	0.96	11 分 46 秒
128	0.9654	0.965	0.9654	0.9615	28 分 12 秒

13.6　K 最近邻算法

K 最近邻算法（K-Nearest Neighbor，KNN），是基于某个样本最接近的 K 个邻居的类别情况来预测该样本的分类情况。该算法遵循的准则是"近朱者赤，近墨者黑"，即判断一个人的好坏，通过已知的与其关系最近的 K 个人的情况来判断，如果大多数为好人，则判断此人为好人；反之，则认为其为坏人。

使用 K 最近邻算法分类器 KnnClassifier 的代码如下：

```
new KnnClassifier()
    .setK(3)
    .setVectorCol(VECTOR_COL_NAME)
    .setLabelCol(LABEL_COL_NAME)
    .setPredictionCol(PREDICTION_COL_NAME)
    .fit(train_data)
    .transform(test_data)
    .link(
        new EvalMultiClassBatchOp()
            .setLabelCol(LABEL_COL_NAME)
            .setPredictionCol(PREDICTION_COL_NAME)
            .lazyPrintMetrics("KnnClassifier - 3 - EUCLIDEAN")
    );
```

运行结果如下，KNN 算法虽然很简单，但对当前问题的分类效果还是很好的。

```
------------------------------- Metrics: -------------------------------
Accuracy:0.9719 Macro F1:0.9718 Micro F1:0.9719 Kappa:0.9688
|Pred\Real|   9|   8|   7|...|   2|   1|   0|
|---------|----|----|----|---|----|----|----|
|        9| 971|   4|  12|...|   1|   0|   0|
|        8|   4| 924|   0|...|   2|   0|   0|
|        7|   8|   4| 991|...|  13|   0|   1|
|      ...| ...| ...| ...|...| ...| ...| ...|
|        2|   1|   3|   4|...| 994|   2|   1|
|        1|   4|   0|  19|...|   9|1133|   1|
|        0|   4|   7|   0|...|  10|   0| 974|
```

KNN 算法默认是使用欧氏（EUCLIDEAN）距离，我们也可以尝试使用其他距离，譬如余弦（COSINE）距离。设置 DistanceType=COSINE，其他设置、流程与使用默认欧氏距离的示例相同，如下代码所示。

```
new KnnClassifier()
    .setDistanceType(DistanceType.COSINE)
    .setK(3)
    .setVectorCol(VECTOR_COL_NAME)
    .setLabelCol(LABEL_COL_NAME)
    .setPredictionCol(PREDICTION_COL_NAME)
    .fit(train_data)
    .transform(test_data)
    .link(
        new EvalMultiClassBatchOp()
            .setLabelCol(LABEL_COL_NAME)
            .setPredictionCol(PREDICTION_COL_NAME)
            .lazyPrintMetrics("KnnClassifier - 3 - COSINE")
    );
```

运行结果如下。

```
------------------------------ Metrics: ------------------------------
Accuracy:0.974  Macro F1:0.9738  Micro F1:0.974  Kappa:0.9711
|Pred\Real|   9|   8|   7|...|   2|   1|   0|
|---------|----|----|----|---|----|----|----|
|        9| 969|   3|  15|...|   0|   0|   0|
|        8|   3| 940|   0|...|   6|   0|   0|
|        7|   5|   3| 993|...|   8|   0|   1|
|      ...| ...| ...| ...|...| ...| ...| ...|
|        2|   2|   2|   5|...|1006|   3|   0|
|        1|   6|   2|  11|...|   2|1130|   1|
|        0|  10|   6|   3|...|   8|   0| 977|
```

精确度为 0.974，相比欧氏距离有少许提高。

在 KNN 的试验中，我们设置参数 K=3，其实也可以使用其他值，譬如尝试一下 K=7，精确度为 0.97，并没有获得比 K=3 更好的分类效果。

下面我们将 KNN 算法与其他算法进行比较，帮助读者理解机器学习的一些概念，并更深入地了解这些算法的本质。

1. KNN算法与K-Means算法；监督学习与非监督学习

- KNN 算法的训练数据标记了每条记录的类别信息，但 K-Means 算法的训练数据不需要标记类别信息。
- 监督学习（Supervised Learning）是对标记的训练数据进行学习。KNN、朴素贝叶斯、逻辑回归、随机森林等分类方法都属于监督学习。非监督学习（Unsupervised Learning）是对没有标记的训练数据进行学习。典型的例子就是 K-Means 等聚类方法。监督学习与非监督学习的本质区别在于训练数据是否必须被标记。

2. KNN算法与朴素贝叶斯、逻辑回归和随机森林算法；惰性学习与迫切学习

- KNN 算法没有产生模型，对每个新数据的预测都是计算该新数据与各训练数据的距离，再根据距离最近的 K 个数据的标记分类情况，预测新数据的类别。而朴素贝叶斯、逻辑回归和随机森林算法都有模型训练过程，产生分类模型，对于新数据的预测只需要有分类模型，不再需要原始的训练数据；一般来说，预测时需要的计算量更小，预测速度更快。
- 迫切学习（Eager Learning）先利用训练数据进行训练得到一个模型，使用模型对新数据进行预测。朴素贝叶斯、逻辑回归和随机森林算法都属于迫切学习。惰性学习（Lazy Learning）是指直接使用已有的训练数据对新数据进行预测，预测之前没有模型训练过程，KNN 属于这种方式。

14

在线学习

在线学习（Online Learning）是机器学习的一种模型训练方法，可以根据线上数据的变化实时调整模型，使模型能够反映线上的变化，从而提高线上预测的准确率。

为了更好地理解在线学习的概念，我们先介绍与之相对应的概念——批量训练（Batch Learning）。先确定一个样本训练集，针对训练集的全体数据进行训练，一般需要使用迭代过程，重复使用数据集，不断调整参数。在线学习不需要事先确定训练数据集，训练数据是在训练过程中逐条产生的，每来一个训练样本，就会根据该样本产生的损失函数值、目标函数值及梯度，对模型进行一次迭代。

FTRL（Follow The Regularized Leader）算法是一种被广泛应用的在线学习算法，由Google于2013年提出并发表，详见文章"Ad Click Prediction: a View from the Trenches"。

这里实现的FTRL算法，可以用于如下场景。

（1）在线训练、在线预测功能。计算流式训练数据，不断更新模型，得到"模型流"，并将其输出到流式Table。通过流式Table实时获取模型，进行在线预测。

（2）支持初始化模型。实际中通常是对历史数据进行批式训练，既可以用FTRL算法，也可以用逻辑回归等线性算法得到初始化模型。

（3）使用该算法训练批式数据得到一个模型，为了区分于前面的"模型流"，这里称之为固定模型（Fixed Model）。基于此固定模型，可以对批式数据或流式数据进行预测。

14.1 整体流程

我们先关注FTRL在线预测组件及FTRL流式预测组件。如图14-1所示，它们之间通过模型数据流连接，即 FtrlTrain 不断产生新的模型，流式地传给 FtrlPredict 组件。每当 FtrlPredict 组件接收到一个完整的模型时，便会替换其旧的模型，切换成新的模型。对于在线学习 FtrlTrain，需要两个输入，一个是初始的模型，避免系统冷启动；另一个是流式的训练数据。FtrlTrain 输出的就是模型数据流。FtrlPredict 组件同样需要初始模型，可以在 FtrlTrain 输出模型前对已来到的数据进行预测。

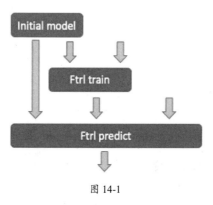

图 14-1

介绍完两个核心组件，我们再看看所需的初始模型、训练数据流和预测数据流是如何准备的。如图 14-2 所示，初始模型是采用传统的离线训练方式，对批式的训练数据进行训练得到的。

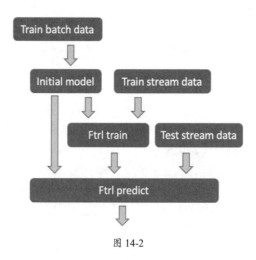

图 14-2

FTRL 算法是线性算法，其输入的数据必须都是数值型的，而原始的数据既有数值型的，也有离散型的，我们需要进行相应的特征工程操作，将原始特征数据变换为向量形式。

如图 14-3 所示，这里我们需要使用特征工程的组件，将批式原始训练数据转化为批式向量训练数据，将流式原始训练数据转化为流式向量训练数据，将流式原始预测数据转化为流式向量预测数据。

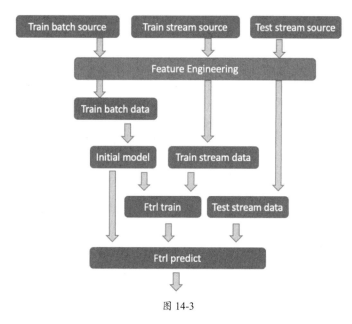

图 14-3

14.2 数据准备

在互联网广告中，点击率（CTR）是衡量广告效果的一个非常重要的指标。因此，点击预测系统在赞助搜索和实时竞价中具有重要的应用价值。

这里使用 Kaggle 比赛的 CTR 数据，原始链接为链接 14-1。Alink 的示例中用到了两个采样数据集（参见链接 14-2、链接 14-3），在国内的访问速度较快。

使用 TextSourceBatchOp 整行读取并输出部分数据，代码如下。

```
new TextSourceBatchOp()
    .setFilePath("http://alink-release.oss-cn-beijing.aliyuncs.com/data-files/avazu-small.csv"
)
    .firstN(10)
    .print();
```

运行结果如图 14-4 所示。

```
text
1000009418151094273,0,14102100,1005,0,1fbe01fe,f3845767,28905ebd,ecad2386,7801e8d9,07d7df22,a99f214a,ddd2926e,44956a24,1,2,15706,320,50,1722,0,35,-1,79
1000016934911786371,0,14102100,1005,0,1fbe01fe,f3845767,28905ebd,ecad2386,7801e8d9,07d7df22,a99f214a,96809ac8,711ee120,1,0,15704,320,50,1722,0,35,100084,79
1000037190421519486,0,14102100,1005,0,1fbe01fe,f3845767,28905ebd,ecad2386,7801e8d9,07d7df22,a99f214a,b3cf8def,8a4875bd,1,0,15704,320,50,1722,0,35,100084,79
1000064072448083876,0,14102100,1005,0,1fbe01fe,f3845767,28905ebd,ecad2386,7801e8d9,07d7df22,a99f214a,e8275b8f,6332421a,1,0,15706,320,50,1722,0,35,100084,79
1000067905641704296,0,14102100,1005,1,fe8cc448,9166c161,0569f928,ecad2386,7801e8d9,07d7df22,a99f214a,9644d0bf,779d90c2,1,0,18993,320,50,2161,0,35,-1,157
1000072075780110386,0,14102100,1005,0,d6137915,bb1ef334,f028772b,ecad2386,7801e8d9,07d7df22,a99f214a,05241af0,8a4875bd,1,0,16920,320,50,1899,0,431,100077,117
1000072472998854911,0,14102100,1005,0,8fda644b,25d4cfcd,f028772b,ecad2386,7801e8d9,07d7df22,a99f214a,b264c159,be6db1d7,1,0,20362,320,50,2333,0,39,-1,157
1000091875574232837,0,14102100,1005,1,e151e245,7e091613,f028772b,ecad2386,7801e8d9,07d7df22,a99f214a,e6f67278,be74e6fe,1,0,20632,320,50,2374,3,39,-1,23
1000094927118602991,1,14102100,1005,0,1fbe01fe,f3845767,28905ebd,ecad2386,7801e8d9,07d7df22,a99f214a,37e8da74,5db079b5,1,2,15707,320,50,1722,0,35,-1,79
1000126448061946736,0,14102100,1002,0,84c7ba46,c4e18dd6,5e219e0,ecad2386,7801e8d9,07d7df22,c357dbff,f1ac7184,373ecbe6,0,0,21689,320,50,2496,3,167,100191,23
```

图 14-4

我们看到每条数据包含多个数据项，以逗号分隔。下面是各数据列的定义。

- id：广告 ID。
- click：0 表示点击，1 表示不点击。
- hour：时间格式为 YYMMDDHH，例如 14091123 表示 2014 年 9 月 11 日 23:00。
- C1：匿名的离散变量。
- banner_pos：标题位置。
- site_id：地点 ID。
- site_domain：地点 Domain。
- site_category：地点类别。
- app_id：应用 ID。
- app_domain：应用 Domain。
- app_category：应用类别。
- device_id：设备 ID。
- device_ip：设备 IP。
- device_model：设备模型。
- device_type：设备类型。
- device_conn_type：设备连接类型。
- C14-C21：匿名的离散变量。

我们根据各列的定义，组装 SCHEMA_STRING 如下。

```
static final String SCHEMA_STRING
    = "id string, click string, dt string, C1 string, banner_pos int, site_id string, site_domain string, "
    + "site_category string, app_id string, app_domain string, app_category string, device_id string, "
    + "device_ip string, device_model string, device_type string, device_conn_type string, C14 int, C15 int, "
    + "C16 int, C17 int, C18 int, C19 int, C20 int, C21 int";
```

接下来，我们就可以通过 CsvSourceBatchOp 读取和显示数据，脚本如下。

```
CsvSourceBatchOp trainBatchData = new CsvSourceBatchOp()
    .setFilePath("http://alink-release.oss-cn-beijing.aliyuncs.com/data-files/avazu-small.csv"
```

```
)
   .setSchemaStr(SCHEMA_STRING);

trainBatchData.firstN(10).print();
```

结果如图 14-5 所示。

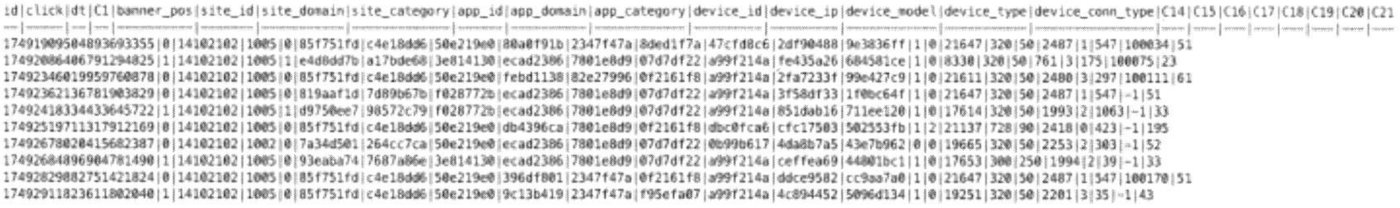

```
id|click|dt|C1|banner_pos|site_id|site_domain|site_category|app_id|app_domain|app_category|device_id|device_ip|device_model|device_type|device_conn_type|C14|C15|C16|C17|C18|C19|C20|C21
--|-----|--|--|----------|-------|-----------|-------------|------|----------|------------|---------|---------|------------|-----------|----------------|---|---|---|---|---|---|---|---
17491909504893693355|0|14102102|1005|0|85f751fd|c4e18dd6|50e219e0|80a0f91b|2347f47a|8ded1f7a|47cfd8c6|2df90488|9e3836ff|1|0|21647|320|50|2487|1|547|100034|51
17492086406791294825|1|14102102|1005|1|e4d8dd7b|a17bde68|3e814130|ecad2386|7801e8d9|07d7df22|a99f214a|fe435a26|684581ce|1|0|8330|320|50|761|3|175|100075|23
17492346019959760878|0|14102102|1005|0|85f751fd|c4e18dd6|50e219e0|febd1138|82e27996|0f2161f8|a99f214a|2fa7233f|99e427c9|1|0|21611|320|50|2480|3|297|100111|61
17492362136781903829|0|14102102|1005|0|819aaf1d|7d89b67b|f028772b|ecad2386|7801e8d9|07d7df22|a99f214a|3f58df33|1f0bc64f|1|0|21647|320|50|2487|1|547|-1|51
17492418334433645722|1|14102102|1005|1|d9750ee7|98572c79|f028772b|ecad2386|7801e8d9|07d7df22|a99f214a|851dab16|711ee120|1|0|17614|320|50|1993|2|1063|-1|33
17492519711317912169|0|14102102|1005|0|85f751fd|c4e18dd6|50e219e0|db4396ca|7801e8d9|0f2161f8|dbc0fca6|cfc17503|502553fb|1|2|21137|728|90|2418|0|423|-1|195
17492678020415682387|0|14102102|1002|0|7a34d501|264cc7ca|50e219e0|ecad2386|7801e8d9|07d7df22|0b99b617|4da8b7a5|43e7b962|0|0|19665|320|50|2253|2|303|-1|52
17492684896904781490|1|14102102|1005|0|93eaba74|7687a86e|3e814130|ecad2386|7801e8d9|07d7df22|a99f214a|ceffea69|44801bc1|1|0|17653|300|250|1994|2|39|-1|33
17492829882751421824|0|14102102|1005|0|85f751fd|c4e18dd6|50e219e0|396df801|2347f47a|0f2161f8|a99f214a|ddce9582|cc9aa7a0|1|0|21647|320|50|2487|1|547|100170|51
17492911823611802040|1|14102102|1005|0|85f751fd|c4e18dd6|50e219e0|9c13b419|2347f47a|f95efa07|a99f214a|4c894452|5096d134|1|0|19251|320|50|2201|3|35|-1|43
```

图 14-5

由于列数较多，我们不易于将数据与列名对应起来。为了更方便地看数据，这里有一个小技巧，输出的文本数据及分隔换行符号正好是 MarkDown 格式，将其复制并粘贴到 MarkDown 编辑器中，即可看到整齐的图片显示，如图　14-6 所示。

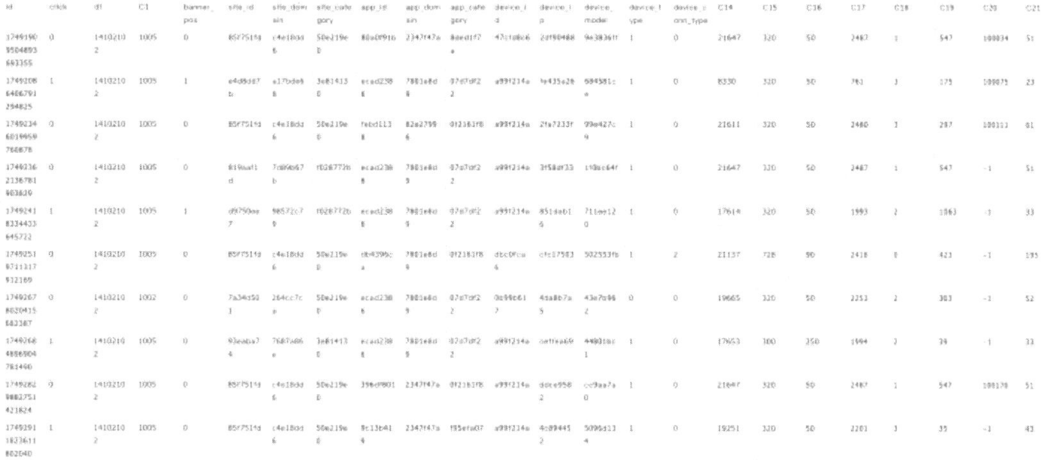

id	click	dt	C1	banner_pos	site_id	site_domain	site_category	app_id	app_domain	app_category	device_id	device_ip	device_model	device_type	device_conn_type	C14	C15	C16	C17	C18	C19	C20	C21
17491909504893693355	0	14102102	1005	0	85f751fd	c4e18dd6	50e219e0	80a0f91b	2347f47a	8ded1f7a	47cfd8c6	2df90488	9e3836ff	1	0	21647	320	50	2487	1	547	100034	51
17492086406791294825	1	14102102	1005	1	e4d8dd7b	a17bde68	3e814130	ecad2386	7801e8d9	07d7df22	a99f214a	fe435a26	684581ce	1	0	8330	320	50	761	3	175	100075	23
17492346019959760878	0	14102102	1005	0	85f751fd	c4e18dd6	50e219e0	febd1138	82e27996	0f2161f8	a99f214a	2fa7233f	99e427c9	1	0	21611	320	50	2480	3	297	100111	61
17492362136781903829	0	14102102	1005	0	819aaf1d	7d89b67b	f028772b	ecad2386	7801e8d9	07d7df22	a99f214a	3f58df33	1f0bc64f	1	0	21647	320	50	2487	1	547	-1	51
17492418334433645722	1	14102102	1005	1	d9750ee7	98572c79	f028772b	ecad2386	7801e8d9	07d7df22	a99f214a	851dab16	711ee120	1	0	17614	320	50	1993	2	1063	-1	33
17492519711317912169	0	14102102	1005	0	85f751fd	c4e18dd6	50e219e0	db4396ca	7801e8d9	0f2161f8	dbc0fca6	cfc17503	502553fb	1	2	21137	728	90	2418	0	423	-1	195
17492678020415682387	0	14102102	1002	0	7a34d501	264cc7ca	50e219e0	ecad2386	7801e8d9	07d7df22	0b99b617	4da8b7a5	43e7b962	0	0	19665	320	50	2253	2	303	-1	52
17492684896904781490	1	14102102	1005	0	93eaba74	7687a86e	3e814130	ecad2386	7801e8d9	07d7df22	a99f214a	ceffea69	44801bc1	1	0	17653	300	250	1994	2	39	-1	33
17492829882751421824	0	14102102	1005	0	85f751fd	c4e18dd6	50e219e0	396df801	2347f47a	0f2161f8	a99f214a	ddce9582	cc9aa7a0	1	0	21647	320	50	2487	1	547	100170	51
17492911823611802040	1	14102102	1005	0	85f751fd	c4e18dd6	50e219e0	9c13b419	2347f47a	f95efa07	a99f214a	4c894452	5096d134	1	0	19251	320	50	2201	3	35	-1	43

图 14-6

14.3　特征工程

14.2 节展示了数据，本节会继续深入了解数据，由数据列的描述信息可知其中含有哪些数值型特征及枚举型特征。具体代码如下所示。

```
static final String[] CATEGORY_COL_NAMES = new String[] {
```

```
    "C1", "banner_pos", "site_category", "app_domain",
    "app_category", "device_type", "device_conn_type",
    "site_id", "site_domain", "device_id", "device_model"};
static final String[] NUMERICAL_COL_NAMES = new String[] {
    "C14", "C15", "C16", "C17", "C18", "C19", "C20", "C21"};
static final String LABEL_COL_NAME = "click";
```

click 列标明了是否被点击,是分类问题的标签列。对于数值型特征,各特征的取值范围差异很大,一般需要进行标准化、归一化等操作。枚举类型的特征不能直接应用到 FTRL 模型,需要将枚举值映射到向量值。后面还需要将各列的变换结果合成为一个向量,即后面模型训练的特征向量。

在此示例中,选择对数值类型进行标准化操作,并使用了 FeatureHash 算法组件,在其参数设置中需要指定处理的各列名称,并需要标明哪些是枚举类型,那么没被标明的列就是数值类型的。FeatureHash 操作会将这些特征通过 hash 的方式映射到一个稀疏向量中,可以设置向量的维度,这里设置为 30000。每个数值列都会被 hash 到一个向量项中,该列的数值就会赋给对应的向量项。而每个枚举特征的不同枚举值也会被 hash 到向量项,并被赋值为 1。其实,FeatureHash 同时完成了枚举类型的映射及汇总为特征向量的工作。因为使用了 hash 的方式,所以会存在不同内容被 hash 到同一项的风险,但是该组件使用起来比较简便,因此在示例中使用或者作为实验开始时的组件,快速得到 baseline 指标,FeatureHash 还是很适合的。相关脚本如下所示。

```
static final String VEC_COL_NAME = "vec";
static final int NUM_HASH_FEATURES = 30000;
... ...

// setup feature enginerring pipeline
Pipeline feature_pipeline = new Pipeline()
    .add(
        new StandardScaler()
            .setSelectedCols(NUMERICAL_COL_NAMES)
    )
    .add(
        new FeatureHasher()
            .setSelectedCols(ArrayUtils.addAll(CATEGORY_COL_NAMES, NUMERICAL_COL_NAMES))
            .setCategoricalCols(CATEGORY_COL_NAMES)
            .setOutputCol(VEC_COL_NAME)
            .setNumFeatures(NUM_HASH_FEATURES)
    );
```

我们定义特征工程处理 Pipeline(管道),其中包括 StandardScaler 和 FeatureHasher,对批式训练数据 trainBatchData 执行 fit 方法,并进行训练,得到 PipelineModel(管道模型)。该管

道模型既可以用在批式数据中，也可以应用在流式数据中，生成特征向量。我们先把这个特征工程处理模型保存到本地，文件名为 feature_model.ak，具体代码如下。

```
static final String FEATURE_MODEL_FILE = "feature_model.ak";
... ...
// fit and save feature pipeline model
feature_pipeline
    .fit(trainBatchData)
    .save(DATA_DIR + FEATURE_MODEL_FILE);
BatchOperator.execute();
```

14.4 特征工程处理数据

在 14.3 节中，我们训练并保存了特征工程处理模型。本节需要使用特征工程处理模型，将批式原始训练数据转化为批式向量训练数据，将流式原始训练数据转化为流式向量训练数据，将流式原始预测数据转化为流式向量预测数据，如图 14-7 所示。

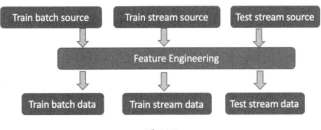

图 14-7

批式原始训练数据如下。

```
CsvSourceBatchOp trainBatchData = new CsvSourceBatchOp()
    .setFilePath("http://alink-release.oss-cn-beijing.aliyuncs.com/data-files/avazu-small.csv"
)
    .setSchemaStr(SCHEMA_STRING);
```

我们可以通过定义一个流式数据源，并按 1:1 的比例实时切分数据，从而得到流式原始训练数据、流式原始预测数据。

```
CsvSourceStreamOp data = new CsvSourceStreamOp()
    .setFilePath("http://alink-release.oss-cn-beijing.aliyuncs.com/data-files/avazu-ctr-train-8M.csv")
    .setSchemaStr(SCHEMA_STRING);

SplitStreamOp spliter = new SplitStreamOp().setFraction(0.5).linkFrom(data);
```

利用 PipelineModel.load 方法，可以载入前面保存的特征工程处理模型。

```
PipelineModel feature_pipelineModel = PipelineModel.load(DATA_DIR + FEATURE_MODEL_FILE);
```

Alink 的 PipelineModel 既能预测批式数据，也可以预测流式数据，而且调用方式相同，使用模型实例的 transform 方法即可。

使用如下代码得到批式向量训练数据。

```
feature_pipelineModel.transform(trainBatchData)
```

使用如下代码得到流式向量训练数据。

```
StreamOperator train_stream_data = feature_pipelineModel.transform(spliter);
```

使用如下代码得到流式向量预测数据。

```
StreamOperator test_stream_data = feature_pipelineModel.transform(spliter.getSideOutput(0));
```

进一步地，我们通过批式向量训练数据，可以训练得到一个线性模型作为后面 FTRL 算法的初始模型。首先定义逻辑回归分类器 lr，然后将批式向量训练数据"连接"到此分类器，输出结果便为逻辑回归模型，并将此模型保存到数据文件中，如下脚本所示。

```
// train initial batch model
LogisticRegressionTrainBatchOp lr = new LogisticRegressionTrainBatchOp()
    .setVectorCol(VEC_COL_NAME)
    .setLabelCol(LABEL_COL_NAME)
    .setWithIntercept(true)
    .setMaxIter(10);

feature_pipelineModel
    .transform(trainBatchData)
    .link(lr)
    .link(
        new AkSinkBatchOp().setFilePath(DATA_DIR + INIT_MODEL_FILE)
    );
BatchOperator.execute();
```

14.5 在线训练

基于前面几节的准备工作，我们已经具备了初始模型、流式向量训练数据、流式向量预测数据。接下来，我们会进入本章的关键内容，演示如何接入 FTRL 在线训练模块及对应的在线预测模块，如图 14-8 所示。

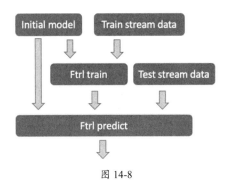

图 14-8

FTRL 在线模型训练的代码如下，在 FtrlTrainStreamOp 的构造函数中输入初始模型 initModel，设置各种参数，并"连接"流式向量训练数据。

```
// ftrl train
FtrlTrainStreamOp model = new FtrlTrainStreamOp(initModel)
  .setVectorCol(VEC_COL_NAME)
  .setLabelCol(LABEL_COL_NAME)
  .setWithIntercept(true)
  .setAlpha(0.1)
  .setBeta(0.1)
  .setL1(0.01)
  .setL2(0.01)
  .setTimeInterval(10)
  .setVectorSize(NUM_HASH_FEATURES)
  .linkFrom(train_stream_data);
```

FTRL 在线预测的代码如下，需要"连接"FTRL 在线模型训练输出的模型流，以及流式向量预测数据。

```
FtrlPredictStreamOp predResult = new FtrlPredictStreamOp(initModel)
  .setVectorCol(VEC_COL_NAME)
  .setPredictionCol(PREDICTION_COL_NAME)
  .setReservedCols(new String[] {LABEL_COL_NAME})
  .setPredictionDetailCol(PRED_DETAIL_COL_NAME)
  .linkFrom(model, test_stream_data);
```

我们可以按照如下方式设置输出流式结果，由于数据较多，输出前先对流式数据进行采样，并在每行预测数据前用 SQL 语句增加标识列，显示"Pred Sample"。注意，对于流式的任务，print 方法不能触发流式任务的执行，必须调用 StreamOperator.execute 方法才能开始执行。

```
predResult
  .sample(0.0001)
  .select("'Pred Sample' AS out_type, *")
  .print();
```

在执行的过程中，会先运行批式的初始模型训练，待批式任务执行结束，再启动流式任务。

最后，我们再将预测结果流 predResult 接入流式二分类评估组件 EvalBinaryClassStreamOp，并设置相应的参数。由于每次评估结果是 Json 格式的，为了便于显示，还可以在后面上 Json 内容提取组件 JsonValueStreamOp，并增加标识列，显示相应行输出的内容为评估指标 Eval Metric，代码如下。

```
predResult
    .link(
        new EvalBinaryClassStreamOp()
            .setPositiveLabelValueString("1")
            .setLabelCol(LABEL_COL_NAME)
            .setPredictionDetailCol(PRED_DETAIL_COL_NAME)
            .setTimeInterval(10)
    )
    .link(
        new JsonValueStreamOp()
            .setSelectedCol("Data")
            .setReservedCols(new String[] {"Statistics"})
            .setOutputCols(new String[] {"Accuracy", "AUC", "ConfusionMatrix"})
            .setJsonPath(new String[] {"$.Accuracy", "$.AUC", "$.ConfusionMatrix"})
    )
    .select("'Eval Metric' AS out_type, *")
    .print();
```

注意：流式的组件"连接"完成后，需要调用流式任务执行命令，即 StreamOperator.execute()，开始执行，显示结果如下。

```
out_type|click|pred|pred_info
--------|-----|----|---------
out_type|Statistics|Accuracy|AUC|ConfusionMatrix
--------|----------|--------|---|---------------
Pred Sample|0|0|{"0":"0.9220358860557992","1":"0.07796411394420077"}
Pred Sample|1|0|{"0":"0.6017776815075736","1":"0.3982223184924264"}
Pred Sample|0|0|{"0":"0.9309052094416868","1":"0.06909479055831325"}
Pred Sample|0|0|{"0":"0.9200218813350143","1":"0.07997811866498572"}
Pred Sample|0|0|{"0":"0.8557440536248255","1":"0.14425594637517447"}
Pred Sample|0|0|{"0":"0.9437830739831977","1":"0.05621692601680228"}
Pred Sample|0|0|{"0":"0.7786343432104281","1":"0.22136565678957187"}
Pred Sample|0|0|{"0":"0.9682519573861983","1":"0.03174804261380171"}
Pred Sample|0|0|{"0":"0.8974879967663411","1":"0.10251200323365894"}
Pred Sample|0|0|{"0":"0.8854488604041583","1":"0.11455113959584173"}
Eval Metric|window|0.8376583064315868|0.6143809440223419|[[53,59],[2556,13440]]
Eval Metric|all|0.8376583064315868|0.6143809440223419|[[53,59],[2556,13440]]
Pred Sample|0|0|{"0":"0.8335324643013561","1":"0.16646753569864392"}
```

```
Pred Sample|0|0|{"0":"0.8423454385153902","1":"0.15765456148460977"}
Pred Sample|0|0|{"0":"0.8452003664043619","1":"0.1547996335956381"}
Pred Sample|0|0|{"0":"0.9368763856950871","1":"0.06312361430491287"}
Pred Sample|0|0|{"0":"0.7963950068001105","1":"0.20360499319988945"}
Eval Metric|window|0.8421089401737267|0.6423637093476301|[[164,180],[6600,35997]]
Eval Metric|all|0.840894850039 7975|0.6343512339990862|[[217,239],[9156,49437]]
Pred Sample|0|0|{"0":"0.8853896240654119","1":"0.11461037593458812"}
Pred Sample|0|0|{"0":"0.9144744898328552","1":"0.08552551016714482"}
Pred Sample|1|0|{"0":"0.8063021776871495","1":"0.1936978223128505"}
Eval Metric|window|0.8424430394846714|0.7265268800218625|[[183,147],[6457,35128]]
Eval Metric|all|0.8415375777504853|0.67178153459723|[[400,386],[15613,84565]]
...  ...
```

其中输出了两种流式数据，一种是预测数据，各列的名称如下。

```
out_type|click|pred|pred_info
--------|-----|----|---------
```

第 1 列为标识信息，第 2 列是原始的 click 信息，第 3 列为预测结果，第 4 列为预测的详细信息。对应的预测结果形式如下。

```
Pred Sample|0|0|{"0":"0.9220358860557992","1":"0.07796411394420077"}
```

另一种是评估指标，各列的名称如下。

```
out_type|Statistics|Accuracy|AUC|ConfusionMatrix
--------|----------|--------|---|---------------
```

其中，Statistics 列有两个值 all 和 window，all 表示从开始运行到现在的所有预测数据的评估结果；window 表示时间窗口（当前设置为 10 秒）的所有预测数据的评估结果。

从输出的信息来看，无论是整体指标，还是窗口指标，Accuracy 和 AUC 都在提升，说明 FTRL 算法是有效的。

14.6 模型过滤

在上节的示例中，在线学习训练器每隔 10 秒就产生一个模型，然后直接推送给流式预测组件。在实际应用中，10 秒内的数据可能有很强的偏向性，从而影响模型的整体预测效果，模型未经验证就直接上线，难免会让人有些担心。本节将介绍模型的过滤机制，对新产生的模型进行验证，只有当评估指标达到了阈值，才将新模型参数传给流式预测组件。

使用 FtrlModelFilterStreamOp 组件 model_filter，设置模型评估精确度的阈值 AccuracyThreshold=

0.83，AUC 的阈值 AucThreshold=0.71，并设置一些模型评估需要用到的参数。将 model_filter 接入在线学习组件输出的模型流，并接入用来验证模型的数据流。这里设置了一个流式输出，将输出过滤后的模型，使用前面的技巧，在每行模型数据前用 SQL 语句增加标识列，显示 "Model"。对于流式预测组件 FtrlPredictStreamOp，接入模型过滤组件 model_filter 输出的模型流，并接入所要预测的数据流 test_stream_data。具体代码如下。

```
// ftrl train
FtrlTrainStreamOp model = new FtrlTrainStreamOp(initModel)
    .setVectorCol(VEC_COL_NAME)
    .setLabelCol(LABEL_COL_NAME)
    .setWithIntercept(true)
    .setAlpha(0.1)
    .setBeta(0.1)
    .setL1(0.01)
    .setL2(0.01)
    .setTimeInterval(10)
    .setVectorSize(NUM_HASH_FEATURES)
    .linkFrom(train_stream_data);

// model filter
FtrlModelFilterStreamOp model_filter = new FtrlModelFilterStreamOp()
    .setPositiveLabelValueString("1")
    .setVectorCol(VEC_COL_NAME)
    .setLabelCol(LABEL_COL_NAME)
    .setAccuracyThreshold(0.83)
    .setAucThreshold(0.71)
    .linkFrom(model, train_stream_data);

model_filter
    .select("'Model' AS out_type, *")
    .print();

// ftrl predict
FtrlPredictStreamOp predResult = new FtrlPredictStreamOp(initModel)
    .setVectorCol(VEC_COL_NAME)
    .setPredictionCol(PREDICTION_COL_NAME)
    .setReservedCols(new String[] {LABEL_COL_NAME})
    .setPredictionDetailCol(PRED_DETAIL_COL_NAME)
    .linkFrom(model_filter, test_stream_data);
```

输出计算结果如图 14-9 所示，我们看到在前两轮评估前，模型都没有更新，这是因为初始模型的评估指标较低，需要使用更多的训练样本才能达到指定的指标阈值。FTRL 组件虽然每 10 秒就产生一个模型，但是由于评估指标没有到阈值，所以都被过滤掉了。

```
out_type|bid|ntab|model_id|model_info|label_value
--------|---|----|--------|----------|-----------
out_type|click|pred|pred_info
--------|-----|----|---------
out_type|Statistics|Accuracy|AUC|ConfusionMatrix
--------|----------|--------|---|---------------
Pred Sample|0|0|{"0":"0.9377529917730091","1":"0.06224700822699092"}
Eval Metric|window|0.8335293287548803|0.6122200800597303|[[61,88],[3451,17659]]
Eval Metric|all|0.8335293287548803|0.6122200800597303|[[61,88],[3451,17659]]
Pred Sample|0|0|{"0":"0.843055646316327","1":"0.156944353683673"}
Pred Sample|0|0|{"0":"0.9129594617918085","1":"0.08704053820819146"}
Pred Sample|0|0|{"0":"0.9532587643017029","1":"0.046741235698297134"}
Pred Sample|0|0|{"0":"0.9357957904044711","1":"0.06420420959552886"}
Pred Sample|0|0|{"0":"0.8641536244526069","1":"0.1358463755473931"}
Pred Sample|0|0|{"0":"0.7865946845445935","1":"0.21340531545540653"}
Pred Sample|0|0|{"0":"0.9234134648219126","1":"0.07658653517808744"}
Eval Metric|window|0.8404184729717774|0.6198697322921425|[[142,157],[7058,37855]]
Eval Metric|all|0.8382151614989996|0.6173660384648244|[[203,245],[10509,55514]]
Pred Sample|0|0|{"0":"0.9186098221320143","1":"0.08139017786798575"}
Pred Sample|0|0|{"0":"0.8029715291034955","1":"0.19702847089650455"}
Model|0|20|0|{"hasInterceptItem":"true","vectorCol":"\"vec\"","modelName":"\"Logistic Regression\"","labelCol":n
Model|0|20|1048576|{"featureColNames":null,"featureColTypes":null,"coefVector":{"data":[-0.6784407860097712,0.0,
Model|0|20|1048577|-1.015492755263354E-4,-3.899382446905679E-5,1.135019436935083E-4,0.0,-6.819390973151679E-5,-0
Model|0|20|1048578|687E-5,-0.04365726078229945,0.0,0.0,-3.273834796168124E-5,-0.05451212263974209,-1.983135132297495
Model|0|20|1048579|0.0,-0.08506139587506867,-8.595045479817747E-5,1.149032894371878E-4,0.0,0.0,0.0,0.0,-2.718098
Model|0|20|1048580|4E-5,0.0,0.0,0.0,-3.274184053093851E-5,-1.5153002429807585E-4,-0.017178455686681304,-3.0386246467
Model|0|20|1048581|25848969E-5,-5.6092900005054916E-5,0.0,0.0,1.4860481330858388E-4,-0.033398008429795685,-1.0855589
Model|0|20|1048582|03,-7.755721408571834E-5,0.0,0.0,-2.600655001784036E-5,0.0,0.0,-2.7342190154608805E-5,1.461146502914
Model|0|20|1048583|-5,0.08310933503473497,-0.05472890571896318,-3.071246336463626E-5,0.0,0.0,0.0,1.3936238214184944
Model|0|20|1048584|876,-2.789863613806454E-5,-0.05620678923208143,-0.0552696873467867,-2.2732915274547903E-5,0.
Model|0|20|1048585|425927,2.79729607179458E-4,-0.03915229482230068,-8.670768826912563E-5,2.2848705357318133E-4,-
Model|0|20|1048586|63405432166148,-3.176819906462355E-5,-3.4106051793327794E-5,0.0,-2.7191147181643166E-5,-2.779
Model|0|20|1048587|836231E-5,0.0,1.4606971848951493E-4,-6.855781569505958E-4,-2.2572380287123274E-5,0.0,0.0,-2.2
Model|0|20|1048588|2345451943E-6,-0.17615794486620295,-1.8779470874865728E-4,-8.415450395865494E-5,-3.4888453510
Model|0|20|1048589|996693038E-4,-3.0812477279027554E-5,-4.784778241234657E-5,0.0,0.00824580250951806,0.0,-0.0786
Model|0|20|1048590|-0.007410252415562256,0.0,1.3567399015933853E-4,-6.798770308079837E-5,-1.0191936079852558E-4,
Model|0|20|1048591|52494017115,0.0,-1.385972170012299E-4,-1.05059711258772E-4,0.0,0.0,0.0,-0.012647716635138372,1.48
Model|0|20|1048592|E-5,0.0,1.605422787428073E-4,-0.121684396839641,0.0,0.0,0.0,1.2745802766813677E-4,-5.199610201294
Model|0|20|2251799812636672|null|1
Model|0|20|2251799812636673|null|0
Pred Sample|0|0|{"0":"0.9835687374543344","1":"0.016431262545665626"}
```

图 14-9

15 回归的由来

前面的章节都在建立分类或聚类的机器学习模型，模型预测输出的结果都属于类别，所有的类别是平等的，编号 1 的类别并不比编号 100 的类别差。本章同样会建立机器学习模型，但模型预测输出的是数值，即回归模型，数值能够反映不同预测样本在数量多少、品质高低、程度深浅上的差异。

“回归”的概念是由高尔顿（Francis Galton）在研究人类遗传问题时提出来的，随着《遗传的身高向平均数方向的回归》[1]一文的发表而得到普及。本节将使用当年的数据，利用Alink的分析工具，再现分析过程，通过数据理解“回归”的含义。

在 Kaggle 上有原始数据，为 1078 条父亲与儿子的身高记录，参见链接 15-2。下载解压后为文本格式数据，如图 15-1 所示，第一行为数据列名称，第一列为 Father（父亲身高，单位：英寸），第二列为 Son（儿子身高，单位：英寸），中间以制表符分隔。

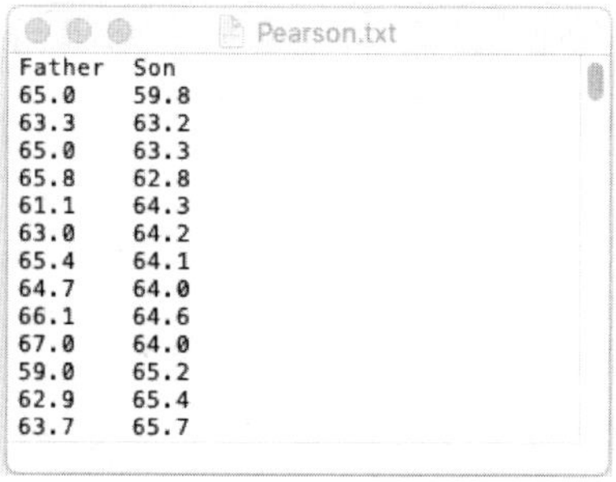

Father	Son
65.0	59.8
63.3	63.2
65.0	63.3
65.8	62.8
61.1	64.3
63.0	64.2
65.4	64.1
64.7	64.0
66.1	64.6
67.0	64.0
59.0	65.2
62.9	65.4
63.7	65.7

图 15-1

1 该论文的英文名称为“Regression towards mediocrity in hereditary stature”，参见链接 15-1。

15.1 平均数

我们先使用 CsvSourceBatchOp 组件读取原始数据，具体代码如下，设列名分别为 father 和 son，都是 DOUBLE 类型；字段分隔符为制表符，即 FieldDelimiter="\t"；原始数据的第一行用于记录数据列名称，在读取数据时应该略过，需要设置参数 IgnoreFirstLine=true。

```
CsvSourceBatchOp source = new CsvSourceBatchOp()
    .setFilePath(DATA_DIR + ORIGIN_FILE)
    .setSchemaStr("father double, son double")
    .setFieldDelimiter("\t")
    .setIgnoreFirstLine(true);

source.firstN(5).print();
```

数据输出结果如下。

```
father|son
------|---
65.0000|59.8000
63.3000|63.2000
65.0000|63.3000
65.8000|62.8000
61.1000|64.3000
```

为了便于整体观察，我们将这 1000 多个数据以散点图的方式显示出来，如图 15-2 所示。

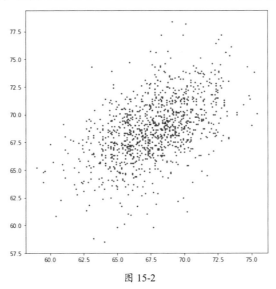

图 15-2

再使用 lazyPrintStatistics 方法，对数据进行统计操作，具体代码如下。

```
source.lazyPrintStatistics();
```

计算得到的统计结果如下。

```
|colName|count|missing|     sum|   mean|variance| min| max|
|-------|-----|-------|--------|-------|--------|----|----|
| father| 1078|      0| 72966.4|67.6868|  7.5396|  59|75.4|
|    son| 1078|      0| 74041.6|68.6842|  7.9309|58.5|78.4|
```

可知，父亲的平均身高是 67.6868 英寸，儿子的平均身高为 68.6842 英寸，之间相差了 0.9974 英寸，约为 1 英寸。

15.2 向平均数方向的回归

基于 15.1 节的统计结果，儿子的平均身高比父亲的平均身高多 0.9974 英寸（约 1 英寸），我们可以自然地猜测，身高 72 英寸的父亲平均会有身高 73 英寸的儿子，身高 65 英寸的父亲平均会有身高 66 英寸的儿子，即可以用直线 $y = x + 1$ 来近似预测儿子的身高。

我们在原散点图的基础上绘制直线 $y = x + 1$，如图 15-3 所示。

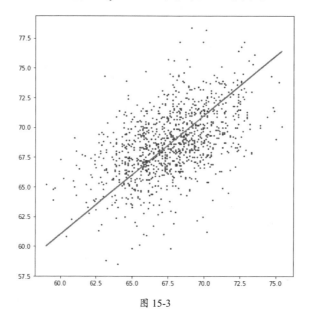

图 15-3

可以发现，在中间区域，数据点平均分布在直线两侧；在左侧区域，似乎直线上方的数据点多一些；在右侧区域，似乎直线下方的数据点多一些。

下面我们选取两个有代表性的父亲身高区域，通过统计进行定量分析。具体代码如下，分别是父亲身高为 72 英寸左右时，儿子身高的统计情况；父亲身高为 65 英寸左右时，儿子身高的统计情况。

```
source.filter("father>=71.5 AND father<72.5").lazyPrintStatistics("father 72");

source.filter("father>=64.5 AND father<65.5").lazyPrintStatistics("father 65");
```

运行结果如下。

```
father 72
|colName|count|missing|    sum|   mean|variance| min| max|
|-------|-----|-------|-------|-------|--------|----|----|
| father|   52|      0| 3745.1|72.0212|  0.1037|71.5|72.4|
|    son|   52|      0| 3675.9|70.6904|  5.3542|64.5|76.4|

father 65
|colName|count|missing|    sum|   mean|variance| min| max|
|-------|-----|-------|-------|-------|--------|----|----|
| father|  104|      0| 6750.5|64.9087|   0.087|64.5|65.4|
|    son|  104|      0| 6986.1| 67.174|  6.3402|59.8|73.9|
```

可以看出，身高为 65 英寸左右的父亲们的平均身高为 64.9087 英寸，其儿子们的平均身高为 67.174 英寸，提高了 2.2653 英寸；身高为 72 英寸左右的父亲们的平均身高为 72.0212 英寸，其儿子们的平均身高为 70.6904 英寸，降低了 1.3308 英寸。这与我们前面猜测的直线规律 $y = x + 1$ 相差很大。

从这些数据中可以看出，对于矮个子的父亲，其儿子们的平均身高会比父辈高一些；对于高个子的父亲，其儿子们的平均身高会比父辈矮一些，即儿子们的身高会向平均值"回归"。

15.3 线性回归

下面使用线性回归模型对此数据进行建模预测，并与前面的结果进行对比，具体代码如下。

```
LinearRegTrainBatchOp linear_model =
  new LinearRegTrainBatchOp()
    .setFeatureCols("father")
```

```
        .setLabelCol("son")
        .linkFrom(source);

linear_model.lazyPrintTrainInfo();
linear_model.lazyPrintModelInfo();
```

输出模型信息如下。

```
-------------------------- model meta info --------------------------
{hasInterception: true, model name: Linear Regression, num feature: 1}
-------------------------- model weight info --------------------------
|intercept|   father|
|---------|---------|
|  33.8928|0.51400662|
```

即，线性回归的表达式为

$$y = 33.8928 + 0.51400662x$$

使用预测组件 LinearRegPredictBatchOp，从 linear_model 获取线性回归模型，对原始数据进行线性回归预测，相关代码如下。

```
LinearRegPredictBatchOp linear_reg =
    new LinearRegPredictBatchOp()
        .setPredictionCol("linear_reg")
        .linkFrom(linear_model, source);

linear_reg.lazyPrint(5);
```

运行结果如下。

```
father|son|linear_reg
------|---|----------
65.0000|59.8000|67.3033
63.3000|63.2000|66.4295
65.0000|63.3000|67.3033
65.8000|62.8000|67.7145
61.1000|64.3000|65.2987
```

最后，将此直线与原始数据及猜测直线放在一张图内，如图 15-4 所示。

两条直线的交点为父亲和儿子身高数据的平均值，即(67.6868, 68.6842)。在其左侧，线性回归值要高于直线 $y = x + 1$；而在其右侧，线性回归值要低于直线 $y = x + 1$。

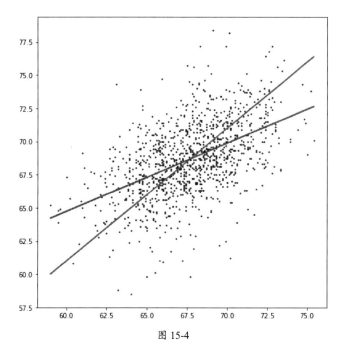

图 15-4

16 常用回归算法

本章以预测葡萄酒品质值为例，演示常用的回归算法。与二分类、多分类模型相似，回归模型也有其评估指标，我们将在 16.1 节中进行介绍。

16.1 回归模型的评估指标

设回归函数为 $f(x)$，回归测试样本共 n 条，第 i 条样本的特征为 $x^{(i)}$，回归标签值为 $y^{(i)}$，并设 $\bar{y}$ 为回归标签值的平均值，即

$$\bar{y} = \frac{1}{n}\sum_{i=1}^{n} y^{(i)}$$

总平方和（Sum of Squared for Total，SST）为

$$\text{SST} = \sum_{i=1}^{n}(y^{(i)} - \bar{y})^2$$

回归平方和（Sum of Squares for Regression，SSR）为

$$\text{SSR} = \sum_{i=1}^{n}(f(x^{(i)}) - \bar{y})^2$$

误差平方和（Sum of Squares for Error，SSE）为

$$\text{SSE} = \sum_{i=1}^{n}\bigl(f(x^{(i)}) - y^{(i)}\bigr)^2$$

R^2 判定系数（Coefficient of Determination）为

$$R^2 = 1 - \frac{\text{SSE}}{\text{SST}}$$

R 多重相关系数（Multiple Correlation Coefficient）为

$$R = \sqrt{R^2}$$

均方误差（Mean Squared Error，MSE）为

$$\text{MSE} = \frac{1}{n}\text{SSE} = \frac{1}{n}\sum_{i=1}^{n}\bigl(f(x^{(i)}) - y^{(i)}\bigr)^2$$

均方根误差（Root Mean Squared Error，RMSE）为

$$\text{RMSE} = \sqrt{\text{MSE}}$$

绝对误差（Sum of Absolute Error/Difference，SAE/SAD）为

$$\text{SAE} = \sum_{i=1}^{n}\bigl|f(x^{(i)}) - y^{(i)}\bigr|$$

平均绝对误差（Mean Absolute Error/Difference，MAE/MAD）为

$$\text{MAE} = \frac{1}{n}\sum_{i=1}^{n}\bigl|f(x^{(i)}) - y^{(i)}\bigr|$$

平均绝对百分误差（Mean Absolute Percentage Error，MAPE）为

$$\text{MAPE} = \frac{100}{n}\sum_{i=1}^{n}\left|\frac{f(x^{(i)}) - y^{(i)}}{y^{(i)}}\right|$$

解释方差（Explained Variance）为

$$\text{ExplainedVariance} = \frac{\text{SSR}}{n}$$

16.2 数据探索

葡萄酒的品质可以用数值表示，数值越大表示品质越好，而葡萄酒本身的理化指标对其品质是有影响的。下面通过建立回归模型对葡萄酒的品质进行预测，可从链接 16-1 下载数据。葡萄牙的绿酒（Vinho Verde），其酒精度中等，口感清淡鲜酸，特别适合夏季饮用。需要注意的是，绿酒中的“绿”指的并非是成熟或陈年的葡萄酒，绿酒可以是白色的，也可以是红色的。原始数据中分别对两种颜色的绿酒提供了不同的数据集，我们只选择了其中的一个（白色的），数据文件的名称为 winequality-white.csv。

该数据集有 11 个特征字段，都是经过理化测试得到的，具体名称及含义如表 16-1 所示。

表 16-1 特征字段说明

编号	特征字段名称	中文含义	编号	特征字段名称	中文含义
1	fixed acidity	固定酸度	7	total sulfur dioxide	总二氧化硫
2	volatile acidity	挥发性酸度	8	density	密度
3	citric acid	柠檬酸	9	pH	pH 值
4	residual sugar	残糖	10	sulphates	硫酸盐
5	Chlorides	氯化物	11	alcohol	酒精
6	free sulfur dioxide	自由二氧化硫			

该数据集的目标变量为葡萄酒的品质，目标变量名称为 quality，分值在 0 和 10 之间，是基于感官的数据。

将数据文件 winequality-white.csv 下载到本地，使用文本编辑器打开，如图 16-1 所示。

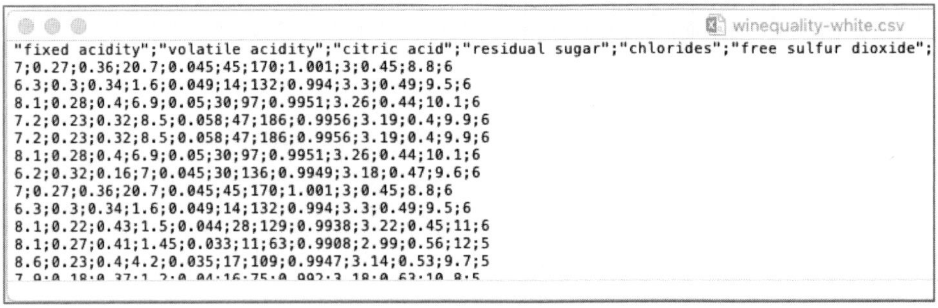
winequality-white.csv

```
"fixed acidity";"volatile acidity";"citric acid";"residual sugar";"chlorides";"free sulfur dioxide";
7;0.27;0.36;20.7;0.045;45;170;1.001;3;0.45;8.8;6
6.3;0.3;0.34;1.6;0.049;14;132;0.994;3.3;0.49;9.5;6
8.1;0.28;0.4;6.9;0.05;30;97;0.9951;3.26;0.44;10.1;6
7.2;0.23;0.32;8.5;0.058;47;186;0.9956;3.19;0.4;9.9;6
7.2;0.23;0.32;8.5;0.058;47;186;0.9956;3.19;0.4;9.9;6
8.1;0.28;0.4;6.9;0.05;30;97;0.9951;3.26;0.44;10.1;6
6.2;0.32;0.16;7;0.045;30;136;0.9949;3.18;0.47;9.6;6
7;0.27;0.36;20.7;0.045;45;170;1.001;3;0.45;8.8;6
6.3;0.3;0.34;1.6;0.049;14;132;0.994;3.3;0.49;9.5;6
8.1;0.22;0.43;1.5;0.044;28;129;0.9938;3.22;0.45;11;6
8.1;0.27;0.41;1.45;0.033;11;63;0.9908;2.99;0.56;12;5
8.6;0.23;0.4;4.2;0.035;17;109;0.9947;3.14;0.53;9.7;5
```

图 16-1

第 1 行为各数据列的名称，从第 2 行开始是具体的数据，每行为一条记录，各数值间使用分号分隔。如下代码所示，定义各列的名称，由于列名中不能有空格，所以我们使用“驼峰”

格式将多个单词连在一起；所有数据都是数值类型的，这里我们定义为双精度浮点类型。根据数据的特点，在使用 CsvSourceBatchOp 组件时，需要设置字段间的分隔符为分号，并忽略第 1 行内容，即设置参数 IgnoreFirstLine=true，最后输出 5 条数据。

```
private static final String[] COL_NAMES = new String[] {
    "fixedAcidity", "volatileAcidity", "citricAcid", "residualSugar", "chlorides",
    "freeSulfurDioxide", "totalSulfurDioxide", "density", "pH", "sulphates",
    "alcohol", "quality"
};

private static final String[] COL_TYPES = new String[] {
    "double", "double", "double", "double", "double",
    "double", "double", "double", "double", "double",
    "double", "double"
};

CsvSourceBatchOp source = new CsvSourceBatchOp()
    .setFilePath(DATA_DIR + ORIGIN_FILE)
    .setSchemaStr(Utils.generateSchemaString(COL_NAMES, COL_TYPES))
    .setFieldDelimiter(";")
    .setIgnoreFirstLine(true);

source.lazyPrint(5);
```

输出数据如表 16-2 所示，quality 列为回归标签值列。

表 16-2 葡萄酒品质数据

fixedAcidity	volatileAcidity	citricAcid	residualSugar	chlorides	freeSulfurDioxide	totalSulfurDioxide	density	pH	sulphates	alcohol	quality
5.6000	0.3500	0.4000	6.3000	0.0220	23.0000	174.0000	0.9822	3.5400	0.5000	11.6000	7.0000
8.8000	0.2400	0.2300	10.3000	0.0320	12.0000	97.0000	0.9957	3.1300	0.4000	10.7000	6.0000
6.0000	0.2900	0.2100	15.5500	0.0430	20.0000	142.0000	0.9966	3.1100	0.5400	10.1000	6.0000
6.1000	0.2700	0.3100	1.5000	0.0350	17.0000	83.0000	0.9908	3.3200	0.4400	11.1000	7.0000
7.4000	0.5600	0.0900	1.5000	0.0710	19.0000	117.0000	0.9950	3.2200	0.5300	9.8000	5.0000

了解变量 quality 与 11 个特征变量间的关系，需要计算相关系数矩阵，具体代码如下。

```
source.link(new CorrelationBatchOp().lazyPrintCorrelation());
```

输出结果如表 16-3 所示。

表 16-3 各列的相关系数

colName	fixedAcidity	volatileAcidity	citricAcid	residualSugar	chlorides	freeSulfurDioxide	totalSulfurDioxide	density	pH	sulphates	alcohol	quality
fixedAcidity	1.0000	-0.0227	0.2892	0.0890	0.0231	-0.0494	0.0911	0.2653	-0.4259	-0.0171	-0.1209	-0.1137
volatileAcidity	-0.0227	1.0000	-0.1495	0.0643	0.0705	-0.0970	0.0893	0.0271	-0.0319	-0.0357	0.0677	-0.1947
citricAcid	0.2892	-0.1495	1.0000	0.0942	0.1144	0.0941	0.1211	0.1495	-0.1637	0.0623	-0.0757	-0.0092
residualSugar	0.0890	0.0643	0.0942	1.0000	0.0887	0.2991	0.4014	0.8390	-0.1941	-0.0267	-0.4506	-0.0976
chlorides	0.0231	0.0705	0.1144	0.0887	1.0000	0.1014	0.1989	0.2572	-0.0904	0.0168	-0.3602	-0.2099
freeSulfurDioxide	-0.0494	-0.0970	0.0941	0.2991	0.1014	1.0000	0.6155	0.2942	-0.0006	0.0592	-0.2501	0.0082
totalSulfurDioxide	0.0911	0.0893	0.1211	0.4014	0.1989	0.6155	1.0000	0.5299	0.0023	0.1346	-0.4489	-0.1747
density	0.2653	0.0271	0.1495	0.8390	0.2572	0.2942	0.5299	1.0000	-0.0936	0.0745	-0.7801	-0.3071
pH	-0.4259	-0.0319	-0.1637	-0.1941	-0.0904	-0.0006	0.0023	-0.0936	1.0000	0.1560	0.1214	0.0994
sulphates	-0.0171	-0.0357	0.0623	-0.0267	0.0168	0.0592	0.1346	0.0745	0.1560	1.0000	-0.0174	0.0537
alcohol	-0.1209	0.0677	-0.0757	-0.4506	-0.3602	-0.2501	-0.4489	-0.7801	0.1214	-0.0174	1.0000	0.4356
quality	-0.1137	-0.1947	-0.0092	-0.0976	-0.2099	0.0082	-0.1747	-0.3071	0.0994	0.0537	0.4356	1.0000

为了便于查看，我们借助可视化工具，各字段间的相关系数情况如图 16-2 所示，除了有具体的数字，还用颜色进行了标识，黑色为+1，白色为-1。颜色越深，说明正相关性越强；颜色越浅，说明负相关性越强。

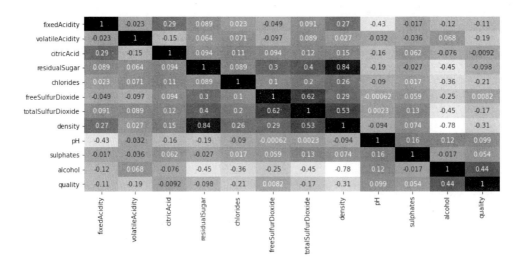

图 16-2

从图 16-2 中可以看出如下内容。
- 特征 alcohol（酒精）与 quality 的相关系数为 0.44，正相关性较强；density（密度）与 quality 的相关系数值为-0.31，负相关性较强。
- 特征 citricAcid（柠檬酸）与 quality 的相关系数为-0.0092，接近 0，是线性无关；特征 freeSulfurDioxide（自由二氧化硫）与 quality 的相关系数为-0.0082，接近 0，是线性无关。
- 特征 density（密度）与 residualSugar（残糖）的正相关性很强，相关系数为 0.84；density（密度）与 alcohol（酒精）的负相关性很强，相关系数为-0.78。

本数据集的标签列（quality 列）都是整数值，可以通过分组计数和排序操作计算其直方图分布，具体代码如下。

```
source
  .groupBy(LABEL_COL_NAME, LABEL_COL_NAME + ", COUNT(*) AS cnt")
  .orderBy(LABEL_COL_NAME, 100)
  .lazyPrint(-1);
```

运行结果如下。

```
quality|cnt
-------|---
3.0000|20
4.0000|163
5.0000|1457
6.0000|2198
7.0000|880
8.0000|175
9.0000|5
```

可以看出 6 的个数最多，其次是 5 和 7，4 和 8 的分布更少，都在 200 以下，比 6 的个数少一个数量级，而 3 和 9 的比例更小。

最后，将原始数据按 8:2 的比例划分为训练集和预测集，保存为 AK 格式文件。本章后面尝试的各种算法都会基于此训练集和测试集，以便于比较效果。

16.3 线性回归

首先尝试使用我们比较熟悉的线性回归方法，相关代码如下。

```
new LinearRegression()
  .setFeatureCols(FEATURE_COL_NAMES)
  .setLabelCol(LABEL_COL_NAME)
```

```
    .setPredictionCol(PREDICTION_COL_NAME)
    .enableLazyPrintTrainInfo()
    .enableLazyPrintModelInfo()
    .fit(train_data)
    .transform(test_data)
    .link(
        new EvalRegressionBatchOp()
            .setLabelCol(LABEL_COL_NAME)
            .setPredictionCol(PREDICTION_COL_NAME)
            .lazyPrintMetrics("LinearRegression")
    );
```

使用线性回归组件，设置特征、标签及预测结果列，其他参数使用默认值；选择使用 Lazy 方式输出训练信息和模型信息；组件 EvalRegressionBatchOp 用于评估回归模型。

训练信息如下。

```
-------------------------- train meta info --------------------------
{model name: Linear Regression, num feature: 11}
-------------------------- train importance info --------------------------
|          colName|importnaceValue|          colName| weightValue|
|-----------------|---------------|-----------------|------------|
|          density|     0.44120155|               pH|   0.64837239|
|     residualSugar|    0.40567559|        sulphates|   0.64099725|
|          alcohol|     0.24270038|          alcohol|   0.19730515|
|          ... ...|        ... ...|          ... ...|      ... ...|
| totalSulfurDioxide|   0.01594283|        chlorides|  -0.32426272|
|        chlorides|     0.00728598|   volatileAcidity|  -1.88292677|
|        citricAcid|     0.00313107|          density|-147.10297328|
-------------------------- train convergence info --------------------------
step:0  loss:14.36145407 gradNorm:5.90856615 learnRate:0.40000000
step:1  loss:5.30536574  gradNorm:5.30978617 learnRate:1.60000000
step:2  loss:1.81822334  gradNorm:3.30213486 learnRate:1.60000000
... ...
step:12 loss:0.28568962 gradNorm:0.00474675 learnRate:4.00000000
step:13 loss:0.28568641 gradNorm:0.00216498 learnRate:4.00000000
step:14 loss:0.28568635 gradNorm:0.00032419 learnRate:4.00000000
```

最重要的 3 个特征为 density、residualSugar 和 alcohol。整个训练过程在第 14 次迭代时达到收敛条件退出。

下面是模型中各个特征的权重值。

```
-------------------------- model meta info --------------------------
{hasInterception: true, model name: Linear Regression, num feature: 11}
-------------------------- model weight info --------------------------
| colName[0,9]| intercept|fixedAcidity|volatileAcidity|citricAcid|residualSugar|chlorides|freeSulfurDioxide|totalSulfurDioxide|    density|         pH|
|  weight[0,9]|  147.2227|   0.05561785|    -1.88292677|-0.02573757|    0.08013539|-0.32426272|       0.00370713|       -0.00037885|-147.10297328| 0.64837239|
|colName[10,11]|  sulphates|     alcohol|               |          |             |         |                 |                  |           |           |
| weight[10,11]| 0.64099725|  0.19730515|               |          |             |         |                 |                  |           |           |
```

回归评估结果如下，我们会把其作为一个基准与后面的方法进行比较。

```
------------------------------ Metrics: ------------------------------
MSE:0.5309 RMSE:0.7286    MAE:0.5748 MAPE:10.0995    R2:0.2655
```

下面我们希望通过 LASSO 算法减少线性模型中非零特征的数量，使用组件，设置参数 Lambda=0.05，其他设置与线性回归一样。

```
new LassoRegression()
   .setLambda(0.05)
   .setFeatureCols(FEATURE_COL_NAMES)
   .setLabelCol(LABEL_COL_NAME)
   .setPredictionCol(PREDICTION_COL_NAME)
   .enableLazyPrintTrainInfo()
   .enableLazyPrintModelInfo("< LASSO model >")
   .fit(train_data)
   .transform(test_data)
   .link(
      new EvalRegressionBatchOp()
          .setLabelCol(LABEL_COL_NAME)
          .setPredictionCol(PREDICTION_COL_NAME)
          .lazyPrintMetrics("LassoRegression")
   );
```

输出训练信息如下。

```
-------------------------- train meta info --------------------------
{model name: LASSO, num feature: 11}
-------------------------- train importance info --------------------------
|          colName|importnaceValue|         colName|weightValue|
|-----------------|---------------|----------------|-----------|
|          alcohol|     0.30556675|         alcohol| 0.24841285|
|   volatileAcidity|     0.14232785|        sulphates| 0.03046873|
|    residualSugar|     0.05538545|    residualSugar| 0.01094060|
|             ... |        ... ...|            ... |       ... |
|        citricAcid|     0.00000000|        chlorides|-0.90947378|
| totalSulfurDioxide|    0.00000000| volatileAcidity|-1.40732785|
|               pH|     0.00000000|         density|-18.19511910|
-------------------------- train convergence info --------------------------
step:0  loss:14.39385147 gradNorm:5.85882684 learnRate:0.40000000
step:1  loss:4.61560283  gradNorm:5.25846666 learnRate:1.60000000
step:2  loss:0.50012696  gradNorm:2.88726519 learnRate:1.60000000
... ...
step:20 loss:0.30128245 gradNorm:0.06450862 learnRate:0.00000000
step:21 loss:0.30128245 gradNorm:0.06450862 learnRate:0.00000000
step:22 loss:0.30128245 gradNorm:0.06450862 learnRate:0.00000000
```

LASSO 算法选出的最重要的 3 个特征为 alcohol、volatileAcidity 和 residualSugar。对比线性回归的 density、residualSugar 和 alcohol，特征 residualSugar 与 alcohol 都保留在前三位，在前面计算相关系数时，volatileAcidity 与 quality 的相关系数也相对较高，所以 volatileAcidity 的出现

也在意料之中；特征 density 与特征 residualSugar 与 alcohol 的相关性较强，权重值下降显著。
模型信息如下。

```
--------------------- model meta info ---------------------
{hasInterception: true, model name: LASSO, num feature: 11}
--------------------- model weight info ---------------------
   |colName[0,9]| intercept|fixedAcidity|volatileAcidity|citricAcid|residualSugar|  chlorides|freeSulfurDioxide|totalSulfurDioxide|    density|         pH|
   |  weight[0,9]|  21.7008|  -0.01094269|    -1.40732785| 0.00000000|   0.01094060| -0.90947378|        0.00132137|         0.00000000|-18.19511910| 0.00000000|
   |colName[10,11]| sulphates|     alcohol|              |          |             |           |                 |                  |           |           |
   |weight[10,11]|0.03046873| 0.24841285|              |          |             |           |                 |                  |           |           |
```

这里出现了 3 个权重为 0 的特征：citricAcid、totalSulfurDioxide 和 pH。

模型的评估指标如下，相比于线性回归模型，指标变化不大。

```
-------------------------- Metrics: --------------------------
MSE:0.5555 RMSE:0.7453    MAE:0.5904 MAPE:10.3725    R2:0.2315
```

16.4 决策树与随机森林

本节将尝试使用决策树和随机森林算法，看看能否获得比线性回归更好的效果。
先看决策树回归器 DecisionTreeRegressor 的实验，具体代码如下。

```
new DecisionTreeRegressor()
    .setFeatureCols(FEATURE_COL_NAMES)
    .setLabelCol(LABEL_COL_NAME)
    .setPredictionCol(PREDICTION_COL_NAME)
    .fit(train_data)
    .transform(test_data)
    .link(
        new EvalRegressionBatchOp()
            .setLabelCol(LABEL_COL_NAME)
            .setPredictionCol(PREDICTION_COL_NAME)
            .lazyPrintMetrics("DecisionTreeRegressor")
    );
BatchOperator.execute();
```

其中设置了特征、标签及预测结果列，其他参数使用默认值。
运行结果如下。

```
-------------------------- Metrics: --------------------------
MSE:0.6215 RMSE:0.7884    MAE:0.494 MAPE:8.7346    R2:0.1401
```

各项指标要落后于线性回归算法。
使用多棵决策树构成的随机森林比较容易获得更好的指标。逐渐调整随机森林中树的棵数，看其对回归效果的影响，如下代码所示。

```
for (int numTrees : new int[] {2, 4, 8, 16, 32, 64, 128}) {
    new RandomForestRegressor()
        .setNumTrees(numTrees)
        .setFeatureCols(FEATURE_COL_NAMES)
        .setLabelCol(LABEL_COL_NAME)
        .setPredictionCol(PREDICTION_COL_NAME)
        .fit(train_data)
        .transform(test_data)
        .link(
            new EvalRegressionBatchOp()
                .setLabelCol(LABEL_COL_NAME)
                .setPredictionCol(PREDICTION_COL_NAME)
                .lazyPrintMetrics("RandomForestRegressor - " + numTrees)
        );
    BatchOperator.execute();
}
```

整理运行结果，如表 16-4 所示，可以看到 64 棵树时的回归效果明显优于线性回归；在 64 棵树之前，随着棵数的增加，各项指标都在提升，但提升的幅度在下降；在 128 棵树时并没有获得更好的效果。因此，在实际应用中，我们需要考虑回归效果和计算代价，选择合适的棵数。

表 16-4 随机森林算法选择不同棵数的参数

棵数	MSE	RMSE	MAE	MAPE	R2
2	0.5126	0.716	0.4945	8.6621	0.2907
4	0.4307	0.6563	0.459	8.0789	0.404
8	0.3866	0.6218	0.4361	7.6881	0.4651
16	0.3705	0.6087	0.428	7.5556	0.4874
32	0.3645	0.6037	0.4256	7.5223	0.4957
64	0.3613	0.601	0.4245	7.4985	0.5002
128	0.3617	0.6014	0.4227	7.4711	0.4996

16.5 GBDT回归

我们再尝试一个非线性的回归模型——GBDT 模型，流程与随机森林模型相同，使用 GBDT 回归组件 GbdtRegressor，设置特征列为 11 个特征，标签列为 quality，并设置预测结果列。为了获取更好的回归模型，我们需要尝试多个参数，这里列出了一组参数，具体设置如下代码所示。

```
for (int numTrees : new int[] {16, 32, 64, 128, 256, 512}) {
```

```
new GbdtRegressor()
    .setLearningRate(0.05)
    .setMaxLeaves(256)
    .setFeatureSubsamplingRatio(0.3)
    .setMinSamplesPerLeaf(2)
    .setMaxDepth(100)
    .setNumTrees(numTrees)
    .setFeatureCols(FEATURE_COL_NAMES)
    .setLabelCol(LABEL_COL_NAME)
    .setPredictionCol(PREDICTION_COL_NAME)
    .fit(train_data)
    .transform(test_data)
    .link(
        new EvalRegressionBatchOp()
            .setLabelCol(LABEL_COL_NAME)
            .setPredictionCol(PREDICTION_COL_NAME)
            .lazyPrintMetrics("GbdtRegressor - " + numTrees)
    );
    BatchOperator.execute();
}
```

汇总运行结果如表 16-5 所示，回归指标要优于随机森林。

表 16-5　GBDT 算法选择不同棵数的参数

棵数	MSE	RMSE	MAE	MAPE	R2
16	0.4906	0.7005	0.5465	9.6852	0.3212
32	0.424	0.6511	0.5095	9.0017	0.4134
64	0.3824	0.6184	0.4691	8.2692	0.4709
128	0.3633	0.6027	0.4356	7.6824	0.4974
256	0.3552	0.596	0.4062	7.176	0.5086
512	0.355	0.5958	0.3926	6.9417	0.5088

17 常用聚类算法

聚类就是将若干个对象的集合分割成几个类,每个类内的对象之间是相似的,但与其他类的对象是不相似的。

例如,前面使用过的 Iris 数据集,每个数据有 4 个特征,可以看作四维空间中的点。我们无法显示数据在四维空间的分布情况,但可以将其 4 个特征进行两两组合,分别绘制散点图,并按类别标以不同的颜色,散点图矩阵如图 17-1 所示。

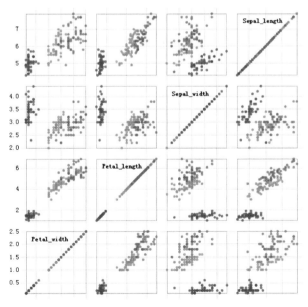

图 17-1

在每个散点图中，这些数据都会明显聚为两簇，根据这些二维平面的投影情况，我们会猜想其在四维空间中应该也会明显地聚合为两类。从类别颜色上看，其中一簇中包含两个类别，它们的界限并不清晰，如果我们的目标是聚为三簇，那么这两个类也会分属于不同的簇，但在每个簇中会以某个类别为主，掺杂其他类别的一些点。

17.1 聚类评估指标

聚类个数为K个，整个数据集的中心为u，样本总数为N，聚类结果为$\Omega = \{\omega_1, \omega_2, \cdots, \omega_K\}$。对于聚类簇$C_k$，$u_k$为聚类中心，包含$n_k$条样本，即$n_k = |\omega_k|$。

17.1.1 基本评估指标

本节介绍 5 个基本的评估指标。

（1）紧密度（Compactness，CP）

$$\mathrm{CP}_k = \frac{1}{|\omega_k|} \sum_{x \in \omega_k} \|x - u_k\|$$

$$\mathrm{CP} = \frac{1}{K} \sum_{k=1}^{K} \mathrm{CP}_k$$

CP 越小，意味着类内聚类距离越小。

（2）分离度（Seperation，SP）

$$\mathrm{SP} = \frac{2}{K^2 - K} \sum_{i=1}^{K} \sum_{j=i+1}^{K} \|u_i - u_j\|$$

SP 越大，意味着类间聚类距离越大。

（3）Davies-Bouldin（DB）指数

$$\mathrm{DB} = \frac{1}{K} \sum_{i=1}^{K} \max_{j \neq i} \left(\frac{\mathrm{CP}_i + \mathrm{CP}_j}{\|u_i - u_j\|} \right)$$

DB 越小，意味着类内距离越小，同时类间距离越大。

（4）方差比准则（Calinski-Harabasz Index，VRC）

$$SSB = \sum_{k=1}^{K} n_k \|u_k - u\|$$

$$SSW = \sum_{k=1}^{K} \sum_{x \in \omega_k} \|x - u_k\|$$

$$VRC = \frac{SSB}{SSW} \cdot \frac{N-K}{K-1}$$

其中，SSB 是组与组之间的平方和误差，SSW 是组内平方和误差。如果 SSW 越小、SSB 越大，那么 VRC 越大，聚类效果越好。

（5）轮廓系数（Silhouette Coefficient）

用于描述聚类中各簇的轮廓清晰度。将所有样本的轮廓系数求平均值，就是该聚类结果总的轮廓系数。

下面主要介绍单个样本轮廓系数的计算。

对于属于簇ω_k的样本i，有以下情况。

- 计算样本i到同簇其他样本的平均距离$a(i)$。$a(i)$越小，说明样本i越应该被聚类到该簇。$a(i)$被称为样本i的**簇内不相似度**。
- 计算样本i到其他某簇ω_j的所有样本的平均距离$b_j(i)$，称为样本i与簇ω_j的不相似度。定义为样本i的**簇间不相似度**为

$$b(i) = \min\{b_1(i), \cdots, b_{k-1}(i), b_{k+1}(i), \cdots, b_K(i)\}$$

- 根据样本i的簇内不相似度$a(i)$和簇间不相似度$b(i)$，定义样本i的**轮廓系数**为

$$S(i) = \frac{b(i) - a(i)}{\max\{a(i), b(i)\}}$$

也可以换一个形式表达，看上去更清晰。

$$S(i) = \begin{cases} 1 - \dfrac{a(i)}{b(i)}, & a(i) < b(i) \\ 0, & a(i) = b(i) \\ \dfrac{b(i)}{a(i)} - 1, & a(i) > b(i) \end{cases}$$

可知轮廓系数的取值范围是$[-1, 1]$。当$S(i)$接近 1 时，样本i聚类合理；$S(i)$接近-1，则说明样本i更应该分类到其他簇；若$S(i)$近似为 0，则说明样本i在两个簇的边界上。

上面介绍的指标计算只需要聚类结果，如果数据集还有标记好的类别信息，则还可以提供更多的指标。

17.1.2　基于标签值的评估指标

设样本数据分属于M个标签类别$C = \{c_1, c_2, \cdots, c_M\}$，$|\omega_k \cap c_m|$为聚类$\omega_k$中的样本属于类别$c_m$的个数，$P_{ki}$为聚类$\omega_k$中的样本属于类别$c_m$的概率，即

$$P_{km} = \frac{|\omega_k \cap c_m|}{|\omega_k|}$$

则每个聚类的熵（Entropy）可以表示为

$$e_k = -\sum_{m=1}^{M} P_{km} \log(P_{km})$$

整个聚类划分的熵表示为

$$e = \sum_{k=1}^{K} \frac{n_k}{N} e_k$$

- 纯度（Purity）

$$\text{purity}(\Omega, C) = \sum_{k=1}^{K} \frac{n_k}{N} \max_m (P_{km}) = \frac{1}{N} \sum_{k=1}^{K} \max_m |\omega_k \cap c_m|$$

Purity 在$[0,1]$区间内，越接近 1，表示聚类结果越好。

- 归一化互信息（Normalized Mutual Information，NMI）

$$H(\Omega) = -\sum_{k=1}^{K} \frac{\omega_k}{N} \log\left(\frac{\omega_k}{N}\right)$$

$$H(C) = -\sum_{m=1}^{M} \frac{c_m}{N} \log\left(\frac{c_m}{N}\right)$$

$$I(\Omega, C) = \sum_{k=1}^{K} \sum_{m=1}^{M} \frac{|\omega_k \cap c_m|}{N} \log\left(\frac{N|\omega_k \cap c_m|}{|\omega_k| \cdot |c_m|}\right)$$

$$\text{NMI} = \frac{2 \cdot I(\Omega, C)}{H(\Omega) + H(C)}$$

NMI 在[0,1]区间内，越接近 1，表示聚类结果越好。

整个数据集会有 $\binom{N}{2} = \frac{N(N-1)}{2}$ 个样本对，考虑如下 4 种情况。

- TP：同一类的样本被分到同一个簇。
- TN：不同类的样本被分到不同簇。
- FP：不同类的样本被分到同一个簇。
- FN：同一类的样本被分到不同簇。

则有

$$\text{TP} + \text{FP} = \sum_{m=1}^{M} \binom{|c_m|}{2}$$

$$\text{TP} + \text{FN} = \sum_{k=1}^{K} \binom{|\omega_k|}{2}$$

$$\text{TP} = \sum_{k=1}^{K} \sum_{m=1}^{M} \binom{|\omega_k \cap c_m|}{2}$$

$$\text{TP} + \text{TN} + \text{FP} + \text{FN} = \binom{N}{2}$$

可以计算得到

$$\text{Precision} = \frac{\text{TP}}{\text{TP} + \text{FP}}$$

$$\text{Recall} = \frac{\text{TP}}{\text{TP} + \text{FN}}$$

$$F_1 = \frac{2 \cdot \text{Recall} \cdot \text{Precision}}{\text{Recall} + \text{Precision}}$$

- 兰德系数（Rand Index，RI）

$$RI = \frac{TP + TN}{TP + TN + FP + FN}$$

RI 在[0,1]区间内,越接近 1,表示聚类结果越好,意味着聚类结果与标签类别情况越吻合。

- 调整兰德系数（Adjusted Rand Index，ARI）

$$Index = TP$$

$$ExpectedIndex = \frac{(TP + FP)(TP + FN)}{TP + TN + FP + FN}$$

$$MaxIndex = \frac{TP + FP + TP + FN}{2}$$

$$ARI = \frac{Index - ExpectedIndex}{MaxIndex - ExpectedIndex}$$

ARI 在[-1,1]区间内，越接近 1，表示聚类结果越好；在聚类结果为随机产生的时候，指标会接近零。

17.2 K-Means聚类

K 均值（K-Means）聚类，用户定义所要的类的个数 K，算法按照距离将数据自动聚合为 K 个类，使类内的对象的距离近，而不同类内的对象的距离远，每个类以该类所有数据的"均值"作为其中心点。

17.2.1 算法简介

算法描述如下。

①在给定的 n 个对象中，随机选取 K 个对象作为每个类的中心（初始均值），或者通过其他方式指定 K 个中心。

②对于全部 n 个对象，分别计算与 K 个中心的距离，将对象指派到最接近的类。

③更新每个类的新均值，得出 K 个新的中心。

④根据 K 个新的中心，重复第②步和第③步，直至满足收敛准则。

下面通过一个例子使读者更直观地理解上述算法。我们生成均匀分布的二维数据，每个维度的取值范围都是[0, 100]，生成 100 亿个这样的数据点。聚类的个数设为 9，我们可以想象到最佳的聚类状态：将整个区域按"井"字等分成9份。

下面通过 K-Means 聚类算法实际计算一下，这里假设最坏的初始点情况为9个点重合且位于区域的边缘，如图 17-2 的左上图所示。随后展现了第5次迭代和第10次迭代后的情况，可以发现，重合的初始中心点被展开了；第30次迭代时，中心点已在整个区域散开了；第50次迭代时，这些中心点按3×3的位置排列；第100次迭代时，中心点已经排列得非常整齐了。

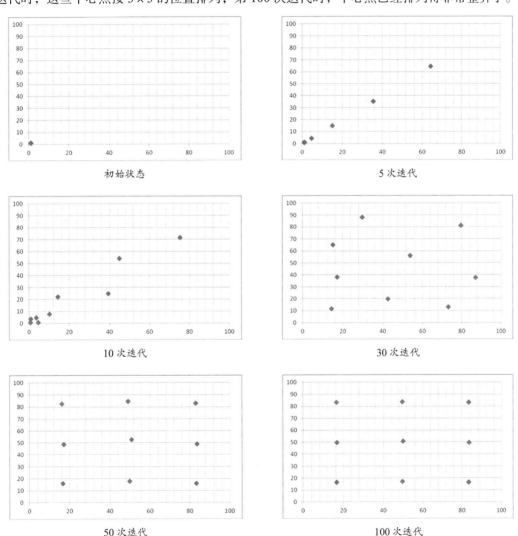

图 17-2

17.2.2　K-Means实例

我们尝试使用K-Means算法对数据集Iris进行自动聚类。由于聚类组件支持的数据为向量格式，我们需要将原始数据集 Iris 中各特征列合并为向量。使用向量装配组件 VectorAssemblerBatchOp，将 Iris 数据集的各特征列作为输入列，输出为一个向量列，并把转换结果保存为 AK 格式文件。具体代码如下。

```
new CsvSourceBatchOp()
    .setFilePath(DATA_DIR + ORIGIN_FILE)
    .setSchemaStr(SCHEMA_STRING)
    .link(
        new VectorAssemblerBatchOp()
            .setSelectedCols(FEATURE_COL_NAMES)
            .setOutputCol(VECTOR_COL_NAME)
            .setReservedCols(LABEL_COL_NAME)
    )
    .link(
        new AkSinkBatchOp().setFilePath(DATA_DIR + VECTOR_FILE)
    );
```

使用 AkSourceBatchOp 读取前面产生的向量数据文件，并输出 5 条数据，具体代码如下。

```
AkSourceBatchOp source = new AkSourceBatchOp().setFilePath(DATA_DIR + VECTOR_FILE);

source.lazyPrint(5);
```

输出数据如下，第 2 列就是向量序列化字符串的数据，各个数值间使用空格分隔。

```
category|vec
--------|---
Iris-versicolor|5.0 2.0 3.5 1.0
Iris-versicolor|5.9 3.0 4.2 1.5
Iris-versicolor|6.0 2.2 4.0 1.0
Iris-versicolor|6.1 2.9 4.7 1.4
Iris-versicolor|5.6 2.9 3.6 1.3
```

下面使用 K-Means 组件进行聚类实验，需要用到 K-Means 训练组件 KMeansTrainBatchOp，设置聚类个数为 2，并设置向量列名称；还用到了 K-Means 预测组件 KMeansPredictBatchOp，需要设置聚类预测结果数据列的名称。数据源连接训练组件，训练组件的输出为 K-Means 模型；再用预测组件接收来自 K-Means 模型和数据源的连接，预测组件的输出即为聚类结果。具体代码如下。

```
KMeansTrainBatchOp kmeans_model = new KMeansTrainBatchOp()
    .setK(2)
    .setVectorCol(VECTOR_COL_NAME);
```

```
KMeansPredictBatchOp kmeans_pred = new KMeansPredictBatchOp()
    .setPredictionCol(PREDICTION_COL_NAME);

source.link(kmeans_model);
kmeans_pred.linkFrom(kmeans_model, source);

kmeans_model.lazyPrintModelInfo();

kmeans_pred.lazyPrint(5);
```

模型信息显示如下。

```
------------------------------ KMeansInfo ------------------------------
KMeans clustering with 2 clusters. Clustering on 150 samples of 4 dimension based on EUCLIDEAN.
========================== ClusterCenters ==========================
{0: [6.301, 2.8866, 4.9588, 1.6959],
 1: [5.0057, 3.3604, 1.5623, 0.2887]}
```

可以看到聚类个数、样本总数及样本中向量的维度,以及采用的是何种距离。下面列出了各个聚类簇中心的向量。

聚类输出结果如下。

```
category|vec|cluster_id
--------|---|----------
Iris-versicolor|5.0 2.0 3.5 1.0|1
Iris-versicolor|5.9 3.0 4.2 1.5|1
Iris-versicolor|6.0 2.2 4.0 1.0|1
Iris-versicolor|6.1 2.9 4.7 1.4|1
Iris-versicolor|5.6 2.9 3.6 1.3|1
```

cluster_id 就是聚类结果预测列,为从 0 开始的聚类索引值。评估聚类结果的好坏不像评估分类问题那样,直接与标签值对比是否一致即可让我们对它有一个大致的印象。在 17.1 节中有详细介绍,其中一部分指标(譬如 VRC、DB、SilhouetteCoefficient 等)是通过样本的向量值和聚类情况计算出来的,还有的指标(譬如 ARI、NMI、Purity 等)是通过样本的标签值与聚类情况计算出来的。当然,如果样本没有标签值,就无法计算这些指标,它们也不会在评估结果中显示。

下面我们使用聚类评估组件 EvalClusterBatchOp 对前面的 K-Means 聚类结果进行评估。

```
kmeans_pred
    .link(
        new EvalClusterBatchOp()
            .setVectorCol(VECTOR_COL_NAME)
            .setLabelCol(LABEL_COL_NAME)
            .setPredictionCol(PREDICTION_COL_NAME)
            .lazyPrintMetrics("KMeans EUCLIDEAN")
    );
```

输出评估结果如下。

```
KMeans EUCLIDEAN
------------------------------- Metrics: -------------------------------
k:2
VRC:513.3038    DB:0.4048    SilhouetteCoefficient:0.8502
ARI:0.5399 NMI:0.6565 Purity:0.6667

| Label\Cluster|  1|  0|
|---------------|---|---|
|   Iris-virginica|  0| 50|
|Iris-versicolor|  3| 47|
|     Iris-setosa| 50|  0|
```

其中显示了一些常用的指标和一个交叉表，对样本的标签值和聚类情况进行了细分。可以看出，标签为 Iris-virginica 的样本都在第 0 个聚类簇；标签为 Iris-setosa 的样本都在第 1 个聚类簇；标签为 Iris-versicolor 的样本在 2 个聚类簇中都有分布，主要分布在第 0 个聚类簇。

下面我们将着重分析 Iris-versicolor 标签的样本，看看其属于不同聚类簇的样本的特点。使用 SQL OrderBy 对样本数据进行排序，先是按聚类结果排序，然后按标签值排序；将输出范围设定为 200，因为 Iris 的总样本数为 150，因此超出 Iris 的总样本数会全部排序输出；将排序是否为升序的参数设为 false，表示使用降序。具体代码如下。

```
kmeans_pred
    .orderBy(PREDICTION_COL_NAME + ", " + LABEL_COL_NAME, 200, false)
    .lazyPrint(-1, "all data");
```

运行结果如下，这里省略了前后端的样本，主要展示标签为 Iris-versicolor，且在第 0 个聚类簇的 3 个样本。

```
... ...
Iris-versicolor|5.7 2.8 4.1 1.3|0
Iris-versicolor|6.0 2.7 5.1 1.6|0
Iris-versicolor|6.1 3.0 4.6 1.4|0
Iris-versicolor|5.8 2.7 4.1 1.0|0
Iris-versicolor|6.6 3.0 4.4 1.4|0
Iris-versicolor|5.5 2.3 4.0 1.3|0
Iris-versicolor|6.2 2.9 4.3 1.3|0
Iris-versicolor|5.5 2.4 3.7 1.0|0
Iris-versicolor|5.5 2.5 4.0 1.3|0
Iris-versicolor|5.1 2.5 3.0 1.1|1
Iris-versicolor|4.9 2.4 3.3 1.0|1
Iris-versicolor|5.0 2.3 3.3 1.0|1
Iris-setosa|4.9 3.0 1.4 0.2|1
Iris-setosa|4.9 3.1 1.5 0.1|1
Iris-setosa|5.1 3.5 1.4 0.3|1
```

```
Iris-setosa|5.0 3.0 1.6 0.2|1
Iris-setosa|5.5 4.2 1.4 0.2|1
Iris-setosa|4.5 2.3 1.3 0.3|1
Iris-setosa|5.0 3.3 1.4 0.2|1
Iris-setosa|4.6 3.1 1.5 0.2|1
... ...
```

可以看到，标签为 Iris-versicolor 且属于第 1 个聚类簇的 3 个样本，与同为 Iris-versicolor 标签的其他样本相比，在"尺寸"上相对较小，在各项指标上较同类小，这样从"距离"的角度看，它和 Iris setosa 类的花更近些。对比下面标签为 Iris-setosa 的数据，会发现其向量值的最后一项都小于 0.3，但在计算欧氏距离时，我们关注的是两个样本间相对的数值。我们是否能换一个度量方式进行聚类呢？

我们可以选择 K-Means 算法使用余弦度量。前面讲了如何使用 K-Means 的训练和预测组件，这里我们使用 Pipeline 的方式，直接使用 PipelineStage K-Means，可以通过 fit 和 transform 方法简化训练和预测流程，使用 Lazy 方式输出模型信息，并连接聚类评估组件。具体代码如下。

```
new KMeans()
    .setK(2)
    .setDistanceType(DistanceType.COSINE)
    .setVectorCol(VECTOR_COL_NAME)
    .setPredictionCol(PREDICTION_COL_NAME)
    .enableLazyPrintModelInfo()
    .fit(source)
    .transform(source)
    .link(
        new EvalClusterBatchOp()
            .setVectorCol(VECTOR_COL_NAME)
            .setPredictionCol(PREDICTION_COL_NAME)
            .setLabelCol(LABEL_COL_NAME)
            .lazyPrintMetrics("KMeans COSINE")
    );
```

输出模型信息如下，聚类个数为 2，使用 COSINE（余弦）度量。

```
------------------------------ KMeansInfo ------------------------------
KMeans clustering with 2 clusters. Clustering on 150 samples of 4 dimension based on COSINE.
========================== ClusterCenters ==========================
{0: [0.8027, 0.5467, 0.2352, 0.0389],
 1: [0.7282, 0.3349, 0.5662, 0.1924]}
```

评估结果如下。

```
------------------------------ Metrics: ------------------------------
k:2
VRC:501.9249    DB:0.3836   SilhouetteCoefficient:0.8462
ARI:0.5681 NMI:0.7337 Purity:0.6667
```

```
| Label\Cluster|  1|  0|
|--------------|---|---|
| Iris-virginica| 50|  0|
|Iris-versicolor| 50|  0|
|    Iris-setosa|  0| 50|
```

可以看到，新的聚类方式可以将 Iris-Setosa 标签的样本与其他两种标签的样本自动聚合为两类，也就是完整地分隔开了。

17.3 高斯混合模型

高斯混合模型（Gaussian Mixed Model，GMM）用多维高斯概率模型来描述每个类，将聚类问题看作多个高斯分布函数的线性组合。

17.3.1 算法介绍

高斯混合模型用多维高斯概率模型来描述每个类，通过计算概率的方式来确定数据点属于哪个类。通过期望最大化（Expectation Maximization，EM）方法进行迭代，求得模型参数的估计值。下面首先介绍多维高斯分布，然后介绍如何使用 EM 方法求解。

所用的概率模型为多维高斯分布，将每条记录的属性集作为一个向量，每个特征作为向量的一个分量，记作 $\boldsymbol{x} = \begin{pmatrix} x_1 \\ \vdots \\ x_m \end{pmatrix}$，各特征列的均值向量为 $\boldsymbol{u} = \begin{pmatrix} \mu_1 \\ \vdots \\ \mu_m \end{pmatrix}$，协方差矩阵为 $\boldsymbol{C} = \begin{pmatrix} c_{11} & \cdots & c_{1m} \\ \vdots & \ddots & \vdots \\ c_{m1} & \cdots & c_{mm} \end{pmatrix}$，则多维高斯分布的概率密度函数定义为

$$N(\boldsymbol{x}\,;\,\boldsymbol{u},\boldsymbol{C}) = \frac{1}{(2\pi)^{m/2}\,|\boldsymbol{C}|^{1/2}} \cdot \exp\left\{-\frac{1}{2}(\boldsymbol{x}-\boldsymbol{u})^{\mathrm{T}}\boldsymbol{C}^{-1}(\boldsymbol{x}-\boldsymbol{u})\right\}$$

观察这个公式，我们可以很容易地理解该方法和 K-Means 算法的关联。从协方差矩阵 $\boldsymbol{C}$ 入手，假设各个特征之间是独立的、各个特征的方差值相等。

由第一个假设可以知道，对于两个不同的特征 i,j，有 $c_{ij} = 0$，所以协方差矩阵 $\boldsymbol{C}$ 只有对角线上的元素不为 0。再由第二个假设，设相等的方差值为 σ^2，则

$$\boldsymbol{C} = \begin{pmatrix} \sigma^2 & \cdots & 0 \\ \vdots & \ddots & \vdots \\ 0 & \cdots & \sigma^2 \end{pmatrix}$$

其逆矩阵为

$$C^{-1} = \begin{pmatrix} \frac{1}{\sigma^2} & \cdots & 0 \\ \vdots & \ddots & \vdots \\ 0 & \cdots & \frac{1}{\sigma^2} \end{pmatrix}$$

m维矩阵C的行列式$|C|$的值为

$$|C| = (\sigma^2)^m$$

则相应的概率密度函数为

$$\begin{aligned} N(\boldsymbol{x};\boldsymbol{u},\boldsymbol{C}) &= \frac{1}{(2\pi)^{m/2}|\boldsymbol{C}|^{1/2}} \cdot \exp\left\{-\frac{1}{2}(\boldsymbol{x}-\boldsymbol{u})^{\mathrm{T}} \boldsymbol{C}^{-1}(\boldsymbol{x}-\boldsymbol{u})\right\} \\ &= \frac{1}{(2\pi)^{m/2}(\sigma^2)^{m/2}} \cdot \exp\left\{-\frac{1}{2}(\boldsymbol{x}-\boldsymbol{u})^{\mathrm{T}} \begin{pmatrix} \frac{1}{\sigma^2} & \cdots & 0 \\ \vdots & \ddots & \vdots \\ 0 & \cdots & \frac{1}{\sigma^2} \end{pmatrix}(\boldsymbol{x}-\boldsymbol{u})\right\} \\ &= \frac{1}{(2\pi\sigma^2)^{m/2}} \cdot \exp\left\{-\frac{1}{2\sigma^2}\sum_{i=1}^{m}(x_i-\mu_i)^2\right\} \end{aligned}$$

$\sum_{i=1}^{m}(x_i - \mu_i)^2$即为 K-Means 算法所使用的向量$\boldsymbol{x}$和$\boldsymbol{u}$的欧氏距离的平方，而由上面的公式可知，该平方与概率密度的值呈负指数关系，即$\boldsymbol{x}$与中心$\boldsymbol{u}$的欧氏距离越小，所对应的概率密度的值越大。

下面详细讲解一下 GMM 算法的过程，我们可以了解到 K-Means 和 GMM 的更多相似之处。对于 GMM 聚类算法，假设在K个聚类的情况下，每个数据$\boldsymbol{x}$都满足如下的概率分布。

$$p(\boldsymbol{x}) = \sum_{k=1}^{K} \pi_k N(\boldsymbol{x};\boldsymbol{u}_k,\boldsymbol{C}_k)$$

对数似然函数为

$$\begin{aligned} \log\left(\prod_{i=1}^{N} p(\boldsymbol{x}_i)\right) &= \log\left(\prod_{i=1}^{N}\sum_{k=1}^{K} \pi_k N(\boldsymbol{x}_k;\boldsymbol{u}_k,\boldsymbol{C}_k)\right) \\ &= \sum_{i=1}^{N} \log\left(\sum_{k=1}^{K} \pi_k N(\boldsymbol{x}_k;\boldsymbol{u}_k,\boldsymbol{C}_k)\right) \end{aligned}$$

对数据进行训练的目的就是得到使对数似然函数达到最大值的参数π_k, u_k, C_k。
算法过程如下。

（1）用随机函数初始化 K 个高斯分布的参数，同时保证

$$\sum_{k=1}^{K} \pi_k = 1$$

（2）（E 步骤）对每个训练数据 x，计算 K 个高斯函数 $\pi_k \cdot N(x; u_k, C_k)$ 中概率的大小，把概率最大的作为其分类。

（3）（M 步骤）用最大似然估计，在 E 步骤计算的分类的基础上，估计相应的参数值，具体公式如下。

$$u_k = \frac{1}{N_k} \sum_{x_k} x_k$$

$$C_k = \frac{1}{N_k} \sum_{x_k} (x_k - u_k)^{\mathrm{T}} (x_k - u_k)$$

$$\pi_k = \frac{N_k}{N}$$

（4）判断是否收敛，若收敛，则返回当前参数；否则返回第 2 步继续计算。

在 M 步骤中，新中心点 u_k 的计算公式与 K-Means 相同，但在 E 步骤判断分类时，还要考虑每个分类占总体的比例 π_k，不像 K-Means 只考虑距离。

17.3.2　GMM实例

下面我们利用 GMM 算法对 17.2 节中的数据进行聚类。使用组件设置向量列名称、设置聚类结果列名称，定义聚类个数为 2，并选择使用 Lazy 方式输出模型信息。具体代码如下。

```
new GaussianMixture()
  .setK(2)
  .setVectorCol(VECTOR_COL_NAME)
  .setPredictionCol(PREDICTION_COL_NAME)
  .enableLazyPrintModelInfo()
  .fit(source)
  .transform(source)
  .link(
    new EvalClusterBatchOp()
      .setVectorCol(VECTOR_COL_NAME)
      .setPredictionCol(PREDICTION_COL_NAME)
```

```
        .setLabelCol(LABEL_COL_NAME)
        .lazyPrintMetrics("GaussianMixture 2")
);
```

显示模型信息如下,包括各聚类中心点的向量及各聚类的协方差矩阵。

```
------------------------- GMMInfo -------------------------
GMM clustering with 2 clusters.
========================= ClusterCenters =========================
{0: [6.262, 2.872, 4.906, 1.676],
 1: [5.006, 3.418, 1.464, 0.244]}
-------------------- CovarianceMatrix of each clusters --------------------
{0: mat[4,4]:
 [[0.435, 0.1209, 0.4489, 0.1655],
  [0.1209, 0.1096, 0.1414, 0.0792],
  [0.4489, 0.1414, 0.6749, 0.2859],
  [0.1655, 0.0792, 0.2859, 0.1786]],
 1: mat[4,4]:
 [[0.1218, 0.0983, 0.0158, 0.0103],
  [0.0983, 0.1423, 0.0114, 0.0112],
  [0.0158, 0.0114, 0.0295, 0.0056],
  [0.0103, 0.0112, 0.0056, 0.0113]]}
```

各原始数据的聚类评估如下。

```
-------------------------- Metrics: --------------------------
k:2
VRC:501.9249    DB:0.3836    SilhouetteCoefficient:0.8462
ARI:0.5681 NMI:0.7337 Purity:0.6667

| Label\Cluster|  1|  0|
|--------------|---|---|
|  Iris-virginica|  0| 50|
|Iris-versicolor|  0| 50|
|     Iris-setosa| 50|  0|
```

显然,Iris setosa 亚属的样本与其他两个亚属的样本被自动聚合为两类,即被完整地分隔开了。其结果要优于使用欧氏距离的 K-Means 算法,与使用夹角余弦作为距离的 K-Means 算法的聚类结果相同。

17.4 二分K-Means聚类

二分 K 均值(Bisecting K-Means)聚类算法是 K-Means 聚类算法的一个变体,采用"自顶向下"方法的层次聚类。首先将全部数据点作为一个簇,使用 K-Means 算法将其分为两个簇,然后在所有簇中选择聚类指标误差平方和最大的簇，继续使用 K-Means 算法将其分为两个簇，持续进行下去，每次一分二的操作都会增加一个簇，直到达到用户的聚类数目为止。

二分 K-Means 聚类组件的使用方式和参数设置与 K-Means 聚类组件完全相同。如下代码所示，使用组件 BisectingKMeans，设置聚类个数为 3。

```
new BisectingKMeans()
    .setK(3)
    .setVectorCol(VECTOR_COL_NAME)
    .setPredictionCol(PREDICTION_COL_NAME)
    .enableLazyPrintModelInfo("BiSecting KMeans EUCLIDEAN")
    .fit(source)
    .transform(source)
    .link(
        new EvalClusterBatchOp()
            .setVectorCol(VECTOR_COL_NAME)
            .setPredictionCol(PREDICTION_COL_NAME)
            .setLabelCol(LABEL_COL_NAME)
            .lazyPrintMetrics("Bisecting KMeans EUCLIDEAN")
    );
BatchOperator.execute();
```

整理运行结果，并与 K-Means 二分类的结果进行对比，如表 17-1 所示。

表 17-1　聚类中心对比

Bisecting K-Means 3 个聚类簇	K-Means 2 个聚类簇
聚类簇中心： {0: [5.0057, 3.3604, 1.5623, 0.2887], 　1: [5.9475, 2.7661, 4.4542, 1.4542], 　2: [6.85, 3.0737, 5.7421, 2.0711]}	聚类簇中心： {0: [6.301, 2.8866, 4.9588, 1.6959], 　1: [5.0057, 3.3604, 1.5623, 0.2887]}
VRC:520.6334　DB:0.6755 SilhouetteCoefficient:0.7232 ARI:0.6809　　NMI:0.6914　　Purity:0.8733 |Label\\Cluster| 2| 1| 0| |--------------|---|---|---| |Iris-virginica| 36| 14| 0| |Iris-versicolor| 2| 45| 3| | Iris-setosa| 0| 0| 50|	VRC:513.3038　DB:0.4048 SilhouetteCoefficient:0.8502 ARI:0.5399　　NMI:0.6565　　Purity:0.6667 |Label\\Cluster| 1| 0| |--------------|---|---| |Iris-virginica| 0| 50| |Iris-versicolor| 3| 47| | Iris-setosa| 50| 0|

可以得到如下信息。

- 3 个聚类簇的第 0 簇中心为 [5.0057, 3.3604, 1.5623, 0.2887]，与 2 个聚类簇的第 1 簇中心的位置相同。

- 左边的第 0 簇与右边的第 1 簇都是由全部标签为 Iris-setosa 的样本和 3 个 Iris-versicolor 标签的样本构成的。
- 左边的第 1、2 簇是将右边的第 0 簇进行一次二分聚类产生的。

K-Means 算法在使用 COSINE 度量时，实现了完美的二分聚类。这里我们使用二分 K-Means 聚类算法，同样也选择 COSINE 度量，看看分类效果如何。相比前面的示例，只需设置距离类型 DistanceType=COSINE。具体代码如下。

```
new BisectingKMeans()
    .setDistanceType(DistanceType.COSINE)
    .setK(3)
    .setVectorCol(VECTOR_COL_NAME)
    .setPredictionCol(PREDICTION_COL_NAME)
    .enableLazyPrintModelInfo("BiSecting KMeans COSINE")
    .fit(source)
    .transform(source)
    .link(
        new EvalClusterBatchOp()
            .setDistanceType("COSINE")
            .setVectorCol(VECTOR_COL_NAME)
            .setPredictionCol(PREDICTION_COL_NAME)
            .setLabelCol(LABEL_COL_NAME)
            .lazyPrintMetrics("Bisecting KMeans COSINE")
    );
BatchOperator.execute();
```

整理运行结果，并与 K-Means 二分类的结果进行对比，如表 17-2 所示。

表 17-2 COSINE 度量的聚类中心对比

Bisecting K-Means 3 个聚类簇	K-Means 2 个聚类簇
聚类簇中心： {0: [0.8027, 0.5467, 0.2352, 0.0389], 1: [0.7536, 0.3495, 0.532, 0.1641], 2: [0.7058, 0.3222, 0.5931, 0.2152]}	聚类簇中心： {0: [0.8027, 0.5467, 0.2352, 0.0389], 1: [0.7282, 0.3349, 0.5662, 0.1924]}
VRC:21836.8509 DB:0.3085 SilhouetteCoefficient:0.8488 ARI:0.9039 NMI:0.8997 Purity:0.9667 | Label\Cluster| 2| 1| 0| |---------------|---|---|---| | Iris-virginica| 50| 0| 0| |Iris-versicolor| 5| 45| 0| | Iris-setosa| 0| 0| 50|	VRC:501.9249 DB:0.3836 SilhouetteCoefficient:0.8462 ARI:0.5681 NMI:0.7337 Purity:0.6667 | Label\Cluster| 1| 0| |---------------|---|---| | Iris-virginica| 50| 0| |Iris-versicolor| 50| 0| | Iris-setosa| 0| 50|

可以得到如下信息。

- 3 个聚类簇的第 0 簇中心为[0.8027, 0.5467, 0.2352, 0.0389]，与 2 个聚类簇的第 0 簇中心的位置相同。
- 3 个聚类簇的各项聚类指标都优于 2 个聚类簇。DB 指标越小越好，VRC、Silhouette-Coefficient、ARI、NMI、Purity 都是越大越好。
- 左右两边的第 0 簇都是完全由标签为 Iris-setosa 的样本构成的，左边的第 1、2 簇是将右边的第 1 簇进行一次二分聚类产生的。

有了 17.4 节的基础，下面我们来挑战难度更高的聚类数为 3 的情形。原始的样本数据本来就属于 3 个分类，从散点图可以看出，Iris virginica 和 Iris versicolor 分类的样本点没有清晰的分界线。如何能较好地聚成 3 簇，还尽量与其原来的分类属性一致呢？

我们先尝试 17.4 节的最佳参数组合，在余弦距离下，选择 2 个特征列与选择全部特征列的实验组件，对于选择两个特征 petal_width 和 petal_length 的情况，散点图中左下角聚集的点被分到 3 个聚类中。

下面尝试在欧氏距离下，分别选择 2 个特征列与选择全部特征列的实验组件，从散点图上不容易看出区别，通过对比原始分类类别与聚类结果的统计数据可知，选择 2 个特征列的情况会更好一些。

综合起来进行对比，我们虽然没有找到一种参数设置，能够使原始的分类类别与聚类结果一致，但是可以使它们之间的差异变小。

17.5 基于经纬度的聚类

我们将使用美国 50 个州首府的经纬度数据，通过 Alink 经纬度聚类组件 GeoKMeans 进行聚类试验。

各州首府的经纬度数据集显示如下。

```
State|Region|Division|longitude|latitude
-----|------|--------|---------|--------
Alabama|South|East South Central|-86.7509|32.5901
Alaska|West|Pacific|-127.2500|49.2500
Arizona|West|Mountain|-111.6250|34.2192
Arkansas|South|West South Central|-92.2992|34.7336
California|West|Pacific|-119.7730|36.5341
```

第 1 列是州名，第 2 列和第 3 列分别是对 50 个州的两种划分方式，Region 划分为 4 个部分，Division 划分为 9 个部分，第 4 列为州首府所在的经度值，第 5 列为州首府所在的纬度值。

为了进一步了解 Region 与 Division 的关系，我们可以使用 SQL groupBy 操作。

```
source
    .groupBy("Region, Division", "Region, Division, COUNT(*) AS numStates")
    .orderBy("Region, Division", 100)
    .lazyPrint(-1);
```

运行结果如下。

```
Region|Division|numStates
------|--------|---------
North Central|East North Central|5
North Central|West North Central|7
Northeast|Middle Atlantic|3
Northeast|New England|6
South|East South Central|4
South|South Atlantic|8
South|West South Central|4
West|Mountain|8
West|Pacific|5
```

Division 的 9 个部分是在 Region 的 4 个部分基础上进行的，除了 Region 的 South 被分为 3 个部分，其他的 Region 都被分为了 2 个部分。

随后，使用经纬度聚类组件 GeoKMeans 分别针对 2 个和 4 个聚类的情况进行实验。具体代码如下。

```
for (int nClusters : new int[] {2, 4}) {
    BatchOperator <?> pred = new GeoKMeans()
        .setLongitudeCol("longitude")
        .setLatitudeCol("latitude")
        .setPredictionCol(PREDICTION_COL_NAME)
        .setK(nClusters)
        .fit(source)
        .transform(source);

    pred.link(
        new EvalClusterBatchOp()
            .setPredictionCol(PREDICTION_COL_NAME)
            .setLabelCol("Region")
            .lazyPrintMetrics(nClusters + " with Region")
    );
    pred.link(
```

```
        new EvalClusterBatchOp()
            .setPredictionCol(PREDICTION_COL_NAME)
            .setLabelCol("Division")
            .lazyPrintMetrics(nClusters + " with Division")
    );
    BatchOperator.execute();
}
```

经纬度聚类组件 GeoKMeans 的输入数据列参数的设置上与聚类组件不同，不是通过一个统一的 Vector 输入，而是需要指定具体的经度列 LongitudeCol 和纬度列 LatitudeCol。在评估时，只需输入聚类结果列 PredictionCol 和标签列 LabelCol，这里分别针对 Region 和 Division 进行了评估。

2 个聚类的评估结果对比如表 17-3 所示。

表 17-3　2 个聚类的评估结果对比

2 with Region	2 with Division
ARI:0.2698　　NMI:0.3849　　Purity:0.54 |Label\Cluster| 1| 0| |--------------|---|---| | West| 0| 13| | South| 14| 2| | Northeast| 9| 0| |North Central | 8| 4|	ARI:0.1428　　NMI:0.3647　　Purity:0.32 | Label\Cluster| 1| 0| |------------------|---|---| |West South Central| 2| 2| |West North Central| 3| 4| | South Atlantic| 8| 0| | Pacific| 0| 5| | New England| 6| 0| | Mountain| 0| 8| | Middle Atlantic| 3| 0| |East South Central| 4| 0| |East North Central| 5| 0|

在左边 Region 的 4 个区域中，West 都属于聚类簇 0，Northeast 都属于聚类簇 1；South 和 North Central 大部分属于聚类簇 1，并分别有 2 个和 4 个州属于聚类簇 0。对应右边 Division 的情况，West South Central 和 West North Central 中分别有 2 个和 4 个州属于聚类簇 0。

4 个聚类的评估结果对比如表 17-4 所示。

表 17-4 4 个聚类的评估结果对比

4 with Region	4 with Division
ARI:0.5149 NMI:0.6297 Purity:0.78	ARI:0.3221 NMI:0.5862 Purity:0.46
<pre>|Label\Cluster| 3| 2| 1| 0|
\|-------------\|---\|---\|---\|---\|
\| West\| 0\| 1\| 12\| 0\|
\| South\| 11\| 0\| 0\| 5\|
\| Northeast\| 0\| 0\| 0\| 9\|
\|North Central\| 3\| 7\| 0\| 2\|</pre> | <pre>\| Label\Cluster\| 3\| 2\| 1\| 0\|
\|-----------------\|---\|---\|---\|---\|
\|West South Central\| 4\| 0\| 0\| 0\|
\|West North Central\| 1\| 6\| 0\| 0\|
\| South Atlantic\| 3\| 0\| 0\| 5\|
\| Pacific\| 0\| 0\| 5\| 0\|
\| New England\| 0\| 0\| 0\| 6\|
\| Mountain\| 0\| 1\| 7\| 0\|
\| Middle Atlantic\| 0\| 0\| 0\| 3\|
\|East South Central\| 4\| 0\| 0\| 0\|
\|East North Central\| 2\| 1\| 0\| 2\|</pre> |

在左边 Region 的 4 个区域中，Northeast 都属于聚类簇 0，而聚类簇 0 也是以 Northeast 为主的；聚类簇 1 都是以 West 的各州构成的；聚类簇 3 以 South 为主，聚类簇 2 以 North Central 为主。可以看到，聚类数为 4 的评估指标 ARI 都要优于聚类数为 2 的 ARI。

18

批式与流式聚类

第 17 章侧重于介绍聚类算法，使用的是小数据集。本章将使用手写数字数据集，模型训练时间较长，各个算法在计算时间上会有明显的差异，输入的向量是稠密格式还是稀疏格式对计算量也有显著的影响。本章还会介绍与流式数据相关的聚类功能，一个是通过批式训练的聚类模型预测流式数据，另一个是直接针对流式数据进行聚类模型训练和预测。

18.1 稠密向量与稀疏向量

第 13 章介绍了如何从原始数据中解析稠密向量数据和稀疏向量数据，并保存为 AK 格式数据文件。我们分别选取其稠密向量训练数据和稀疏向量训练数据作为聚类实验的稠密向量数据源与稀疏向量数据源，并定义一个停表对象，用来对各次试验进行计时。具体代码如下。

```
AkSourceBatchOp dense_source = new AkSourceBatchOp().setFilePath(DATA_DIR + DENSE_TRAIN_FILE);
AkSourceBatchOp sparse_source = new AkSourceBatchOp().setFilePath(DATA_DIR + SPARSE_TRAIN_FILE);
Stopwatch sw = new Stopwatch();
```

由于各个试验流程相似，而且都要分别使用稠密向量和稀疏向量进行试验，使用 Pipeline 会使代码更简洁。定义 Pipeline 和相关的标识字符串作为一个二元组，以及使用默认距离的 KMeans、余弦距离的 KMeans 和 BisectingKMeans。具体代码如下。

```
ArrayList <Tuple2 <String, Pipeline>> pipelineList = new ArrayList <>();
pipelineList.add(new Tuple2 <>("KMeans EUCLIDEAN",
   new Pipeline()
      .add(
         new KMeans()
```

```
                    .setK(10)
                    .setVectorCol(VECTOR_COL_NAME)
                    .setPredictionCol(PREDICTION_COL_NAME)
            )
));
pipelineList.add(new Tuple2 <>("KMeans COSINE",
    new Pipeline()
        .add(
            new KMeans()
                    .setDistanceType(DistanceType.COSINE)
                    .setK(10)
                    .setVectorCol(VECTOR_COL_NAME)
                    .setPredictionCol(PREDICTION_COL_NAME)
            )
));
pipelineList.add(new Tuple2 <>("BisectingKMeans",
    new Pipeline()
        .add(
            new BisectingKMeans()
                    .setK(10)
                    .setVectorCol(VECTOR_COL_NAME)
                    .setPredictionCol(PREDICTION_COL_NAME)
            )
));
```

整体的代码如下。

```
For (Tuple2 <String, Pipeline> pipelineTuple2 : pipelineList) {
    sw.reset();
    sw.start();
    pipelineTuple2.f1
        .fit(dense_source)
        .transform(dense_source)
        .link(
            new EvalClusterBatchOp()
                .setVectorCol(VECTOR_COL_NAME)
                .setPredictionCol(PREDICTION_COL_NAME)
                .setLabelCol(LABEL_COL_NAME)
                .lazyPrintMetrics(pipelineTuple2.f0 + " DENSE")
        );
    BatchOperator.execute();
    sw.stop();
    System.out.println(sw.getElapsedTimeSpan());

    sw.reset();
    sw.start();
    pipelineTuple2.f1
        .fit(sparse_source)
        .transform(sparse_source)
        .link(
            new EvalClusterBatchOp()
                .setVectorCol(VECTOR_COL_NAME)
                .setPredictionCol(PREDICTION_COL_NAME)
```

```
        .setLabelCol(LABEL_COL_NAME)
        .lazyPrintMetrics(pipelineTuple2.f0 + " SPARSE")
    );
    BatchOperator.execute();
    sw.stop();
    System.out.println(sw.getElapsedTimeSpan());
}
```

针对每个 Pipeline 及其标识字符串构成的二元组 pipelineTuple2，选择其 Pipeline 分量 pipelineTuple2.f1，先对稠密向量数据源进行训练、预测及评估，然后再对稀疏向量数据源进行训练、预测及评估。在输出评估结果信息时，加上标识字符串 pipelineTuple2.f0，便于我们查看结果。

运行结果汇总如表 18-1 所示。

表 18-1 稠密向量与稀疏向量聚类对比

距离	稠密向量	稀疏向量
KMeans（EUCLIDEAN）	VRC:2278.5669 DB:2.8836 SilhouetteCoefficient:0.0931 ARI:0.4048 NMI:0.5184 Purity:0.618 34 seconds 763.0 milliseconds	VRC:2236.2382 DB:2.911 SilhouetteCoefficient:0.1182 ARI:0.4063 NMI:0.5045 Purity:0.6121 15 seconds 658.0 milliseconds
KMeans（COSINE）	VRC:2221.7892 DB:2.8565 SilhouetteCoefficient:0.0763 ARI:0.4064 NMI:0.5395 Purity:0.613 24 seconds 567.0 milliseconds	VRC:2248.4341 DB:2.8205 SilhouetteCoefficient:0.0815 ARI:0.3962 NMI:0.5188 Purity:0.6075 12 seconds 75.0 milliseconds
BisectingKMeans	VRC:1984.6809 DB:3.2848 SilhouetteCoefficient:0.0836 ARI:0.226 NMI:0.3646 Purity:0.4626 47 seconds 263.0 milliseconds	VRC:1898.6387 DB:3.562 SilhouetteCoefficient:0.0736 ARI:0.2546 NMI:0.372 Purity:0.4601 25 seconds 227.0 milliseconds

从表中能明显看到，使用稠密向量的计算时间是使用稀疏向量的 2 倍。

18.2 使用聚类模型预测流式数据

本节的内容与 18.3 节是有关联的。本节演示如何使用已有的聚类模型预测流式的数据，在整个预测过程中，聚类模型是不变的；18.3 节会将已有的聚类模型作为初始模型，随着流式数据不断调整聚类模型，并基于新的聚类模型对数据进行聚类预测。

首先获取数据源，使用批式 AK 数据源组件 AkSourceBatchOp 和流式 AK 数据源组件 AkSourceStreamOp，分别以批和流的方式从 AK 格式数据文件中获取稀疏向量格式的数据源。

```
AkSourceBatchOp batch_source = new AkSourceBatchOp().setFilePath(DATA_DIR + SPARSE_TRAIN_FILE);
AkSourceStreamOp stream_source = new AkSourceStreamOp().setFilePath(DATA_DIR + SPARSE_TRAIN_FILE);
```

随后，我们构建一个初始模型。从批式数据源中采样 100 条样本，训练 K-Means 模型，并将模型保存到文件 INIT_MODEL_FILE 中。具体代码如下。

```
batch_source
    .sampleWithSize(100)
    .link(
        new KMeansTrainBatchOp()
            .setVectorCol(VECTOR_COL_NAME)
            .setK(10)
    )
    .link(
        new AkSinkBatchOp()
            .setFilePath(DATA_DIR + INIT_MODEL_FILE)
    );
```

基于这个聚类模型，我们可以对批式数据进行聚类预测。新建一个批式数据源读取聚类模型，然后使用 K-Means 批式预测组件，接入聚类模型和待预测的批式数据，将预测结果与聚类评估组件连接。相关代码如下。

```
AkSourceBatchOp init_model = new AkSourceBatchOp().setFilePath(DATA_DIR + INIT_MODEL_FILE);

new KMeansPredictBatchOp()
    .setPredictionCol(PREDICTION_COL_NAME)
    .linkFrom(init_model, batch_source)
    .link(
        new EvalClusterBatchOp()
            .setVectorCol(VECTOR_COL_NAME)
            .setPredictionCol(PREDICTION_COL_NAME)
            .setLabelCol(LABEL_COL_NAME)
            .lazyPrintMetrics("Batch Prediction")
    );
```

评估结果如下。

```
-------------------------------- Metrics: --------------------------------
k:10
VRC:1743.4614   DB:3.3561   SilhouetteCoefficient:0.0821
ARI:0.3064 NMI:0.407  Purity:0.5202
```

```
|Label\Cluster|    9|    8|    7|    6|    5|    4|    3|    2|    1|    0|
|-------------|-----|-----|-----|-----|-----|-----|-----|-----|-----|-----|
|            9| 1294| 1581|   74|  576|    1| 1689|  322|  187|    7|  218|
|            8|   34|   64|  378| 1665|  686|  361|   40| 1749|   48|  826|
|            7| 4698|  378|    0|   89|    9|  460|   67|  222|   11|  331|
|            6|    1|    2|  218|  122|    1| 1138| 3659|   45|   21|  711|
|            5|   13|   47| 1480| 1521|   17|  986|   42| 1094|    7|  214|
|            4|   86| 1080|    2|  350|    0| 2931| 1231|    3|    7|  152|
|            3|   73|   15| 2346| 2351|  594|   92|   15|  141|   96|  408|
|            2|  145|   30|   96|  126|  596|  554|  369|  180| 2730| 1132|
|            1|    5|    0|   10|   17|    3|   77|    4|    8|    1| 6617|
|            0|   17|  222|  402|  600|   28|  354|  403| 3614|  268|   15|
```

下面使用流式预测组件 KMeansPredictStreamOp 对流式数据进行预测，并把结果保存到数据文件中，便于后面使用批式聚类评估组件进行详细评估。将批式聚类模型数据源 init_model 作为 KMeansPredictStreamOp 构造函数的输入，预测结果写到 AK 格式数据文件中，如下代码所示。

```
stream_source
    .link(
        new KMeansPredictStreamOp(init_model)
            .setPredictionCol(PREDICTION_COL_NAME)
    )
    .link(
        new AkSinkStreamOp()
            .setFilePath(DATA_DIR + TEMP_STREAM_FILE)
            .setOverwriteSink(true)
    );
StreamOperator.execute();
```

接下来，使用批式数据源打开刚才流式预测保存的数据文件，然后使用批式聚类评估组件进行详细评估，具体代码如下。

```
new AkSourceBatchOp()
    .setFilePath(DATA_DIR + TEMP_STREAM_FILE)
    .link(
        new EvalClusterBatchOp()
            .setVectorCol(VECTOR_COL_NAME)
            .setPredictionCol(PREDICTION_COL_NAME)
            .setLabelCol(LABEL_COL_NAME)
            .lazyPrintMetrics("Stream Prediction")
    );
```

评估结果如下。

```
------------------------------- Metrics: -------------------------------
k:10
VRC:1743.4614   DB:3.3561   SilhouetteCoefficient:0.0821
ARI:0.3064 NMI:0.407 Purity:0.5202
```

```
|Label\Cluster|    9|    8|    7|    6|    5|    4|    3|    2|    1|    0|
|-------------|-----|-----|-----|-----|-----|-----|-----|-----|-----|-----|
|            9| 1294| 1581|   74|  576|    1| 1689|  322|  187|    7|  218|
|            8|   34|   64|  378| 1665|  686|  361|   40| 1749|   48|  826|
|            7| 4698|  378|    0|   89|    9|  460|   67|  222|   11|  331|
|            6|    1|    2|  218|  122|    1| 1138| 3659|   45|   21|  711|
|            5|   13|   47| 1480| 1521|   17|  986|   42| 1094|    7|  214|
|            4|   86| 1080|    2|  350|    0| 2931| 1231|    3|    7|  152|
|            3|   73|   15| 2346| 2351|  594|   92|   15|  141|   96|  408|
|            2|  145|   30|   96|  126|  596|  554|  369|  180| 2730| 1132|
|            1|    5|    0|   10|   17|    3|   77|    4|    8|    1| 6617|
|            0|   17|  222|  402|  600|   28|  354|  403| 3614|  268|   15|
```

与批式聚类预测结果的评估结果完全一致。

18.3 流式聚类

Alink 的流式聚类组件需要输入初始聚类模型，这样可以保证最初的流式数据也可以有比较好的预测结果。注意，流式聚类个数已经由初始模型决定了。流式聚类以一个时间窗口为周期，每个周期产生一个新的聚类模型，并用此新模型预测后续数据。随着新数据的不断流入，我们用来训练聚类模型的数据也有一个逐步退出的过程。这里使用半衰期（HalfLife）的概念，含义是经过几个周期后，数据量减半。

先获取流式数据源 stream_source 和初始模型 init_model，然后由初始模型 init_model 构建流式聚类组件 StreamingKMeansStreamOp，并将流式数据源 stream_source 连接到流式聚类组件。流式聚类组件需要设置时间窗口间隔参数 TimeInterval=1，设置半衰期参数 HalfLife=1，随后使用 SQL Select 方法选择数据，以便于查看。对于预测结果数据，一方面，我们使用采样方式输出部分结果，便于看到计算的过程和结果；另一方面，将预测结果导出到文件中，便于后续分析。具体代码如下。

```
AkSourceStreamOp stream_source
    = new AkSourceStreamOp().setFilePath(DATA_DIR + SPARSE_TRAIN_FILE);

AkSourceBatchOp init_model
    = new AkSourceBatchOp().setFilePath(DATA_DIR + INIT_MODEL_FILE);

StreamOperator<?> stream_pred = stream_source
    .link(
        new StreamingKMeansStreamOp(init_model)
            .setTimeInterval(1L)
            .setHalfLife(1)
```

```
        .setPredictionCol(PREDICTION_COL_NAME)
)
.select(PREDICTION_COL_NAME + ", " + LABEL_COL_NAME +", " + VECTOR_COL_NAME);

stream_pred.sample(0.001).print();

stream_pred
  .link(
    new AkSinkStreamOp()
        .setFilePath(DATA_DIR + TEMP_STREAM_FILE)
        .setOverwriteSink(true)
  );
StreamOperator.execute();
```

在计算过程中不断输出聚类预测数据，整体的预测结果也被保存到了文件中。我们再以批式数据的方式打开此文件，就可以使用上批式聚类评估组件了，具体代码如下。

```
new AkSourceBatchOp()
    .setFilePath(DATA_DIR + TEMP_STREAM_FILE)
    .link(
        new EvalClusterBatchOp()
            .setVectorCol(VECTOR_COL_NAME)
            .setPredictionCol(PREDICTION_COL_NAME)
            .setLabelCol(LABEL_COL_NAME)
            .lazyPrintMetrics("StreamingKMeans")
    );
BatchOperator.execute();
```

输出的结果如下。

```
-------------------------- Metrics: --------------------------
k:10
VRC:1970.4495   DB:3.1637   SilhouetteCoefficient:0.1009
ARI:0.3549 NMI:0.4549 Purity:0.5651
```

Label\Cluster	9	8	7	6	5	4	3	2	1	0
9	1624	1869	36	312	13	1746	87	56	9	197
8	87	93	314	1263	1440	705	40	983	59	867
7	4738	573	1	44	13	408	17	45	33	393
6	1	14	227	52	5	884	4078	70	35	552
5	63	87	1334	1602	21	1160	50	771	7	326
4	221	2015	1	110	1	2869	478	3	8	136
3	92	20	2373	2213	767	98	29	58	114	367
2	113	27	84	97	868	379	272	87	3135	896
1	8	0	12	18	13	50	5	1	8	6627
0	4	57	288	546	33	212	290	4419	66	8

对比前面直接使用初始模型预测的聚类评估结果，可以发现，各项指标都得到了提升，这也说明了流式聚类算法的有效性。

19 主成分分析

主成分分析（Principal Component Analysis，PCA）是一种多元统计分析方法，将多个变量通过线性变换组合为少数几个新的变量，使它们尽可能多地保留原始变量的信息，且彼此互不相关。

对于一个变量，不同个体间的差异越大，变量包含的信息量越大，我们将首选"差异大"的那些新变量作为主成分。借助主成分分析得到的新变量要能概括诸多信息的主要方面，我们也希望新变量之间能够互相独立，每个新变量能独立代表某一方面的性质。

设 m 个变量分别为 $X^{(1)}, \cdots, X^{(m)}$，由 n 个样本形成的数据为

$$X = \begin{pmatrix} X_1^{(1)} & \cdots & X_1^{(m)} \\ \vdots & \ddots & \vdots \\ X_n^{(1)} & \cdots & X_n^{(m)} \end{pmatrix}$$

设单位正交矩阵为

$$A = \begin{pmatrix} A_1^{(1)} & \cdots & A_1^{(m)} \\ \vdots & \ddots & \vdots \\ A_m^{(1)} & \cdots & A_m^{(m)} \end{pmatrix}$$

并设

$$A^{(i)} = \begin{pmatrix} A_1^{(i)} \\ \vdots \\ A_m^{(i)} \end{pmatrix}$$

则经过线性变换得到新变量 $Z^{(1)}, \cdots, Z^{(m)}$，并满足

$$\begin{pmatrix} Z_1^{(1)} & \cdots & Z_1^{(m)} \\ \vdots & \ddots & \vdots \\ Z_n^{(1)} & \cdots & Z_n^{(m)} \end{pmatrix} = \begin{pmatrix} X_1^{(1)} & \cdots & X_1^{(m)} \\ \vdots & \ddots & \vdots \\ X_n^{(1)} & \cdots & X_n^{(m)} \end{pmatrix} \begin{pmatrix} A_1^{(1)} & \cdots & A_1^{(m)} \\ \vdots & \ddots & \vdots \\ A_m^{(1)} & \cdots & A_m^{(m)} \end{pmatrix}$$

则 $\boldsymbol{Z}^{(i)}$ 与 $\boldsymbol{Z}^{(j)}$ 的协方差为

$$\begin{aligned} \mathrm{cov}\big(\boldsymbol{Z}^{(i)}, \boldsymbol{Z}^{(j)}\big) &= \frac{1}{n-1}\big(\boldsymbol{Z}^{(i)} - \overline{\boldsymbol{Z}^{(i)}}\big)\big(\boldsymbol{Z}^{(j)} - \overline{\boldsymbol{Z}^{(j)}}\big) \\ &= \frac{1}{n-1}\big(\boldsymbol{X}\boldsymbol{A}^{(i)} - \overline{\boldsymbol{X}\boldsymbol{A}^{(i)}}\big)^{\mathrm{T}}\big(\boldsymbol{X}\boldsymbol{A}^{(j)} - \overline{\boldsymbol{X}\boldsymbol{A}^{(j)}}\big) \\ &= \frac{1}{n-1}\boldsymbol{A}^{(i)\mathrm{T}}(\boldsymbol{X} - \overline{\boldsymbol{X}})^{\mathrm{T}}(\boldsymbol{X} - \overline{\boldsymbol{X}})\boldsymbol{A}^{(j)} \\ &= \boldsymbol{A}^{(i)\mathrm{T}}\left[\frac{1}{n-1}(\boldsymbol{X} - \overline{\boldsymbol{X}})^{\mathrm{T}}(\boldsymbol{X} - \overline{\boldsymbol{X}})\right]\boldsymbol{A}^{(j)} \end{aligned}$$

即

$$\mathrm{cov}\big(\boldsymbol{Z}^{(i)}, \boldsymbol{Z}^{(j)}\big) = \boldsymbol{A}^{(i)\mathrm{T}}\mathrm{cov}(\boldsymbol{X})\boldsymbol{A}^{(j)}$$

当 $i = j$ 时，$\boldsymbol{Z}^{(i)}$ 的方差为

$$\mathrm{var}\big(\boldsymbol{Z}^{(i)}\big) = \boldsymbol{A}^{(i)\mathrm{T}}\mathrm{cov}(\boldsymbol{X})\boldsymbol{A}^{(i)}$$

另外，由于 $\mathrm{cov}(\boldsymbol{X})$ 是对称的非负定矩阵，对其进行特征值分解，可得

$$\mathrm{cov}(\boldsymbol{X})\boldsymbol{V} = \begin{pmatrix} \lambda_1 & & 0 \\ & \ddots & \\ 0 & & \lambda_m \end{pmatrix}\boldsymbol{V}$$

其中，$\lambda_1 \geqslant \lambda_2 \geqslant \cdots \geqslant \lambda_m$，$\boldsymbol{V}^{\mathrm{T}}\boldsymbol{V} = \boldsymbol{I}$。
令 $\boldsymbol{A}^{(i)} = \boldsymbol{V}^{(i)}$，则

$$\mathrm{var}\big(\boldsymbol{Z}^{(i)}\big) = \lambda_i$$

方差满足如下大小关系

$$\mathrm{var}\big(\boldsymbol{Z}^{(1)}\big) \geqslant \mathrm{var}\big(\boldsymbol{Z}^{(2)}\big) \geqslant \cdots \geqslant \mathrm{var}\big(\boldsymbol{Z}^{(m)}\big)$$

方差的和为

$$\sum_{i=1}^{m} \text{var}(\boldsymbol{Z}^{(i)}) = \sum_{i=1}^{m} \lambda_i$$

且协方差满足

$$\text{cov}(\boldsymbol{Z}^{(i)}, \boldsymbol{Z}^{(j)}) = 0$$

综合前面的分析过程，$\boldsymbol{X}^{(1)}, \cdots, \boldsymbol{X}^{(m)}$ 通过单位正交的矩阵 $\boldsymbol{A}$ 在高维空间进行了旋转变换，得到 $\boldsymbol{Z}^{(1)}, \cdots, \boldsymbol{Z}^{(m)}$，满足 $\text{var}(\boldsymbol{Z}^{(1)}) \geqslant \text{var}(\boldsymbol{Z}^{(2)}) \geqslant \cdots \geqslant \text{var}(\boldsymbol{Z}^{(m)})$。显然，索引号较小的 $\boldsymbol{Z}^{(i)}$ 更重要，可以看作数据的"主成分"。

$\lambda_1, \lambda_2, \cdots, \lambda_m$ 是由矩阵进行特征值分解得到的，由矩阵的迹（trace）的定义和性质，可知

$$\sum_{i=1}^{m} \lambda_i = \text{trace}(\text{cov}(\boldsymbol{X})) = \sum_{i=1}^{m} \text{var}(\boldsymbol{X}^{(i)})$$

$\sum_{i=1}^{m} \lambda_i$ 可以看作矩阵的总能量，λ_i 可以看作每个特征向量所对应的能量，分解前后的总能量相等。从 λ_1 开始，依次加入 $\lambda_2, \cdots, \lambda_m$，当加入到第 p 个时，若这 p 个成分的能量和 $\sum_{i=1}^{p} \lambda_i$ 与总能量 $\sum_{i=1}^{m} \lambda_i$ 的比值大于 w（默认值为 0.9），即

$$\frac{\sum_{i=1}^{p} \lambda_i}{\sum_{i=1}^{m} \lambda_i} > w$$

我们就可以确定主成分的个数为 p。

在实际使用中，有时会先将各个变量进行标准化，此时的协方差矩阵就相当于原始数据的相关系数矩阵。所以 Alink 的主成分分析组件提供了两种计算选择，参数 CalculationType 可以设置为相关系数矩阵（CORR）或者协方差矩阵（COV），默认为相关系数矩阵，即对标准化后的数据计算其主成分。

19.1 主成分的含义

本节的案例是调查美国 50 个州的 7 种犯罪率。数据是美国 50 个州每 10 万人中 7 种犯罪的概率数据。这 7 种犯罪分别是 murder（杀人罪）、rape（强奸罪）、robbery（抢劫罪）、assault（斗殴罪）、burglary（夜盗罪）、larceny（偷盗罪）和 auto（汽车犯罪）。下面我们来做主成

分分析，直接从这 7 个变量出发来评价各州的治安和犯罪情况是很难的，而使用主成分分析则可以把这些变量概括为 2 个或 3 个综合变量（即主成分），以便帮助我们更简便地分析这些数据。

数据只有 50 条，将其写在代码中，定义为数组 CRIME_ROWS_DATA，数据源组件 MemSourceBatchOp 会在构造函数中使用该数组，如下代码所示。

```
private static final Row[] CRIME_ROWS_DATA = new Row[] {
    Row.of("ALABAMA", 14.2, 25.2, 96.8, 278.3, 1135.5, 1881.9, 280.7),
    Row.of("ALASKA", 10.8, 51.6, 96.8, 284.0, 1331.7, 3369.8, 753.3),
    Row.of("ARIZONA", 9.5, 34.2, 138.2, 312.3, 2346.1, 4467.4, 439.5),
    ... ...
    Row.of("WEST VIRGINIA", 6.0, 13.2, 42.2, 90.9, 597.4, 1341.7, 163.3),
    Row.of("WISCONSIN", 2.8, 12.9, 52.2, 63.7, 846.9, 2614.2, 220.7),
    Row.of("WYOMING", 5.4, 21.9, 39.7, 173.9, 811.6, 2772.2, 282.0)
};
static final String[] CRIME_COL_NAMES =
    new String[] {"state", "murder", "rape", "robbery", "assault", "burglary", "larceny", "auto"};
... ...
MemSourceBatchOp source = new MemSourceBatchOp(CRIME_ROWS_DATA, CRIME_COL_NAMES);

source.lazyPrint(10, "Origin data");
```

输出部分原始数据如表 19-1 所示。

表 19-1 原始数据

state	murder	rape	robbery	assault	burglary	larceny	auto
ALABAMA	14.2000	25.2000	96.8000	278.3000	1135.5000	1881.9000	280.7000
ALASKA	10.8000	51.6000	96.8000	284.0000	1331.7000	3369.8000	753.3000
ARIZONA	9.5000	34.2000	138.2000	312.3000	2346.1000	4467.4000	439.5000
ARKANSAS	8.8000	27.6000	83.2000	203.4000	972.6000	1862.1000	183.4000
CALIFORNIA	11.5000	49.4000	287.0000	358.0000	2139.4000	3499.8000	663.5000
COLORADO	6.3000	42.0000	170.7000	292.9000	1935.2000	3903.2000	477.1000
CONNECTICUT	4.2000	16.8000	129.5000	131.8000	1346.0000	2620.7000	593.2000
DELAWARE	6.0000	24.9000	157.0000	194.2000	1682.6000	3678.4000	467.0000
FLORIDA	10.2000	39.6000	187.9000	449.1000	1859.9000	3840.5000	351.4000
GEORGIA	11.7000	31.1000	140.5000	256.5000	1351.1000	2170.2000	297.9000

主成分分析组件为 PCA，设置主成分的个数为 K=4；设置参与计算的数据列为除了州名称

的其他列；主成分分量的输出形式为向量，设置输出列的名称为 VECTOR_COL_NAME；设置使用 Lazy 的方式输出模型信息。为了方便后面的分析，使用格式转换组件将向量列转化为 4 个数据列，名称分别为 "pc_1、pc_2、pc_3、pc_4"，数据类型都为 double 类型。计算主成分的代码如下。

```
BatchOperator <?> pca_result = new PCA()
    .setK(4)
    .setSelectedCols("murder", "rape", "robbery", "assault", "burglary", "larceny", "auto")
    .setPredictionCol(VECTOR_COL_NAME)
    .enableLazyPrintModelInfo()
    .fit(source)
    .transform(source)
    .link(
        new VectorToColumnsBatchOp()
            .setVectorCol(VECTOR_COL_NAME)
            .setSchemaStr("pc_1 double, pc_2 double, pc_3 double, pc_4 double")
            .setReservedCols("state")
    )
    .lazyPrint(10, "state with principle components");
```

我们可以得到如下模型信息。

```
------------------------------------- PCA -------------------------------------
CalculationType: CORR
Number of Principal Component: 4

EigenValues:
| Prin|Eigenvalue|Proportion|Cumulative|
|-----|----------|----------|----------|
|Prin1|    4.115 |   0.5879 |   0.5879 |
|Prin2|   1.2387 |   0.177  |   0.7648 |
|Prin3|   0.7258 |   0.1037 |   0.8685 |
|Prin4|   0.3164 |   0.0452 |   0.9137 |

EigenVectors:
| colName|   Prin1|   Prin2|   Prin3|   Prin4|
|--------|--------|--------|--------|--------|
|  murder| -0.3003| -0.6292| -0.1782|  0.2321|
|    rape| -0.4318| -0.1694|  0.2442| -0.0622|
| robbery| -0.3969|  0.0422| -0.4959|  0.558 |
| assault| -0.3967| -0.3435|  0.0695| -0.6298|
|burglary| -0.4402|  0.2033|  0.2099|  0.0576|
| larceny| -0.3574|  0.4023|  0.5392|  0.2349|
|    auto| -0.2952|  0.5024| -0.5684| -0.4192|
```

由 EigenValues 可知各主成分的贡献率。在结果表中，第一列为主成分所对应的奇异值大小，第二列为各主成分的贡献率。可以看出，第一个主成分的贡献率非常突出，达到了 58.8%；最后一列为累计贡献率，可以看到前两个主成分的累计贡献率达 76.5%，前三个主成分的累计贡献率达 86.8%，前四个主成分的累计贡献率达 91.4%。在分析数据时，我们可以根据实际需要考虑选取几个主成分。

由 EigenVectors 可知，主成分的计算表达式如下。

Prin1 = -0.3003 * murder - 0.4318 * rape - 0.3969 * robbery - 0.3967 * assault - 0.4402 * burglary - 0.3574 * larceny - 0.2952 * auto
Prin2 = -0.6292 * murder - 0.1694 * rape + 0.0422 * robbery - 0.3435 * assault + 0.2033 * burglary + 0.4023 * larceny + 0.5024 * auto

> **注意**：这里的 murder、rape 等变量是原变量标准化后的变量，即原变量减去均值，再除以标准差得到的变量。第一主成分 Prin1 对所有的变量都有近似相等的权重，可认为第一主成分是对所有犯罪率的总度量取相反数；第二主成分 Prin2 可以看作抢、盗罪（robbery、burglary、larceny 和 auto 权重为正）与杀、淫罪（murder、rape 和 assault 权重为负）的对比，Prin2 值较小的州的暴力犯罪比重较大。

输出的部分 PCA 结果如表 19-2 所示。

表 19-2　主成分分析结果

state	prin1	prin2	prin3	prin4
ALABAMA	0.0499	-2.0961	-0.5016	-0.2510
ALASKA	-2.4215	0.1665	0.0697	-1.1605
ARIZONA	-3.0141	0.8449	1.7520	0.1162
ARKANSAS	1.0544	-1.3454	0.0183	-0.0215
CALIFORNIA	-4.2838	0.1432	-0.2762	-0.0251
COLORADO	-2.5093	0.9166	1.1516	-0.1126
CONNECTICUT	0.5413	1.5012	-0.7839	-0.0862
DELAWARE	-0.9646	1.2967	0.5259	0.4173
FLORIDA	-3.1118	-0.6039	1.2154	-0.4951
GEORGIA	-0.4904	-1.3808	-0.2446	0.0625

除了关注每个州的主成分分量，我们也希望了解每个分量下最大值和最小值对应的是哪个州？下面分别对 Prin1 和 Prin2 的数值进行排序，具体代码如下。需要注意的是，orderBy 函数的第二参数为输出排序结果的前多少项，当数据量很大时，可以根据这个参数大幅减少计算量，

避免全量排序；orderBy 函数的第三参数为是否升序，这里设置为 false，则会按降序排序，输出前 100 个数据，因为我们整体的数据量为 50，所以会排序输出全部数据。

```
pca_result
   .select("state, prin1")
   .orderBy("prin1", 100, false)
   .lazyPrint(-1, "Order by prin1");

pca_result
   .select("state, prin2")
   .orderBy("prin2", 100, false)
   .lazyPrint(-1, "Order by prin2");
```

运行结果汇总在表 19-3 中，左边为按第一主成分值排序的结果，右边为按第二主成分值排序的结果。这里只列出了排名在前 3 名和后 3 名的各州。

表 19-3　排名在前 3 名和后 3 名的各州

Order by prin1	Order by prin2
state|prin1	state|prin2
-----|-----	-----|-----
NORTH DAKOTA|3.9641	MASSACHUSETTS|2.6311
SOUTH DAKOTA|3.1720	RHODE ISLAND|2.1466
WEST VIRGINIA|3.1477	HAWAII|1.8239
... ...	... ...
NEW YORK|-3.4525	ALABAMA|-2.0961
CALIFORNIA|-4.2838	SOUTH CAROLINA|-2.1621
NEVADA|-5.2670	MISSISSIPPI|-2.5467

第一主成分 Prin1 被看作所有犯罪率的总度量的相反数，Prin1 值较大的州的犯罪率较低，其中，NORTH DAKOTA 州的犯罪率最低（Prin1 = 3.9641）；Prin1 值较小的州的犯罪率较高，NEVADA 州的犯罪率最高（Prin1 = −5.2670）。第二主成分 Prin2 值较小的州的暴力犯罪性质比重较大，MISSISSIPPI 州的暴力犯罪性质比重最大（Prin2= −2.5467）；Prin2 值较大的州的暴力犯罪性质比重较小，MASSACHUSETTS 州的暴力犯罪性质比重最小（Prin2=2.6311）。

19.2　两种计算方式

Alink 的主成分分析组件提供了两种计算方式，参数 CalculationType 可以设置为相关系数矩阵或者协方差矩阵，默认为相关系数矩阵，即计算标准化后的数据的主成分。

构建实验流程如下。

```
Pipeline std_pca = new Pipeline()
    .add(
        new StandardScaler()
            .setSelectedCols("murder", "rape", "robbery", "assault", "burglary", "larceny", "auto")
    )
    .add(
        new PCA()
            .setCalculationType(CalculationType.COV)
            .setK(4)
            .setSelectedCols("murder", "rape", "robbery", "assault", "burglary", "larceny", "auto")
            .setPredictionCol(VECTOR_COL_NAME)
            .enableLazyPrintModelInfo()
    );

std_pca
    .fit(source)
    .transform(source)
    .link(
        new VectorToColumnsBatchOp()
            .setVectorCol(VECTOR_COL_NAME)
            .setSchemaStr("prin1 double, prin2 double, prin3 double, prin4 double")
            .setReservedCols("state")
    )
    .lazyPrint(10, "state with principle components");
BatchOperator.execute();
```

在 Pipeline 中先使用标准化变换组件 StandardScaler 对各数据列进行标准化，然后调用主成分组件 PCA。注意：这里特别设置了参数 CalculationType=COV。

首先看 PCA 阶段输出的模型信息，对比 19.1 节的模型信息，除了"CalculationType: COV"与 19.1 节模型信息中的"CalculationType: CORR"不同，其他数据值都一样。

```
-------------------------------- PCA --------------------------------
CalculationType: COV
Number of Principal Component: 4

EigenValues:
    |Prin|Eigenvalue|Proportion|Cumulative|
    |----|----------|----------|----------|
    |Prin1|    4.115|    0.5879|    0.5879|
    |Prin2|   1.2387|     0.177|    0.7648|
    |Prin3|   0.7258|    0.1037|    0.8685|
    |Prin4|   0.3164|    0.0452|    0.9137|

EigenVectors:
```

```
| colName | Prin1  | Prin2  | Prin3  | Prin4  |
|---------|--------|--------|--------|--------|
|  murder |-0.3003 |-0.6292 |-0.1782 | 0.2321 |
|    rape |-0.4318 |-0.1694 | 0.2442 |-0.0622 |
| robbery |-0.3969 | 0.0422 |-0.4959 | 0.558  |
| assault |-0.3967 |-0.3435 | 0.0695 |-0.6298 |
|burglary |-0.4402 | 0.2033 | 0.2099 | 0.0576 |
| larceny |-0.3574 | 0.4023 | 0.5392 | 0.2349 |
|    auto |-0.2952 | 0.5024 |-0.5684 |-0.4192 |
```

接下来看 PCA 输出的部分计算结果，如表 19-4 所示，也与 19.1 节 PCA 输出的结果一样。

表 19-4　新的 PCA 输出结果

state	prin1	prin2	prin3	prin4
ALABAMA	0.0499	-2.0961	-0.5016	-0.2510
ALASKA	-2.4215	0.1665	0.0697	-1.1605
ARIZONA	-3.0141	0.8449	1.7520	0.1162
ARKANSAS	1.0544	-1.3454	0.0183	-0.0215
CALIFORNIA	-4.2838	0.1432	-0.2762	-0.0251
COLORADO	-2.5093	0.9166	1.1516	-0.1126
CONNECTICUT	0.5413	1.5012	-0.7839	-0.0862
DELAWARE	-0.9646	1.2967	0.5259	0.4173
FLORIDA	-3.1118	-0.6039	1.2154	-0.4951
GEORGIA	-0.4904	-1.3808	-0.2446	0.0625

综合上面的PCA模型信息和输出结果，证实了两种计算方式间的差距只在于数据标准化变换。选择相关系数矩阵（CORR）的计算方式，通过标准化变换，统一了各数据列的尺度；而选择协方差矩阵（COV）的计算方式，输出的主成分实质上是原数据的降维数据，是原始数据在高维空间进行了旋转变换后[1]提取了几个重要维度而形成的数据。

19.3　在聚类方面的应用

本节及 19.4 节将使用手写数字数据集来演示如何通过 PCA 获取降维数据，以便更高效地

1　注意：变换保持了各点间的距离。

处理一些聚类和分类问题。

我们将只使用原始数据 5% 的维度，即 784×5% ≈ 39（维）。使用稀疏格式的训练数据作为数据源，训练出来的 PCA 模型会被多次用到，并会将模型保存到 AK 格式文件（本地路径为 DATA_DIR + PCA_MODEL_FILE）中。具体代码如下。

```
AkSourceBatchOp source = new AkSourceBatchOp().setFilePath(DATA_DIR + SPARSE_TRAIN_FILE);
source
    .link(
        new PcaTrainBatchOp()
            .setK(39)
            .setCalculationType(CalculationType.COV)
            .setVectorCol(VECTOR_COL_NAME)
            .lazyPrintModelInfo()
    )
    .link(
        new AkSinkBatchOp()
            .setFilePath(DATA_DIR + PCA_MODEL_FILE)
            .setOverwriteSink(true)
    );
BatchOperator.execute();
```

运行结束后，生成了 PCA 模型文件，同时输出了如下模型信息，这里略去了后面的 EigenVectors 部分。

```
------------------------------------ PCA ------------------------------------
CalculationType: COV
Number of Principal Component: 39

EigenValues:
|   Prin|Eigenvalue|Proportion|Cumulative|
|-------|----------|----------|----------|
|  Prin1|332724.6674|    0.097|     0.097|
|  Prin2|243283.9391|    0.071|     0.168|
|  Prin3|211507.3671|   0.0617|    0.2297|
|  Prin4|184776.3859|   0.0539|    0.2836|
|  Prin5|166926.8313|   0.0487|    0.3323|
|  Prin6|147844.9617|   0.0431|    0.3754|
|  Prin7|112178.2027|   0.0327|    0.4081|
|  Prin8| 98874.4296|   0.0288|     0.437|
|  Prin9| 94696.2491|   0.0276|    0.4646|
| Prin10| 80809.8245|   0.0236|    0.4881|
|    ...|       ...|      ...|       ...|
| Prin30| 23686.123|   0.0069|    0.7305|
| Prin31| 22562.7619|   0.0066|    0.7371|
| Prin32| 22221.7664|   0.0065|    0.7436|
| Prin33| 20660.6718|    0.006|    0.7496|
| Prin34| 20110.9854|   0.0059|    0.7555|
```

```
|Prin35| 19543.2009|    0.0057|    0.7612|
|Prin36| 18638.2921|    0.0054|    0.7666|
|Prin37| 17340.9003|    0.0051|    0.7717|
|Prin38| 16726.2448|    0.0049|    0.7766|
|Prin39| 16505.8174|    0.0048|    0.7814|
```

... ...

可以看到，前 39 个主成分占了总能量的 78%。

载入 PCA 模型，并对原始数据进行 PCA 变换，结果为 pca_result。

```
BatchOperator <?> pca_result = new PcaPredictBatchOp()
    .setVectorCol(VECTOR_COL_NAME)
    .setPredictionCol(VECTOR_COL_NAME)
    .linkFrom(
        new AkSourceBatchOp().setFilePath(DATA_DIR + PCA_MODEL_FILE),
        source
    );
```

接下来，针对原始数据及 PCA 结果数据 pca_result 分别进行 K-Means 聚类操作，并对比聚类评估结果，具体代码如下。

```
KMeans kmeans = new KMeans()
    .setK(10)
    .setVectorCol(VECTOR_COL_NAME)
    .setPredictionCol(PREDICTION_COL_NAME);

sw.reset();
sw.start();
kmeans
    .fit(source)
    .transform(source)
    .link(
        new EvalClusterBatchOp()
            .setVectorCol(VECTOR_COL_NAME)
            .setPredictionCol(PREDICTION_COL_NAME)
            .setLabelCol(LABEL_COL_NAME)
            .lazyPrintMetrics("KMeans")
    );
BatchOperator.execute();
sw.stop();
System.out.println(sw.getElapsedTimeSpan());

sw.reset();
sw.start();
kmeans
```

```
        .fit(pca_result)
        .transform(pca_result)
        .link(
            new EvalClusterBatchOp()
                .setVectorCol(VECTOR_COL_NAME)
                .setPredictionCol(PREDICTION_COL_NAME)
                .setLabelCol(LABEL_COL_NAME)
                .lazyPrintMetrics("KMeans + PCA")
        );
BatchOperator.execute();
sw.stop();
System.out.println(sw.getElapsedTimeSpan());
```

运行结果如下。

```
KMeans
-------------------------------- Metrics: --------------------------------
k:10
VRC:2293.0127   DB:2.8759   SilhouetteCoefficient:0.0925
ARI:0.3661 NMI:0.4848 Purity:0.5818

|Label\Cluster|    9|    8|    7|    6|    5|    4|    3|    2|    1|    0|
|-------------|-----|-----|-----|-----|-----|-----|-----|-----|-----|-----|
|            9|    9| 1478|  145|   32|    9| 1713|   40| 2385|   87|   51|
|            8|   49|  296|  314|   28|   53|  157| 3171|  171| 1319|  293|
|            7|   29| 2431|  223|   13|    3|  701|    3| 2674|    4|  184|
|            6|  132|    3|  312|   68| 4866|  210|  159|    0|   32|  136|
|            5|   17|  612|  224|   64|  132|  248| 1778|  216| 1739|  391|
|            4|   25| 1559|  119|    7|  115| 2349|   10| 1515|    0|  143|
|            3|  221|   41|  425|   32|   62|  100| 1263|  121| 3817|   49|
|            2| 4208|   68|  357|   58|  181|  204|  144|   39|  371|  328|
|            1|    9|    5| 3697|    0|    8|    4|    8|   10|    5| 2996|
|            0|   47|   61|    6| 4701|  217|   53|  568|   12|  248|   10|

17 seconds 7.0 milliseconds.
KMeans + PCA
-------------------------------- Metrics: --------------------------------
k:10
VRC:3205.3665   DB:2.4339   SilhouetteCoefficient:0.146
ARI:0.3992 NMI:0.5131 Purity:0.6146

|Label\Cluster|    9|    8|    7|    6|    5|    4|    3|    2|    1|    0|
|-------------|-----|-----|-----|-----|-----|-----|-----|-----|-----|-----|
|            9|   86|   52|  100| 2743|  223|  175| 2498|   28|   35|    9|
|            8| 1265| 3202|  363|  312|  340|   21|  214|   55|   29|   50|
|            7|    5|    7|  306|  320|  313| 4129| 1127|   40|   15|    3|
|            6|   31|   92|  165|   12|  247|    0|  135|  571|   60| 4605|
|            5| 1762| 1648|  625|  628|  177|   15|  330|   13|   66|  157|
```

```
|   4|    0|    7| 152|2713| 127|   5|2660|  66|   6| 106|
|   3| 3835| 1275|  83|  31| 442|  40| 163| 161|  31|  70|
|   2|  356|  181| 347|  28| 338|  96| 148|4260|  64| 140|
|   1|    5|    6|3023|   5|3672|   5|   6|  10|   0|  10|
|   0|  243|  496|  26|  29|   4|   4|  43|  53|4749| 276|
```

9 seconds 627.0 milliseconds.

二者在计算时间上有明显差异，提取的主成分在向量维度上要远小于原始向量，所以在计算量上有优势；计算时间相差一倍，这是因为我们选择的原始数据为稀疏向量格式，其计算量也远小于同等维度的稠密向量。从聚类指标上看，基于主成分聚类并不等价于基于原始数据的聚类，而是各有千秋。

19.4 在分类方面的应用

提取特征向量的主成分可以显著减少向量长度，从而大幅减少计算量，缩短分类模型的训练时间。

我们以手写数字分类为例，看看不同的训练数据格式（784 维稠密向量、784 维稀疏向量、39 维主成分向量）对模型训练时间的影响。首先获取稠密向量和稀疏向量的数据源，并定义一个停表对象，用来对各次试验进行计时。具体代码如下。

```
AkSourceBatchOp dense_train_data = new AkSourceBatchOp().setFilePath(DATA_DIR + DENSE_TRAIN_FILE);
AkSourceBatchOp dense_test_data = new AkSourceBatchOp().setFilePath(DATA_DIR + DENSE_TEST_FILE);
AkSourceBatchOp sparse_train_data = new AkSourceBatchOp().setFilePath(DATA_DIR + SPARSE_TRAIN_FILE);
AkSourceBatchOp sparse_test_data = new AkSourceBatchOp().setFilePath(DATA_DIR + SPARSE_TEST_FILE);

Stopwatch sw = new Stopwatch();
```

设置 K 最近邻分类器 KnnClassifier，并分别对稠密向量数据和稀疏向量数据进行训练、预测，并评估分类效果。左边使用稠密向量，右边使用稀疏向量，具体代码如下。

```
new KnnClassifier()                                new KnnClassifier()
    .setK(3)                                           .setK(3)
    .setVectorCol(VECTOR_COL_NAME)                     .setVectorCol(VECTOR_COL_NAME)
    .setLabelCol(LABEL_COL_NAME)                       .setLabelCol(LABEL_COL_NAME)
    .setPredictionCol(PREDICTION_COL_NAME)             .setPredictionCol(PREDICTION_COL_NAME)
    .fit(dense_train_data)                             .fit(sparse_train_data)
    .transform(dense_test_data)                        .transform(sparse_test_data)
    .link(                                             .link(
        new EvalMultiClassBatchOp()                        new EvalMultiClassBatchOp()
            .setLabelCol(LABEL_COL_NAME)                       .setLabelCol(LABEL_COL_NAME)
            .setPredictionCol(PREDICTION_COL_NAME)             .setPredictionCol(PREDICTION_COL_NAME)
            .lazyPrintMetrics("KnnClassifier Dense")           .lazyPrintMetrics("KnnClassifier Sparse")
    );                                                 );
```

左边使用稠密向量，右边使用稀疏向量，分类效果对比显示如下。

```
KnnClassifier Dense                              KnnClassifier Sparse
-------------------------------                  -------------------------------
Metrics:                                         Metrics:
-------------------------------                  -------------------------------
Accuracy:0.9719 Macro F1:0.9718                  Accuracy:0.9719 Macro F1:0.9718
Micro F1:0.9719 Kappa:0.9688                     Micro F1:0.9719 Kappa:0.9688
|Pred\Real|   9|   8|   7|...|   2|   1|   0|    |Pred\Real|   9|   8|   7|...|   2|   1|   0|
|---------|---|---|---|---|---|---|---|          |---------|---|---|---|---|---|---|---|
|        9|971|   4|  12|...|   1|   0|   0|     |        9|971|   4|  12|...|   1|   0|   0|
|        8|  4|924|   0|...|   2|   0|   0|      |        8|  4|924|   0|...|   2|   0|   0|
|        7|  8|   4|991|...|  13|   0|   1|      |        7|  8|   4|991|...|  13|   0|   1|
|      ...|...|...|...|...|...|...|...|          |      ...|...|...|...|...|...|...|...|
|        2|  1|   3|   4|...|994|   2|   1|      |        2|  1|   3|   4|...|994|   2|   1|
|        1|  4|   0|  19|...|   9|1133|  1|      |        1|  4|   0|  19|...|   9|1133|  1|
|        0|  4|   7|   0|...|  10|   0|974|      |        0|  4|   7|   0|...|  10|   0|974|

5 minutes 17 seconds 179.0 milliseconds.         1 minutes 1 seconds 528.0 milliseconds.
```

可见，稠密向量数据和稀疏向量数据在评估指标上是完全一致的，但计算时间的比例大约是 5:1。

接下来，我们将主成分分析和 K 最近邻分类器组合为 Pipeline，再分别对稠密向量数据和稀疏向量数据进行训练、预测，并评估分类效果，左边使用稠密向量，右边使用稀疏向量，具体代码如下。

```
new Pipeline()                                   new Pipeline()
  .add(                                            .add(
    new PCA()                                        new PCA()
      .setK(39)                                        .setK(39)
      .setCalculationType(CalculationType.COV)          .setCalculationType(CalculationType.COV)
      .setVectorCol(VECTOR_COL_NAME)                   .setVectorCol(VECTOR_COL_NAME)
      .setPredictionCol(VECTOR_COL_NAME)              .setPredictionCol(VECTOR_COL_NAME)
  )                                                )
  .add(                                            .add(
    new KnnClassifier()                              new KnnClassifier()
      .setK(3)                                         .setK(3)
      .setVectorCol(VECTOR_COL_NAME)                   .setVectorCol(VECTOR_COL_NAME)
      .setLabelCol(LABEL_COL_NAME)                     .setLabelCol(LABEL_COL_NAME)
      .setPredictionCol(PREDICTION_COL_NAME)           .setPredictionCol(PREDICTION_COL_NAME)
  )                                                )
  .fit(dense_train_data)                           .fit(sparse_train_data)
  .transform(dense_test_data)                      .transform(sparse_test_data)
  .link(                                           .link(
    new EvalMultiClassBatchOp()                      new EvalMultiClassBatchOp()
      .setLabelCol(LABEL_COL_NAME)                     .setLabelCol(LABEL_COL_NAME)
      .setPredictionCol(PREDICTION_COL_NAME)           .setPredictionCol(PREDICTION_COL_NAME)
      .lazyPrintMetrics("Knn with PCA Dense")           .lazyPrintMetrics("Knn with PCA Sparse")
);                                               );
```

注意，在 PCA 中设置参数 CalculationType=COV。

汇总计算评估结果，左边使用稠密向量，右边使用稀疏向量，对比显示如下。

```
Knn with PCA Dense                          Knn with PCA Sparse
-------------------------------             -------------------------------
Metrics:                                    Metrics:
-------------------------------             -------------------------------
Accuracy:0.9755 Macro F1:0.9753             Accuracy:0.9755 Macro F1:0.9753
Micro F1:0.9755 Kappa:0.9728                Micro F1:0.9755 Kappa:0.9728
|Pred\Real|   9|   8|   7|...|   2|   1|   0|    |Pred\Real|   9|   8|   7|...|   2|   1|   0|
|---------|---|---|---|---|----|----|---|       |---------|---|---|---|---|----|----|---|
|        9|972|   4|  10|...|   0|   1|   0|    |        9|972|   4|  10|...|   0|   1|   0|
|        8|  7| 937|   0|...|   5|   0|   0|    |        8|  7| 937|   0|...|   5|   0|   0|
|        7|  3|   3| 994|...|  10|   0|   1|    |        7|  3|   3| 994|...|  10|   0|   1|
|      ...|...|...|...|...| ...| ...|...|      |      ...|...|...|...|...| ...| ...|...|
|        2|  2|   4|   7|...|1008|   3|   1|    |        2|  2|   4|   7|...|1008|   3|   1|
|        1|  3|   0|  13|...|   1|1131|   1|    |        1|  3|   0|  13|...|   1|1131|   1|
|        0|  6|   5|   0|...|   5|   0| 975|    |        0|  6|   5|   0|...|   5|   0| 975|

1 minutes  33 seconds  676.0 milliseconds.      27 seconds  183.0 milliseconds.
```

可以观察到如下信息。
- 稠密向量数据和稀疏向量数据在评估指标上是完全一致的。
- 使用了 PCA 后，放入分类器的特征个数减少了，但评估指标非但没有下降，还有少许上升，这可能是由于提取主成分后减少了一些噪声干扰。
- 稠密向量使用 PCA 前后，计算时间由 5 分 17 秒降到 1 分 33 秒。
- 稀疏向量使用 PCA 前后，计算时间由 1 分 1 秒降到 27 秒。
- 对于 PCA+KNN 的 Pipeline，稠密向量数据与稀疏向量数据的计算时间比例大约是 3:1。

如果有已训练好的 PCA 模型，我们可以直接将 PCAModel 放入 Pipeline 中，PCAModel 不需要进行训练，直接通过 setModelData 方法载入训练好的模型。左边使用稠密向量，右边使用稀疏向量，具体代码如下。

```
new Pipeline()                                  new Pipeline()
    .add(                                           .add(
        new PCAModel()                                  new PCAModel()
            .setVectorCol(VECTOR_COL_NAME)                  .setVectorCol(VECTOR_COL_NAME)
            .setPredictionCol(VECTOR_COL_NAME)              .setPredictionCol(VECTOR_COL_NAME)
            .setModelData(                                  .setModelData(
                new AkSourceBatchOp()                           new AkSourceBatchOp()
                    .setFilePath(DATA_DIR + PCA_MODEL_FILE)        .setFilePath(DATA_DIR + PCA_MODEL_FILE)
            )                                               )
    )                                               )
    .add(                                           .add(
        new KnnClassifier()                             new KnnClassifier()
            .setK(3)                                        .setK(3)
            .setVectorCol(VECTOR_COL_NAME)                  .setVectorCol(VECTOR_COL_NAME)
            .setLabelCol(LABEL_COL_NAME)                    .setLabelCol(LABEL_COL_NAME)
```

```
        .setPredictionCol(PREDICTION_COL_NAME)              .setPredictionCol(PREDICTION_COL_NAME)
)                                                     )
.fit(dense_train_data)                                .fit(sparse_train_data)
.transform(dense_test_data)                           .transform(sparse_test_data)
.link(                                                .link(
   new EvalMultiClassBatchOp()                           new EvalMultiClassBatchOp()
       .setLabelCol(LABEL_COL_NAME)                          .setLabelCol(LABEL_COL_NAME)
       .setPredictionCol(PREDICTION_COL_NAME)                .setPredictionCol(PREDICTION_COL_NAME)
       .lazyPrintMetrics("Knn PCAModel Dense")               .lazyPrintMetrics("Knn PCAModel Sparse")
);                                                    );
```

汇总计算评估结果，左边使用稠密向量，右边使用稀疏向量，对比显示如下。

```
Knn PCAModel Dense                                    Knn PCAModel Sparse
-------------------------------                       -------------------------------
Metrics:                                              Metrics:
-------------------------------                       -------------------------------
Accuracy:0.9755 Macro F1:0.9753                       Accuracy:0.9755 Macro F1:0.9753
Micro F1:0.9755 Kappa:0.9728                          Micro F1:0.9755 Kappa:0.9728
|Pred\Real|   9|   8|   7|...|   2|   1|   0|         |Pred\Real|   9|   8|   7|...|   2|   1|   0|
|---------|---|---|---|---|---|---|---|               |---------|---|---|---|---|---|---|---|
|        9|972|   4|  10|...|   0|   1|   0|          |        9|972|   4|  10|...|   0|   1|   0|
|        8|  7|937|   0|...|   5|   0|   0|           |        8|  7|937|   0|...|   5|   0|   0|
|        7|  3|  3|994|...|  10|   0|   1|            |        7|  3|  3|994|...|  10|   0|   1|
|      ...|...|...|...|...|...|...|...|               |      ...|...|...|...|...|...|...|...|
|        2|  2|  4|  7|...|1008|   3|   1|            |        2|  2|  4|  7|...|1008|   3|   1|
|        1|  3|  0| 13|...|   1|1131|   1|            |        1|  3|  0| 13|...|   1|1131|   1|
|        0|  6|  5|  0|...|   5|   0|975|             |        0|  6|  5|  0|...|   5|   0|975|

22 seconds  951.0 milliseconds.                       20 seconds  425.0 milliseconds.
```

可以观察到如下信息。

- 直接加载模型的方式，对评估指标没有影响。
- 计算时间明显缩短，稠密向量由 1 分 33 秒降到 22 秒，稀疏向量由 27 秒降到 20 秒。
- PCA 的模型是由稀疏向量计算的，但在本实验中应用到稠密向量，仍然可以正常使用。

20 超参数搜索

在机器学习领域，超参数（Hyper-parameter）是在模型训练过程开始之前设置的参数。超参数一般是根据经验确定的变量（譬如学习速率、迭代次数、神经网络的层数）、随机森林中决策树的棵数、K-Means 聚类中的聚类数目，等等。

超参数值的选择会影响模型的效果。例如决策树的树深度参数，如果参数值太小，则不能对一些普遍场景进行细分，分类器太简单，即机器学习常说的欠拟合（Under-fitting），对测试集的分类效果差；如果参数值太大，那么模型对训练数据有很好的细分能力，但是影响模型泛化能力，即过拟合（Over-fitting），在测试数据集上分类效果也会比较差。

分类评估和回归评估有很多评估指标，各有侧重，用户可以根据具体的问题选择适合的评估指标，再针对目标评估指标选择适合的超参数值。

Alink 的模型选择组件是将一个基本的估算器（Estimator）或一个管道（Pipeline）作为输入，针对具体的训练数据选择适合的评估指标，在超参数空间内找到最佳组合超参数。

我们提供了两种常用的模型选择方式。

（1）网格搜索（Grid Search）：针对所要搜索的每个参数，指定其可以选择的数值列表。需要搜索的超参数组合就是各参数可选值的"网格"（笛卡儿积）组合。

（2）随机搜索（Random Search）：针对所要搜索的每个参数，指定其分布范围（可以是离散值数组，也可以是服从某种分布的连续区间），然后在各超参数的分布范围内抽样固定次数，得到需要进行搜索的超参数组合。该方法用来避免网格搜索中的组合爆炸。更多内容可以参考 Bergstra 和 Bengio 的文章"Random Search for Hyper-Parameter Optimization"。

评估指标验证的方式有如下两种。

（1）K 折交叉验证（Cross Validation，CV）：机器学习领域经典的验证方式，对一个超参数组合需要计算 K 次，取 K 次实验的平均值。

（2）训练集和验证集拆分（Train-Validation Split）：简单指定训练数据与验证数据的比例，其好处是对一个超参数组合需要计算 1 次，虽然在评估指标计算的稳定性方面不如 K 折交叉验证，但更适合注重计算时间和资源的场景。

针对不同的问题提供了 4 个调参评估器。
- 二分类调参评估器（BinaryClassificationTuningEvaluator）。
- 多分类调参评估器（MultiClassClassificationTuningEvaluator）。
- 回归调参评估器（RegressionTuningEvaluator）。
- 聚类调参评估器（ClusterTuningEvaluator）。

20.1 示例一：尝试正则系数

使用网格搜索的代码如下所示。

```
LogisticRegression lr = new LogisticRegression()
    .setFeatureCols(new_features)
    .setLabelCol(Chap04.LABEL_COL_NAME)
    .setPredictionCol(Chap04.PREDICTION_COL_NAME)
    .setPredictionDetailCol(Chap04.PRED_DETAIL_COL_NAME);

Pipeline pipeline = new Pipeline().add(lr);

GridSearchCV gridSearch = new GridSearchCV()
    .setNumFolds(5)
    .setEstimator(pipeline)
    .setParamGrid(
        new ParamGrid()
            .addGrid(lr, LogisticRegression.L_1,
                new Double[] {0.0000001, 0.000001, 0.00001, 0.0001, 0.001, 0.01, 0.1, 1.0, 10.0})
    )
    .setTuningEvaluator(
        new BinaryClassificationTuningEvaluator()
            .setLabelCol(Chap04.LABEL_COL_NAME)
            .setPredictionDetailCol(Chap04.PRED_DETAIL_COL_NAME)
            .setTuningBinaryClassMetric(TuningBinaryClassMetric.AUC)
    )
    .enableLazyPrintTrainInfo();
```

```
GridSearchCVModel bestModel = gridSearch.fit(train_data);
```

这里使用的是 5 折交叉验证，设置要评估的估算器为 lr，并列举全部候选正则系数的取值。GridSearchCV 执行 fit 方法后得到 GridSearchCVModel，即使用搜索出的最优超参数对全部训练数据进行训练而得到的模型。

运行结果如下，前面是超参数搜索的输出结果，显示了各个参数对应的指标值；后面是应用当前最优超参数对整个数据集进行训练得到的模型评估结果。

```
Metric information:
  Metric name: AUC
  Larger is better: true
Tuning information:
|                AUC|              stage|param| value|
|-------------------|-------------------|-----|------|
|0.7952618317852187 |LogisticRegression |   11| 1.0E-5|
| 0.793731422403357 |LogisticRegression |   11| 1.0E-6|
|0.7819673037230401 |LogisticRegression |   11|  0.001|
|0.7796277279311166 |LogisticRegression |   11|   0.01|
|0.7715346390692316 |LogisticRegression |   11| 1.0E-7|
|0.7632430058789119 |LogisticRegression |   11| 1.0E-4|
|0.7441272544405644 |LogisticRegression |   11|    0.1|
|                0.5|LogisticRegression |   11|    1.0|
|                0.5|LogisticRegression |   11|   10.0|
GridSearchCV
-------------------------------- Metrics: --------------------------------
AUC:0.7905  Accuracy:0.78   Precision:0.5882    Recall:0.566    F1:0.5769  LogLoss:0.4799
|Pred\Real|  2|  1|
|---------|---|---|
|        2| 30| 21|
|        1| 23|126|
```

20.2　示例二：搜索GBDT超参数

这里使用随机搜索，同时为了减少计算次数，因此没有使用 K 折交叉验证，而是直接按指定的比例（TrainRatio=0.8）划分训练集与验证集，随机搜索中的搜索参数可以设置为在指定数组中选择或者在指定取值区间内随机取值。具体代码如下。

```
GbdtClassifier gbdt = new GbdtClassifier()
    .setFeatureCols(featuresColNames)
    .setLabelCol(Chap05.LABEL_COL_NAME)
    .setPredictionCol(Chap05.PREDICTION_COL_NAME)
    .setPredictionDetailCol(Chap05.PRED_DETAIL_COL_NAME);
```

```
RandomSearchTVSplit randomSearch = new RandomSearchTVSplit()
    .setNumIter(20)
    .setTrainRatio(0.8)
    .setEstimator(gbdt)
    .setParamDist(
        new ParamDist()
            .addDist(gbdt, GbdtClassifier.NUM_TREES, ValueDist.randArray(new Integer[] {50, 100}))
            .addDist(gbdt, GbdtClassifier.MAX_DEPTH, ValueDist.randInteger(4, 10))
            .addDist(gbdt, GbdtClassifier.MAX_BINS, ValueDist.randArray(new Integer[] {64, 128, 256, 512}))
            .addDist(gbdt, GbdtClassifier.LEARNING_RATE, ValueDist.randArray(new Double[] {0.3, 0.1, 0.01}))
    )
    .setTuningEvaluator(
        new BinaryClassificationTuningEvaluator()
            .setLabelCol(Chap05.LABEL_COL_NAME)
            .setPredictionDetailCol(Chap05.PRED_DETAIL_COL_NAME)
            .setTuningBinaryClassMetric(TuningBinaryClassMetric.F1)
    )
    .enableLazyPrintTrainInfo();

RandomSearchTVSplitModel bestModel = randomSearch.fit(train_sample);
```

运行结果如下，可以看出最优指标所对应的各个参数值。

```
Metric information:
  Metric name: F1
  Larger is better: true
Tuning information:
|                 F1|          stage|   param|value|         stage 2|  param 2|value 2|         stage 3|  param 3|value 3|         stage 4|     param 4|value 4|
|-------------------|---------------|--------|-----|----------------|---------|-------|----------------|---------|-------|----------------|------------|-------|
| 0.44776119402985076|GbdtClassifier|numTrees|  100|GbdtClassifier|maxDepth|      4|GbdtClassifier|maxBins|    512|GbdtClassifier|learningRate|    0.3|
| 0.410958904109589 |GbdtClassifier|numTrees|  100|GbdtClassifier|maxDepth|      8|GbdtClassifier|maxBins|    128|GbdtClassifier|learningRate|    0.3|
| 0.39285714285714285|GbdtClassifier|numTrees|   50|GbdtClassifier|maxDepth|      4|GbdtClassifier|maxBins|    512|GbdtClassifier|learningRate|    0.1|
| 0.38095238095238093|GbdtClassifier|numTrees|  100|GbdtClassifier|maxDepth|      6|GbdtClassifier|maxBins|    256|GbdtClassifier|learningRate|    0.1|
| 0.3561643835616438|GbdtClassifier|numTrees|  100|GbdtClassifier|maxDepth|      6|GbdtClassifier|maxBins|    512|GbdtClassifier|learningRate|    0.3|
| 0.3492063492063492|GbdtClassifier|numTrees|  100|GbdtClassifier|maxDepth|      6|GbdtClassifier|maxBins|     64|GbdtClassifier|learningRate|    0.3|
| 0.3384615384615385|GbdtClassifier|numTrees|  100|GbdtClassifier|maxDepth|      8|GbdtClassifier|maxBins|    256|GbdtClassifier|learningRate|    0.3|
| 0.3333333333333333|GbdtClassifier|numTrees|   50|GbdtClassifier|maxDepth|     10|GbdtClassifier|maxBins|     64|GbdtClassifier|learningRate|    0.3|
| 0.3283582089552239|GbdtClassifier|numTrees|   50|GbdtClassifier|maxDepth|      4|GbdtClassifier|maxBins|    128|GbdtClassifier|learningRate|    0.1|
| 0.2933333333333333|GbdtClassifier|numTrees|   50|GbdtClassifier|maxDepth|      5|GbdtClassifier|maxBins|    256|GbdtClassifier|learningRate|    0.3|
| 0.2909090909090909|GbdtClassifier|numTrees|  100|GbdtClassifier|maxDepth|      9|GbdtClassifier|maxBins|    128|GbdtClassifier|learningRate|   0.01|
| 0.2807017543859649|GbdtClassifier|numTrees|  100|GbdtClassifier|maxDepth|      7|GbdtClassifier|maxBins|    256|GbdtClassifier|learningRate|    0.3|
| 0.26666666666666666|GbdtClassifier|numTrees|   50|GbdtClassifier|maxDepth|      4|GbdtClassifier|maxBins|    512|GbdtClassifier|learningRate|    0.1|
| 0.25806451612903225|GbdtClassifier|numTrees|  100|GbdtClassifier|maxDepth|      5|GbdtClassifier|maxBins|    512|GbdtClassifier|learningRate|    0.1|
| 0.22641509433962265|GbdtClassifier|numTrees|  100|GbdtClassifier|maxDepth|      7|GbdtClassifier|maxBins|    128|GbdtClassifier|learningRate|   0.01|
|                0.0|GbdtClassifier|numTrees|   50|GbdtClassifier|maxDepth|     10|GbdtClassifier|maxBins|    512|GbdtClassifier|learningRate|   0.01|
|                0.0|GbdtClassifier|numTrees|   50|GbdtClassifier|maxDepth|      6|GbdtClassifier|maxBins|     64|GbdtClassifier|learningRate|   0.01|
|                0.0|GbdtClassifier|numTrees|   50|GbdtClassifier|maxDepth|      8|GbdtClassifier|maxBins|    256|GbdtClassifier|learningRate|   0.01|
|                0.0|GbdtClassifier|numTrees|   50|GbdtClassifier|maxDepth|      6|GbdtClassifier|maxBins|    256|GbdtClassifier|learningRate|   0.01|
|                0.0|GbdtClassifier|numTrees|   50|GbdtClassifier|maxDepth|      5|GbdtClassifier|maxBins|    512|GbdtClassifier|learningRate|   0.01|
```

20.3 示例三：最佳聚类个数

对于聚类问题，需要使用聚类调参评估器（ClusterTuningEvaluator），其他方面的设置与

20.1 节和 20.2 节相似，具体代码如下。

```java
KMeans kmeans = new KMeans()
    .setVectorCol(VECTOR_COL_NAME)
    .setPredictionCol(PREDICTION_COL_NAME);

GridSearchCV cv = new GridSearchCV()
    .setNumFolds(4)
    .setEstimator(kmeans)
    .setParamGrid(
        new ParamGrid()
            .addGrid(kmeans, KMeans.K, new Integer[] {2, 3, 4, 5, 6})
            .addGrid(kmeans, KMeans.DISTANCE_TYPE,
                new DistanceType[] {DistanceType.EUCLIDEAN, DistanceType.COSINE})
    )
    .setTuningEvaluator(
        new ClusterTuningEvaluator()
            .setVectorCol(VECTOR_COL_NAME)
            .setPredictionCol(PREDICTION_COL_NAME)
            .setLabelCol(LABEL_COL_NAME)
            .setTuningClusterMetric(TuningClusterMetric.RI)
    )
    .enableLazyPrintTrainInfo();

GridSearchCVModel bestModel = cv.fit(source);
```

搜索结果如下，最优的参数为使用余弦距离，选择聚类个数为 3。

```
Metric information:
  Metric name: ri
  Larger is better: true
Tuning information:
  |                  ri|stage|        param|   value|stage 2|param 2|value 2|
  |--------------------|-----|-------------|--------|-------|-------|-------|
  |  0.8994191559981034|KMeans|distanceType|  COSINE|KMeans |      k|      3|
  |  0.8700608503240082|KMeans|distanceType|EUCLIDEAN|KMeans|      k|      4|
  |  0.8698237711395606|KMeans|distanceType|  COSINE|KMeans |      k|      5|
  |  0.8688952110004742|KMeans|distanceType|EUCLIDEAN|KMeans|      k|      5|
  |  0.8517662399241347|KMeans|distanceType|  COSINE|KMeans |      k|      4|
  |  0.8423028291449345|KMeans|distanceType|  COSINE|KMeans |      k|      6|
  |  0.8401296032874981|KMeans|distanceType|EUCLIDEAN|KMeans|      k|      6|
  |  0.8264185237869449|KMeans|distanceType|EUCLIDEAN|KMeans|      k|      3|
  |  0.7933854907539117|KMeans|distanceType|  COSINE|KMeans |      k|      2|
  |  0.7813734787418998|KMeans|distanceType|EUCLIDEAN|KMeans|      k|      2|
```

利用最佳参数所得模型的聚类评估结果如下。

```
-------------------------------- Metrics: --------------------------------
k:3
VRC:469.9545    DB:0.7682   SilhouetteCoefficient:0.6398
ARI:0.9222 NMI:0.9144 Purity:0.9733

|   Label\Cluster|  2|  1|  0|
|----------------|---|---|---|
|   Iris-virginica|  0| 50|  0|
|Iris-versicolor|  0|  4| 46|
|     Iris-setosa| 50|  0|  0|
```

21 文本分析

我们接触到的很多数据是文本形式的，譬如新闻的标题和内容、电影评论、商品描述等。如何从这些文本数据中寻找相似的内容？新闻中的关键词是什么？用户评论了一段话，表达的意思是喜欢还是不喜欢？本章会讨论如何解决这样的问题。

本章主要介绍文本分析中的基本内容，譬如分词、词频统计、TF-IDF 等。后面的两章会基于文本内容构造特征，使用机器模型进行情感分析和预测。

21.1 数据探索

本章使用的数据来自今日头条中的新闻，参见链接 21-1。下载压缩文件，解压后得到 57MB 的文本文件，用文本编辑器打开，如图 21-1 所示。

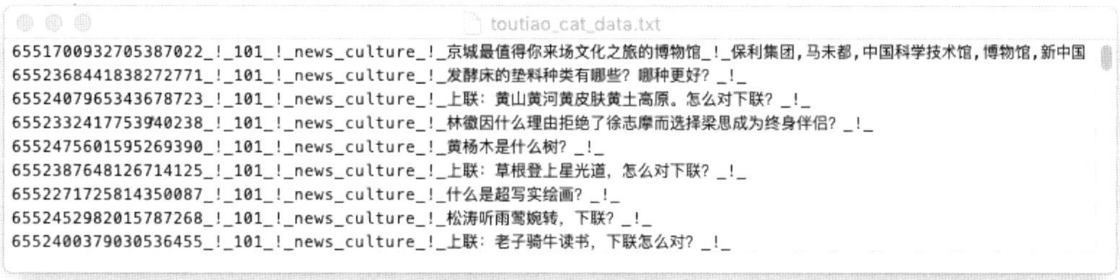

图 21-1

可以看到，每条记录为一行，各数据列间使用 "_!_" 进行分隔。使用 Alink CSV 格式数据源组件 CsvSourceBatchOp，经过如下配置即可读取。

```
static final String DATA_DIR = "/Users/yangxu/alink/data/news_toutiao/";
static final String ORIGIN_TRAIN_FILE = "toutiao_cat_data.txt";
static final String FIELD_DELIMITER = "_!_";
static final String SCHEMA_STRING =
    "id string, category_code int, category_name string, news_title string, keywords string";
private static BatchOperator getSource() {
    return new CsvSourceBatchOp()
        .setFilePath(DATA_DIR + ORIGIN_TRAIN_FILE)
        .setSchemaStr(SCHEMA_STRING)
        .setFieldDelimiter(FIELD_DELIMITER);
}
```

我们使用两个函数获取数据的基本信息，具体代码如下。

```
getSource()
  .lazyPrint(10)
  .lazyPrintStatistics();
```

获取前 10 条数据，输出如下。

```
id|category_code|category_name|news_title|keywords
--|-------------|-------------|----------|--------
6551700932705387022|101|news_culture|京城最值得你来场文化之旅的博物馆|保利集团,马未都,中国科学技术馆,博物馆,新中国
6552368441838272771|101|news_culture|发酵床的垫料种类有哪些？哪种更好？|null
6552407965343678723|101|news_culture|上联：黄山黄河黄皮肤黄土高原。怎么对下联？|null
6552332417753940238|101|news_culture|林徽因什么理由拒绝了徐志摩而选择梁思成为终身伴侣？|null
6552475601595269390|101|news_culture|黄杨木是什么树？|null
6552387648126714125|101|news_culture|上联：草根登上星光道，怎么对下联？|null
6552271725814350087|101|news_culture|什么是超写实绘画？|null
6552452982015787268|101|news_culture|松涛听雨莺婉转，下联？|null
6552400379030536455|101|news_culture|上联：老子骑牛读书，下联怎么对？|null
6552339283632455939|101|news_culture|上联：山水醉人何须酒。如何对下联？|null
```

输出的全表基本统计结果如下。

```
|      colName|  count|missing|      sum|    mean|variance|min|max|
|-------------|-------|-------|---------|--------|--------|---|---|
|           id| 382688|      0|      NaN|     NaN|     NaN|NaN|NaN|
|category_code| 382688|      0|  41163656| 107.5645| 22.4489|100|116|
|category_name| 382688|      0|      NaN|     NaN|     NaN|NaN|NaN|
|   news_title| 382688|      0|      NaN|     NaN|     NaN|NaN|NaN|
|     keywords| 382688| 122453|      NaN|     NaN|     NaN|NaN|NaN|
```

总记录数为 382688 条，keywords 列有缺失值，缺失比例约为 1/3。

我们再看一下类别分布情况，数据中提供了 category_code 和 category_name 两项，分别表示分类代码和分类名称。一般来说，它们应该是一致的，保险起见，我们也一并进行验证。同

时对 category_code、category_name 两列使用 groupBy，并进行计数操作，具体代码如下。

```
getSource()
    .groupBy("category_code, category_name", "category_code, category_name, COUNT(category_name) AS cnt")
    .orderBy("category_code", 100)
    .lazyPrint(-1);
```

结果如表 21-1 所示。

表 21-1 Category 代码与名称

category_code	category_name	cnt
100	news_story	6273
101	news_culture	28031
102	news_entertainment	39396
103	news_sports	37568
104	news_finance	27085
106	news_house	17672
107	news_car	35785
108	news_edu	27058
109	news_tech	41543
110	news_military	24984
112	news_travel	21422
113	news_world	26909
114	stock	340
115	news_agriculture	19322
116	news_game	29300

显然，category_code 列与 category_name 列的值是一一对应的；类别分布并不平均，最大的类别 news_tech 有 41543 条数据，最小的类别只有 340 条数据，相差 100 多倍。

21.2 分词

相比于单个的字符，单词可以表达更明确的意义，一些数字和成语经过分词后还是作为一个整体存在的。

21.2.1　中文分词

中文分词是指基于词法分析系统，对指定的文本内容进行分词，分词后的各个词语间以空格作为分隔符。

我们先构造 3 个句子，然后使用分词组件 SegmentBatchOp，设置选择进行分词操作的文本列 SelectedCol，并设置分词结果输出列 OutputCol，如下代码所示。

```
String[] strings = new String[] {
    "大家好！我在学习、使用 Alink。",
    "【流式计算和批式计算】、(Alink)",
    "《人工智能》，"机器学习"？2020"
};

MemSourceBatchOp source = new MemSourceBatchOp(strings, "sentence");

source.link(
    new SegmentBatchOp()
        .setSelectedCol("sentence")
        .setOutputCol("words")
).lazyPrint(-1, "< Segment >");
```

输出结果如表 21-2 所示。

表 21-2　分词结果

sentence	words
大家好！我在学习、使用 Alink。	大家 好 ！ 我 在 学习 、 使用 alink 。
【流式计算和批式计算】、(Alink)	【 流式 计算 和 批式 计算 】 、 (alink)
《人工智能》，"机器学习"？2020	《 人工智能 》 ， " 机器 学习 " ？ 2020

左边是原始文本，右边是分词结果，显然，单词之间、单词与标点符号之间都以空格分隔，英文字母变成了小写的，"机器学习"被拆成了两个单词——"机器"与"学习"，但我们希望把其当作整体看待，这就需要使用分词组件的自定义词典功能。使用方法 UserDefinedDict，输入需要被当作整体看待的单词，具体代码如下。

```
source.link(
    new SegmentBatchOp()
        .setSelectedCol("sentence")
        .setOutputCol("words")
        .setUserDefinedDict("流式计算", "机器学习")
).print();
```

运行结果如表 21-3 所示，右边是分词后的结果，我们发现"流式计算"和"机器学习"没有被继续拆分，已经被当作单词看待。

表 21-3　使用自定义词典的分词结果

sentence	words
大家好！我在学习、使用 Alink。	大家 好 ！ 我 在 学习 、 使用 alink 。
【流式计算和批式计算】、(Alink)	【 流式计算 和 批式 计算 】 、 (alink)
《人工智能》，"机器学习"？2020	《 人工智能 》 ， " 机器学习 " ？ 2020

接下来，演示停用词过滤组件 StopWordsRemoverBatchOp，它经常与分词组件配合使用，基于分词的结果，过滤掉常用的标点符号，还要把一些助词、副词、代词等经常出现且不能单独体现意义的词过滤掉。分词组件 SegmentBatchOp 后面连接停用词过滤组件 StopWordsRemoverBatchOp，并设置 SelectedCol，选择输入分词结果列，并设置输出列，具体代码如下。

```
source.link(
    new SegmentBatchOp()
        .setSelectedCol("sentence")
        .setOutputCol("words")
        .setUserDefinedDict("流式计算", "机器学习")
).link(
    new StopWordsRemoverBatchOp()
        .setSelectedCol("words")
        .setOutputCol("left_words")
).print();
```

运行结果如表 21-4 所示。

表 21-4　停用词过滤的结果

sentence	words	left_words
大家好！我在学习、使用 Alink。	大家 好 ！ 我 在 学习 、 使用 alink 。	大家 好 学习 使用 alink
【流式计算和批式计算】、(Alink)	【 流式计算 和 批式 计算 】 、 (alink)	流式计算 批式 计算 alink
《人工智能》，"机器学习"？2020	《 人工智能 》 ， " 机器学习 " ？ 2020	人工智能 机器学习 2020

停用词过滤结果在最右边一列，过滤后的内容没有了标点符号，"我""在""和"这些

单词也被过滤掉了。

我们同样也可以定义新的停用词，使用方法 setStopWords，新增了两个停用词——"计算"和"2020"，如下代码所示。

```
source.link(
  new SegmentBatchOp()
    .setSelectedCol("sentence")
    .setOutputCol("words")
    .setUserDefinedDict("流式计算", "机器学习")
).link(
  new StopWordsRemoverBatchOp()
    .setSelectedCol("words")
    .setOutputCol("left_words")
    .setStopWords("计算", "2020")
).print();
```

运行结果如表 21-5 所示，可以看到"计算"和"2020"这两个词都被过滤掉了。

表 21-5 扩展停用词

sentence	words	left_words
大家好！我在学习、使用 Alink。	大家 好 ！ 我 在 学习 、 使用 alink 。	大家 好 学习 使用 alink
【流式计算和批式计算】、(Alink)	【 流式计算 和 批式 计算 】 、 (alink)	流式计算 批式 alink
《人工智能》，"机器学习"？2020	《 人工智能 》 ， " 机器学习 " ？ 2020	人工智能 机器学习

我们再对本章的数据集进行分词操作，具体代码如下。

```
getSource()
  .select("news_title")
  .link(
    new SegmentBatchOp()
      .setSelectedCol("news_title")
      .setOutputCol("segmented_title")
  )
  .firstN(10)
  .print();
```

运行结果如表 21-6 所示。

表 21-6 新闻标题的分词结果

news_title	segmented_title
京城最值得你来场文化之旅的博物馆	京城 最 值得 你 来场 文化 之旅 的 博物馆
发酵床的垫料种类有哪些？哪种更好？	发酵 床 的 垫料 种类 有 哪些 ？ 哪种 更好 ？
上联：黄山黄河黄皮肤黄土高原。怎么对下联？	上联 ： 黄山 黄河 黄皮肤 黄土高原 。 怎么 对 下联 ？
林徽因什么理由拒绝了徐志摩而选择梁思成为终身伴侣？	林徽因 什么 理由 拒绝 了 徐志摩 而 选择 梁思成 为 终身 伴侣 ？
黄杨木是什么树？	黄杨木 是 什么 树 ？
上联：草根登上星光道，怎么对下联？	上联 ： 草根 登上 星光 道 ， 怎么 对 下联 ？
什么是超写实绘画？	什么 是 超 写实 绘画 ？
松涛听雨莺婉转，下联？	松涛 听雨莺 婉转 ， 下联 ？
上联：老子骑牛读书，下联怎么对？	上联 ： 老子 骑牛 读书 ， 下联 怎么 对 ？
上联：山水醉人何须酒。如何对下联？	上联 ： 山水 醉人 何须 酒 。 如何 对 下联 ？

21.2.2 Tokenizer和RegexTokenizer

Tokenizer 针对英文文本，将文档（譬如句子）分解为单元个体（譬如单词、数字、标点符号等）。使用时只需指定输入列和输出列，不需要设置其他参数。Tokenizer 会将文档内容转化为小写的，并以\s+为分隔符（即以单个或连续的多个空格作为分隔符）将文档分割为多个单元个体，并用空格符再将各单元个体连接为一个大字符串。

RegexTokenizer 是 Tokenizer 的升级版，可以通过正则表达式匹配完成更高级的操作。其中有两个参数非常重要——gaps 和 pattern。参数 gaps 是布尔型变量，表明正则匹配针对的对象，默认值为 true，即对分割符进行正则匹配，具体的正则表达式为参数 pattern 的内容；参数 pattern 的默认值为\s+，所以在默认参数情况下，RegexTokenizer 是 Tokenizer 的输出是一致的。将参数 gaps 设置为 false 时，则是对要提取的内容进行正则匹配，找到所有出现的匹配并以空格符连接，作为输出结果。

常用的正则表达式字符如表 21-7 所示。

表 21-7 常用的正则表达式字符

字符	含义
+	规定其前导字符必须在目标对象中出现一次或连续多次
*	规定其前导字符必须在目标对象中出现零次或连续多次
?	规定其前导对象必须在目标对象中出现零次或一次

续表

字符	含义
\s	用于匹配单个空格符，包括 Tab 键和换行符
\S	用于匹配除单个空格符之外的所有字符
\d	用于匹配从 0 到 9 的数字
\w	用于匹配字母、数字或下画线字符
\W	用于匹配所有与\w 不匹配的字符
.	用于匹配除换行符之外的所有字符

容易看出，正则表达式\s+可以用于匹配目标对象中的一个或多个空格字符。而\s、\S、\w、\W 都可以看作相反的操作，再考虑到参数 gaps 取值 true 和 false 时会影响正则匹配的对象是分隔符还是分隔出的内容，所以表 21-8 列出了等效的参数设置。

表 21-8 等效的参数设置

参数设置	等效的参数设置
gaps=true;pattern="\W"	gaps=false;pattern="\w"
gaps=true;pattern="\s"	gaps=false;pattern="\S"

下面通过实验看看各种组件和参数组合的结果。

首先来看 TokenizerBatchOp 组件和 RegexTokenizerBatchOp 组件在默认参数情况下的处理结果。这里构造了两个字符串，包含了多种元素，具体代码如下。

```
String[] strings = new String[] {
   "Hello!    This is Alink!",
   "Flink,Alink..AI#ML@2020"
};

MemSourceBatchOp source = new MemSourceBatchOp(strings, "sentence");

source
   .link(
      new TokenizerBatchOp()
         .setSelectedCol("sentence")
         .setOutputCol("tokens")
   )
   .link(
      new RegexTokenizerBatchOp()
         .setSelectedCol("sentence")
         .setOutputCol("regex_tokens")
   ).lazyPrint(-1);
```

运行结果如表 21-9 所示，显然，在默认参数的情况下，两个组件的处理结果一致。我们也注意到，第一个字符串中有连续的多个空格，在分词结果中没有出现；在默认参数情况下，标点符号没有与单词分离开。

表 21-9　英文 Tokenizer 的结果

sentence	tokens	regex_tokens
Hello!　　This is Alink!	hello! this is alink!	hello! this is alink!
Flink,Alink..AI#ML@2020	flink,alink..ai#ml@2020	flink,alink..ai#ml@2020

下面的实验基于 RegexTokenizerBatchOp 组件，看看参数 gaps=true;pattern="\W"与 gaps=false; pattern="\w"的结果是否一致。注意，组件的默认 gaps 参数值为 true，具体代码如下。

```
source
  .link(
    new RegexTokenizerBatchOp()
      .setSelectedCol("sentence")
      .setOutputCol("tokens_1")
      .setPattern("\\W+")
  )
  .link(
    new RegexTokenizerBatchOp()
      .setSelectedCol("sentence")
      .setOutputCol("tokens_2")
      .setGaps(false)
      .setPattern("\\w+")
  ).lazyPrint(-1);
```

运行结果如表 21-10 所示，两种参数设置得到的结果是一致的。结果中将单词与标点符号分离了，并去除了标点符号。

表 21-10　不同参数对 Tokenizer 的影响

sentence	tokens_1	tokens_2
Hello!　　This is Alink!	hello this is alink	hello this is alink
Flink,Alink..AI#ML@2020	flink alink ai ml 2020	flink alink ai ml 2020

我们注意到，结果中的单词都是小写的，这是因为 RegexTokenizerBatchOp 组件的参数 ToLowerCase 的默认值为 true。如下代码所示，进行对比实验。

```
source
  .link(
```

```
    new RegexTokenizerBatchOp()
        .setSelectedCol("sentence")
        .setOutputCol("tokens_1")
        .setPattern("\\W+")
)
.link(
    new RegexTokenizerBatchOp()
        .setSelectedCol("sentence")
        .setOutputCol("tokens_2")
        .setPattern("\\W+")
        .setToLowerCase(false)
).lazyPrint(-1);
```

运行结果如表 21-11 所示，最右边一列中的单词没有被转换为小写形式。

表 21-11 小写形式转换

sentence	tokens_1	tokens_2
Hello!　　This is Alink!	hello this is alink	Hello This is Alink
Flink,Alink..AI#ML@2020	flink alink ai ml 2020	Flink Alink AI ML 2020

对于 Tokenize 的结果，我们同样可以使用停用词过滤组件 StopWordsRemoverBatchOp，如下代码所示。

```
source
    .link(
        new RegexTokenizerBatchOp()
            .setSelectedCol("sentence")
            .setOutputCol("tokens")
            .setPattern("\\W+")
    )
    .link(
        new StopWordsRemoverBatchOp()
            .setSelectedCol("tokens")
            .setOutputCol("left_tokens")
    ).lazyPrint(-1);
```

运行结果如表 21-12 所示，可以看到，单词"this"和"is"都被过滤掉了。

表 21-12 英文停用词过滤

sentence	tokens	left_tokens
Hello!　　This is Alink!	hello this is alink	hello alink
Flink,Alink..AI#ML@2020	flink alink ai ml 2020	flink alink ai ml 2020

21.3 词频统计

Alink 提供了 WordCountBatchOp 组件，统计各个单词在数据集中出现的总数。为了便于比对，这里从数据集中选出了 10 条数据，先进行分词，然后连接 WordCountBatchOp 组件，组件中只需要设置文本数据所在列的名称，其输出的格式是固定的，共 2 列（word 和 cnt），随后按个数进行降序排序并输出。具体代码如下。

```
BatchOperator titles = getSource()
   .firstN(10)
   .select("news_title")
   .link(
      new SegmentBatchOp()
         .setSelectedCol("news_title")
         .setOutputCol("segmented_title")
         .setReservedCols(new String[] {})
   );
titles
   .link(
      new WordCountBatchOp()
         .setSelectedCol("segmented_title")
   )
   .orderBy("cnt", 100, false)
   .lazyPrint(-1, "WordCount");
```

运行结果如表 21-13 所示，其中显示了出现次数较多的单词。问号出现的次数最多，"下联"在单词中出现的次数最多。

表 21-13　单词出现的次数

word	cnt
?	10
下联	5
对	4
上联	4
:	4
,	3
怎么	3
什么	3

续表

word	cnt
。	2
是	2
的	2
写实	1
博物馆	1
发酵	1
……	……

进一步，如果想知道各单词在每个文本数据中的出现次数，可以使用组件 DocWordCount-BatchOp。将每个文本数据看作一个 Doc，这里直接将文本列当作 ID 列，相关代码如下。

```
titles
  .link(
    new DocWordCountBatchOp()
      .setDocIdCol("segmented_title")
      .setContentCol("segmented_title")
  )
  .lazyPrint(-1, "DocWordCount");
```

运行结果如表 21-14 所示，此时的 cnt 列是左边文本中出现相应单词的次数。

表 21-14 DocWordCount 组件的运行结果

segmented_title	word	cnt
……	……	……
发酵 床 的 垫料 种类 有 哪些 ？ 哪种 更好 ？	种类	1
发酵 床 的 垫料 种类 有 哪些 ？ 哪种 更好 ？	的	1
发酵 床 的 垫料 种类 有 哪些 ？ 哪种 更好 ？	更好	1
发酵 床 的 垫料 种类 有 哪些 ？ 哪种 更好 ？	垫料	1
发酵 床 的 垫料 种类 有 哪些 ？ 哪种 更好 ？	哪种	1
发酵 床 的 垫料 种类 有 哪些 ？ 哪种 更好 ？	有	1
发酵 床 的 垫料 种类 有 哪些 ？ 哪种 更好 ？	床	1
发酵 床 的 垫料 种类 有 哪些 ？ 哪种 更好 ？	哪些	1

续表

segmented_title	word	cnt
发酵 床 的 垫料 种类 有 哪些 ？ 哪种 更好 ？	发酵	1
发酵 床 的 垫料 种类 有 哪些 ？ 哪种 更好 ？	？	2
……	……	……

21.4　单词的区分度

下面首先介绍几个重要概念。

词频（Term Frequency，TF），即单词在该文档中出现的频率。

$$TF = \frac{\text{单词在该文档出现的次数}}{\text{该文档中单词的总数}}$$

显然，TF 的值越大，表示这个单词越重要。

逆文本频率指数（Inverse Document Frequency，IDF），是指在一个文档库中，一个分词出现在的文档数越少，那么它越能和其他文档区别开来。

$$IDF = \ln\left(\frac{\text{总文档数}}{\text{出现该单词的文档数}}\right)$$

TF-IDF（Term Frequency-Inverse Document Frequency）通过计算单词在文档中的频率，进而得到单词的权重，可以用来评估单词对于一个文档集或一个语料库中的其中一份文档的重要程度。单词的重要性与它在文档中出现的次数成正比，但同时会与它在语料库中出现的频率成反比。

计算 TF-IDF 的值就是将 TF 和 IDF 的值相乘。对于某一特定文档内的高频（即 TF 的值较高）单词，并且包含该单词的文档数目在整个文档集合中所占的比例较低（即 IDF 的值较高），则可以产生高权重的 TF-IDF。所以，使用 TF-IDF 可以过滤掉常见的单词，保留能区分该文档的单词。

Alink 提供了两个组件 DocCountVectorizer 和 DocHashCountVectorizer，都可以用于计算上述的指标。二者的区别在于，DocCountVectorizer 是把每个单词对应一个向量分量，单词的总个数就是向量的维度；DocHashCountVectorizer 是指定向量维度，示例中的 setNumFeatures 方法就限定了向量维度为 100，各个单词通过 hash 的方式映射到向量的相应分量。这里会存在多个单词映射到同一个分量的问题，使用大的向量维度可以降低产生这种问题的概率。

```
for (String featureType : new String[] {"WORD_COUNT", "Binary", "TF", "IDF", "TF_IDF"}) {
    new DocCountVectorizer()
        .setFeatureType(featureType)
        .setSelectedCol("segmented_title")
        .setOutputCol("vec")
        .fit(titles)
        .transform(titles)
        .lazyPrint(-1, "DocCountVectorizer + " + featureType);
}
for (String featureType : new String[] {"WORD_COUNT", "Binary", "TF", "IDF", "TF_IDF"}) {
    new DocHashCountVectorizer()
        .setFeatureType(featureType)
        .setSelectedCol("segmented_title")
        .setOutputCol("vec")
        .setNumFeatures(100)
        .fit(titles)
        .transform(titles)
        .lazyPrint(-1, "DocHashCountVectorizer + " + featureType);
}
```

输出内容较多，我们选择一条经过分词操作的文本数据（"发酵 床 的 垫料 种类 有 哪些 ？ 哪种 更好 ？"），使用组件 DocCountVectorizer 在各种方式下生成的结果向量汇总见表 21-15。

表 21-15　向量生成方式及结果

向量生成方式	发酵 床 的 垫料 种类 有 哪些 ？ 哪种 更好 ？
WORD_COUNT	$61$0:2.0 10:1.0 20:1.0 22:1.0 23:1.0 24:1.0 28:1.0 33:1.0 35:1.0 43:1.0
BINARY	$61$0:1.0 10:1.0 20:1.0 22:1.0 23:1.0 24:1.0 28:1.0 33:1.0 35:1.0 43:1.0
TF	$61$0:0.18181818181818182 10:0.09090909090909091 20:0.09090909090909091 22:0.09090909090909091 23:0.09090909090909091 24:0.09090909090909091 28:0.09090909090909091 33:0.09090909090909091 35:0.09090909090909091 43:0.09090909090909091
IDF	$61$0:0.09531017980432493 10:1.2992829841302609 20:1.7047480922384253 22:1.7047480922384253 23:1.7047480922384253 24:1.7047480922384253 28:1.7047480922384253 33:1.7047480922384253 35:1.7047480922384253 43:1.7047480922384253
TF_IDF	$61$0:0.017329123600786353 10:0.11811663492093281 20:0.15497709929440232 22:0.15497709929440232 23:0.15497709929440232 24:0.15497709929440232 28:0.15497709929440232 33:0.15497709929440232 35:0.15497709929440232 43:0.15497709929440232

选择另一条文本数据("京城 最 值得 你 来场 文化 之旅 的 博物馆"),使用 DocHashCountVectorizer 组件,并设置参数 NumFeatures=100,在各种方式下生成的结果向量汇总见表 21-16。注意 WORD_COUNT 方法生成的结果向量,索引号 43 对应的值为 2.0,而该文本中并没有相同的单词,应该是某两个单词被 hash 到同一个向量分量相同导致的。

表 21-16 有向量维度限制的向量生成结果

向量生成方式	京城 最 值得 你 来场 文化 之旅 的 博物馆
WORD_COUNT	$100$8:1.0 21:1.0 26:1.0 38:1.0 42:1.0 43:2.0 56:1.0 93:1.0
BINARY	$100$8:1.0 21:1.0 26:1.0 38:1.0 42:1.0 43:1.0 56:1.0 93:1.0
TF	$100$8:0.1111111111111111 21:0.1111111111111111 26:0.1111111111111111 38:0.1111111111111111 42:0.1111111111111111 43:0.2222222222222222 56:0.1111111111111111 93:0.1111111111111111
IDF	$100$8:1.2992829841302609 21:1.7047480922384253 26:1.0116009116784799 38:1.7047480922384253 42:1.2992829841302609 43:1.2992829841302609 56:1.7047480922384253 93:0.6061358035703155
TF_IDF	$100$8:0.14436477601447342 21:0.18941645469315835 26:0.11240010129760887 38:0.18941645469315835 42:0.14436477601447342 43:0.28872955202894685 56:0.18941645469315835 93:0.06734842261892394

21.5 抽取关键词

文档中的关键词是对详细文档内容的提炼和总结,可以体现文档的基本内容,帮助用户迅速了解文档所关注的领域和所持的基本观点。本小节会从常用的关键内容计算原理入手,帮助用户更深刻地理解"关键"的含义,然后使用关键词抽取组件对示例数据进行操作。

21.5.1 原理简介

关键词和关键句的抽取常用到 TextRank 算法,该算法可以看作著名的网页排名 PageRank 算法在文本领域的应用和改进。本节会首先简要地介绍 PageRank 算法,然后把文本的特点融入其中,以便于我们很自然地理解 TextRank 算法。

在网页排名方面,Google 的 PageRank 算法很好地体现了网页的相关性和重要性,一个网页的重要性不但和有多少个网页链接它相关,还与每个链接它的网页的重要性相关。图 21-2 所示的 PageRank 卡通演示图（来自维基百科）比较形象地说明了这一点，图标个头越大，表示它越重要。

图 21-2

PageRank 的计算公式如下。

$$PR(P_i) = (1-d) + d \sum_{P_k \in \text{In}(P_i)} \frac{1}{|\text{Out}(P_k)|} PR(P_k)$$

其中，$PR(P_i)$ 是网页 P_i 的 PageRank 值；d 是阻尼系数，一般设置为 0.85；$\text{In}(P_i)$ 是存在指向网页 P_k 的链接的网页集合；$\text{Out}(P_k)$ 是网页 P_k 中存在的链接指向的网页的集合。

TextRank 算法可以被看作 PageRank 在自然语言处理领域的应用，主要用于为文本生成关键字和摘要。TextRank 在 PageRank 的基础上引入了边的权值的概念，代表两个单词或句子的相似度，公式如下。

$$TR(T_i) = (1-d) + d \cdot \sum_{T_k \in \text{In}(T_i)} \frac{w_{ki}}{\sum_{T_j \in \text{Out}(T_k)} w_{kj}} TR(T_k)$$

其中，$TR(P_i)$ 是单词或句子 T_i 的 TextRank 值；d 是阻尼系数，一般设置为 0.85；$\text{In}(T_i)$ 是指向 T_i 的单词或句子集合；$\text{Out}(T_k)$ 是被 T_k 指向的单词或句子集合。

在计算关键词时，首先要将文档进行分词，并采取停用词过滤等操作。单词之间的联系是通过是否在句子中相邻出现来定义的，具体做法是，定义一个窗口长度 k，对于每个句子的单词

$$w_1, w_2, w_3, \cdots, w_{m-1}, w_m$$

考查各个窗口

$$w_1, w_2, \cdots, w_k$$

$$w_2, w_3, \cdots, w_{k+1}$$

$$\vdots$$

$$w_{m-k+1}, w_{m-k+2}, \cdots, w_m$$

在每个窗口中出现的任意两个单词间定义一个双向的边，每条边出现的次数即为其权重。

在定义了单词和联系后，就可以使用 TextRank 算法进行计算，并根据 rank 值确定单词的重要性。

21.5.2 示例

很多人都听过相声《报菜名》，那么菜单中的关键词是什么呢？下面我们就用关键词提取组件计算一下，相关代码如下。

```
String[] strings = new String[] {
"蒸羊羔、蒸熊掌、蒸鹿尾儿、烧花鸭、烧雏鸡、烧子鹅、卤猪、卤鸭、酱鸡、腊肉、松花小肚儿、晾肉、香肠儿、什锦苏盘、熏鸡白肚儿、清蒸八宝猪、江米酿鸭子、罐儿野鸡、罐儿鹌鹑。"
+ "卤什件儿、卤子鹅、山鸡、兔脯、菜蟒、银鱼、清蒸哈什蚂、烩鸭丝、烩鸭腰、烩鸭条、清拌鸭丝、黄心管儿、焖白鳝、焖黄鳝、豆豉鲇鱼、锅烧鲤鱼、炉烂甲鱼、抓炒鲤鱼、抓炒对儿虾。"
+ "软炸里脊、软炸鸡、什锦套肠儿、卤煮寒鸦儿、麻酥油卷儿、熘鲜蘑、熘鱼脯、熘鱼肚、熘鱼片儿、醋熘肉片儿、烩三鲜、烩白蘑、烩鸽子蛋、炒银丝、烩鳗鱼、炒白虾、炝青蛤、炒面鱼。"
+ "炒竹笋、芙蓉燕菜、炒虾仁儿、烩虾仁儿、烩腰花儿、烩海参、炒蹄筋儿、锅烧海参、锅烧白菜、炸木耳、炒肝尖儿、桂花翅子、清蒸翅子、炸飞禽、炸汁儿、炸排骨、清蒸江瑶柱。"
+ "糖熘芡仁米、拌鸡丝、拌肚丝、什锦豆腐、什锦丁儿、糟鸭、糟熘鱼片儿、熘蟹肉、炒蟹肉、烩蟹肉、清拌蟹肉、蒸南瓜、酿倭瓜、炒丝瓜、酿冬瓜、烟鸭掌儿、焖鸭掌儿、焖笋、炝茭白。"
+ "茄子晒炉肉、鸭羹、蟹肉羹、鸡血汤、三鲜木樨汤、红丸子、白丸子、南煎丸子、四喜丸子、三鲜丸子、氽丸子、鲜虾丸子、鱼脯丸子、烙炸丸子、豆腐丸子、樱桃肉、马牙肉、米粉肉。"
+ "一品肉、栗子肉、坛子肉、红焖肉、黄焖肉、酱豆腐肉、晒炉肉、炖肉、黏糊肉、炉肉、扣肉、松肉、罐儿肉、烧肉、大肉、烤肉、白肉、红肘子、白肘子、熏肘子、水晶肘子、蜜蜡肘子。"
+ "锅烧肘子、扒肘条、炖羊肉、酱羊肉、烧羊肉、烤羊肉、清羔羊肉、五香羊肉、氽三样儿、爆三样儿、炸卷果儿、烩散丹、烩酸燕儿、烩银丝、烩杂碎、氽节子、烩节子、炸绣球。"
+ "三鲜鱼翅、栗子鸡、氽鲤鱼、酱汁鲫鱼、活钻鲤鱼、板鸭、筒子鸡、烩脐肚、烩南荠、爆肚仁儿、盐水肘花儿、锅烧猪蹄儿、拌粳条、炖吊子、烧肝尖儿、烧肥肠儿、烧心、烧肺。"
+ "烧紫盖儿、烧连帖、烧宝盖儿、油炸肺、酱瓜丝儿、山鸡丁儿、拌海蜇、龙须菜、炝冬笋、玉兰片、烧鸳鸯、烧鱼头、烧槟子、烧百合、炸豆腐、炸面筋、炸软巾、糖熘饹儿。"
+ "拔丝山药、糖焖莲子、酿山药、杏仁儿酪、小炒螃蟹、氽大甲、炒荤素儿、什锦葛仙米、鳎目鱼、八代鱼、海鲫鱼、黄花鱼、鲥鱼、带鱼、扒海参、扒燕窝、扒鸡腿儿、扒鸡块儿。"
+ "扒肉、扒面筋、扒三样儿、油泼肉、酱泼肉、炒虾黄、熘蟹黄、炒子蟹、炸子蟹、佛手海参、炸烹儿、炒芡子米、奶汤、翅子汤、三丝汤、熏斑鸠、卤斑鸠、海白米、烩腰丁儿。"
+ "火烧茨菰、炸鹿尾儿、焖鱼头、拌皮渣儿、氽肥肠儿、炸紫盖儿、鸡丝豆苗、十二台菜、汤羊、鹿肉、驼峰、
```

鹿大哈、插根儿、炸花件儿、清拌粉皮儿、炝莴笋、烹芽韭、木樨菜。"
    + "烹丁香、烹大肉、烹白肉、麻辣野鸡、烩酸蕾、熘脊髓、咸肉丝儿、白肉丝儿、荸荠一品锅、素炝春不老、清焖莲子、酸黄菜、烧萝卜、脂油雪花儿菜、烩银耳、炒银枝儿。"
    + "八宝榛子酱、黄鱼锅子、白菜锅子、什锦锅子、汤圆锅子、菊花锅子、杂烩锅子、煮饽饽锅子、肉丁辣酱、炒肉丝、炒肉片儿、烩酸菜、烩白菜、烩豌豆、焖扁豆、氽毛豆、炒豇豆、外加腌苤蓝丝儿。",
};

```
new MemSourceBatchOp(strings, "doc")
    .link(
        new SegmentBatchOp()
            .setSelectedCol("doc")
            .setOutputCol("words")
    )
    .link(
        new StopWordsRemoverBatchOp()
            .setSelectedCol("words")
    )
    .link(
        new KeywordsExtractionBatchOp()
            .setMethod(Method.TEXT_RANK)
            .setSelectedCol("words")
            .setOutputCol("extract_keywords")
    )
    .select("extract_keywords")
    .print();
```

菜名文档较长，只保留了头尾；对文档进行分词、停用词过滤处理，然后设置 KeywordsExtractionBatchOp 组件，选择使用 TEXT_RANK 方法，选择输入的文本数据列（空格分隔的分词结果形式），并设置输出数据列名称。

计算结果如下。

```
extract_keywords
----------------
烩 儿 炒 肉 烧 熘 炸 氽 焖 丸子
```

关键词以烹饪方法为主，有"烩、炒、烧、熘、炸、氽、焖"，排在第 2 位的是儿化音"儿"；作为重要材料的"肉"排在第 4 位；还有一个关键词"丸子"，也在菜谱中多次与其他词组合出现。

TextRank 算法更适合用在长文本中，比如本章中的新闻标题数据，使用 TF-IDF 指标进行判断更为有效。与前面的示例类似，对文档进行分词、停用词过滤处理，然后设置 KeywordsExtractionBatchOp 组件，选择使用 TF_IDF 方法，设置最多输出关键词的个数为 5，选择输入的文本数据列（空格分隔的分词结果形式），并设置输出数据列的名称。

```
getSource()
    .link(
        new SegmentBatchOp()
            .setSelectedCol("news_title")
```

```
      .setOutputCol("segmented_title")
  )
  .link(
    new StopWordsRemoverBatchOp()
      .setSelectedCol("segmented_title")
  )
  .link(
    new KeywordsExtractionBatchOp()
      .setTopN(5)
      .setMethod(Method.TF_IDF)
      .setSelectedCol("segmented_title")
      .setOutputCol("extract_keywords")
  )
  .select("news_title, extract_keywords")
  .firstN(10)
  .print();
```

运行结果如表 21-17 所示，右边一列为提取出来的关键词。

表 21-17　关键词提取

news_title	extract_keywords
京城最值得你来场文化之旅的博物馆	来场 京城 博物馆 之旅 文化
发酵床的垫料种类有哪些？哪种更好？	垫料 种类 床 发酵 哪种
上联：黄山黄河黄皮肤黄土高原。怎么对下联？	黄皮肤 黄土高原 黄河 黄山 下联
林徽因什么理由拒绝了徐志摩而选择梁思成为终身伴侣？	终身伴侣 梁思成 徐志摩 林徽因 理由
黄杨木是什么树？	黄杨木 树
上联：草根登上星光道，怎么对下联？	草根 星光 登上 道 下联
什么是超写实绘画？	写实 绘画 超
松涛听雨莺婉转，下联？	听雨莺 婉转 松涛 下联
上联：老子骑牛读书，下联怎么对？	骑牛 老子 读书 下联 上联
上联：山水醉人何须酒。如何对下联？	何须 醉人 山水 酒 下联

21.6　文本相似度

文本不能像数字那样被定义大小，但是我们可以看出哪些文本更接近，这种"接近"是通过文本间的距离和相似度来衡量的。常用的有编辑距离（Levenshtein 距离）、最长公共子串（Longest Common Substring，LCS）距离、余弦相似度、Jaccard 相似度等。

本节会使用编辑距离进行文本比较。编辑距离是由苏联科学家 Vladimir Levenshtein 在 1965 年提出的，故又叫 Levenshtein 距离，指的是在两个字符串之间，由一个字符串转换成另一个字符串所需要的最少的编辑操作次数。

将编辑操作定义为以下 3 种类型。

（1）插入一个字符。

（2）删除一个字符。

（3）将一个字符替换成另一个字符。

为了加深对此距离的理解，举例如下。

- ab 和 ab 需要有 0 个操作。字符串同理。
- ab 和 a 需要有 1 个删除操作。
- a 和 ab 需要有 1 个插入操作。
- ab 和 ac 需要有 1 个替换操作。

编辑距离相似度的公式如下，即单位 1 减去编辑距离与最大字符串长度的比值。

$$\text{Simi}_{\text{Edit}} = 1 - \frac{\text{EditDistance}}{\max(\text{Length1}, \text{Length2})}$$

21.6.1 文本成对比较

本小节使用的数据为构造的 5 对文本数据，如表 21-18 所示。

表 21-18 构造的 5 对文本数据

col1	col2
机器学习	机器学习
批式计算	流式计算
Machine Learning	ML
Flink	Alink
Good Morning!	Good Evening!

如下代码所示，将要比较的两列数据分别命名为 col1 和 col2，并分别构造批式数据源和流式数据源。

```
Row[] rows = new Row[] {
    Row.of("机器学习", "机器学习"),
    Row.of("批式计算", "流式计算"),
```

```
    Row.of("Machine Learning", "ML"),
    Row.of("Flink", "Alink"),
    Row.of("Good Morning!", "Good Evening!")
};
MemSourceBatchOp source = new MemSourceBatchOp(rows, new String[] {"col1", "col2"});
MemSourceStreamOp source_stream = new MemSourceStreamOp(rows, new String[] {"col1", "col2"});
```

以字符为单位来比较字符串，可以使用字符串相似度成对比较组件 StringSimilarityPairwiseBatchOp，就是将文本看作字符串，以字符为单位进行比较，如下代码所示。

```
source
    .link(
        new StringSimilarityPairwiseBatchOp()
            .setSelectedCols("col1", "col2")
            .setMetric("LEVENSHTEIN")
            .setOutputCol("LEVENSHTEIN")
    )
... ...
    .link(
        new StringSimilarityPairwiseBatchOp()
            .setSelectedCols("col1", "col2")
            .setMetric("JACCARD_SIM")
            .setOutputCol("JACCARD_SIM")
    )
    .lazyPrint(-1, "\n## StringSimilarityPairwiseBatchOp ##");
```

调用该组件 5 次，每次都换一种度量方式，并根据选择的度量方式命名结果数据列。计算结果如表 21-19 所示。

表 21-19 各种句子的相似度指标

col1	col2	LEVENSHTEIN	LEVENSHTEIN_SIM	LCS	LCS_SIM	JACCARD_SIM
机器学习	机器学习	0.0000	1.0000	4.0000	1.0000	1.0000
批式计算	流式计算	1.0000	0.7500	3.0000	0.7500	0.6000
Machine Learning	ML	14.0000	0.1250	2.0000	0.1250	0.1250
Flink	Alink	1.0000	0.8000	4.0000	0.8000	0.6667
Good Morning!	Good Evening!	3.0000	0.7692	10.0000	0.7692	0.6250

可以观察到如下信息。
- LEVENSHTEIN 和 LCS 计算的是距离。
- LEVENSHTEIN_SIM、LCS_SIM 和 JACCARD_SIM，即以"_SIM"结尾的为相似度指标，取值范围为[0, 1]。

以单词为单位来比较字符串，可以使用文本相似度成对比较组件 TextSimilarityPairwiseBatchOp，就是将文本看作由单词构成且以单词为单位的，如下代码所示。

```
source
    .link(
        new SegmentBatchOp()
            .setSelectedCol("col1")
    )
    .link(
        new SegmentBatchOp()
            .setSelectedCol("col2")
    )
    .link(
        new TextSimilarityPairwiseBatchOp()
            .setSelectedCols("col1", "col2")
            .setMetric("LEVENSHTEIN")
            .setOutputCol("LEVENSHTEIN")
    )
... ...
    .link(
        new TextSimilarityPairwiseBatchOp()
            .setSelectedCols("col1", "col2")
            .setMetric("JACCARD_SIM")
            .setOutputCol("JACCARD_SIM")
    )
    .lazyPrint(-1, "\n## TextSimilarityPairwiseBatchOp ##");
```

首先进行分词操作，由于分词组件每次只能处理一列，所以分两次进行调用，分别处理 col1 和 col2。调用该组件 5 次，每次都换一种度量方式，并根据选择的度量方式命名结果数据列。

计算结果如表 21-20 所示。

表 21-20　各种文本的相似度指标

col1	col2	LEVENSHTEIN	LEVENSHTEIN_SIM	LCS	LCS_SIM	JACCARD_SIM
机器 学习	机器 学习	0.0000	1.0000	2.0000	1.0000	1.0000
批式 计算	流式 计算	1.0000	0.5000	1.0000	0.5000	0.3333
machine learning	ml	4.0000	0.0000	0.0000	0.0000	0.0000
flink	alink	1.0000	0.0000	0.0000	0.0000	0.0000
good morning !	good evening !	1.0000	0.8000	4.0000	0.8000	0.6667

可以观察到如下信息。

- 由于以单词为基本单位进行比较,每对文本包含的元素个数大幅下降。
- 从单词角度来看,fline 与 alink 是不同的,3 个相似度指标都为 0,但是它们的大部分字符是相同的,所以在以字符为单位进行计算时,3 个相似度指标都是比较高的。

流式数据的处理与批式类似,要使用相应的流式字符串相似度成对比较组件 StringSimilarityPairwiseStreamOp,如下代码所示。

```
source_stream
  .link(
    new StringSimilarityPairwiseStreamOp()
      .setSelectedCols("col1", "col2")
      .setMetric("LEVENSHTEIN")
      .setOutputCol("LEVENSHTEIN")
  )
... ...
  .link(
    new StringSimilarityPairwiseStreamOp()
      .setSelectedCols("col1", "col2")
      .setMetric("JACCARD_SIM")
      .setOutputCol("JACCARD_SIM")
  )
  .print();

StreamOperator.execute();
```

调用该组件 5 次,每次都换一种度量方式,并根据选择的度量方式命名结果数据列。

计算结果输出如表 21-21 所示,对比批式计算结果,在数值上是一样的。

表 21-21 流式相似度的计算结果

col1	col2	LEVENSHTEIN	LEVENSHTEIN_SIM	LCS	LCS_SIM	JACCARD_SIM
Machine Learning	ML	14.0000	0.1250	2.0000	0.1250	0.1250
Flink	Alink	1.0000	0.8000	4.0000	0.8000	0.6667
批式计算	流式计算	1.0000	0.7500	3.0000	0.7500	0.6000
Good Morning!	Good Evening!	3.0000	0.7692	10.0000	0.7692	0.6250
机器学习	机器学习	0.0000	1.0000	4.0000	1.0000	1.0000

21.6.2 最相似的TopN

给定一个文本,如何从若干文本中找到 N 个最相近的那些文本,就是本小节关注的问题。Alink 提供了字符串最相似 TopN 计算组件 StringNearestNeighbor(以字符为单位,将文本数据看作字符串)和 TextNearestNeighbor(以单词为单位,将文本数据看作文档)。为了提高计算效率，会对作为对比的原始数据进行一些预计算操作，并保留预计算的结果，我们也可将此看作对原始数据的"训练"，而预计算结果就是"模型"。

从原始数据中取出 4 条新闻标题，构造目标数据集，具体代码如下。

```
Row[] rows = new Row[] {
    Row.of("林徽因什么理由拒绝了徐志摩而选择梁思成为终身伴侣？"),
    Row.of("发酵床的垫料种类有哪些？哪种更好？"),
    Row.of("京城最值得你来场文化之旅的博物馆"),
    Row.of("什么是超写实绘画？")
};
MemSourceBatchOp target = new MemSourceBatchOp(rows, new String[] {TXT_COL_NAME});
BatchOperator <?> source = getSource();
```

1. 基本算法

我们以字符为单位，将文本数据看作字符串，使用 StringNearestNeighbor 组件进行计算。该组件需要输入两个数据，一个是文本数据所在的列，另一个是用来显示结果的，直接罗列各个相似文本。如果每个文本较长，结果会比较臃肿，所以单独提供了参数 IdCol，用于设置文本对应的 ID 列，计算结果会显示 ID 信息。鉴于本章的文本数据为新闻标题信息，比较短小，所以 IdCol 列仍然设置为标题列，便于我们查看结果。

```
for (String metric : new String[] {"LEVENSHTEIN", "LCS", "SSK", "COSINE"}) {
    new StringNearestNeighbor()
        .setMetric(metric)
        .setSelectedCol(TXT_COL_NAME)
        .setIdCol(TXT_COL_NAME)
        .setTopN(5)
        .setOutputCol("similar_titles")
        .fit(source)
        .transform(target)
        .lazyPrint(-1, "StringNearestNeighbor + " + metric.toString());
    BatchOperator.execute();
}
```

计算结果如表 21-22 所示，左边一列为 StringNearestNeighbor 组件的度量参数 metric 值。

表 21-22　字符串最近邻计算结果

metric 值	林徽因什么理由拒绝了徐志摩而选择梁思成为终身伴侣？
String : LEVENSHTEIN	{"ID":"[\"林徽因什么理由拒绝了徐志摩而选择梁思成为终身伴侣？\",\"林徽因拒绝徐志摩而选择梁思成为终身伴侣的原因是什么？\",……]","METRIC":"[1.0,12.0,19.0,20.0,20.0]"}

续表

metric 值	林徽因什么理由拒绝了徐志摩而选择梁思成为终身伴侣？
String：LCS	{"ID":"[\"林徽因什么理由拒绝了徐志摩而选择梁思成为终身伴侣？\",\"林徽因拒绝徐志摩而选择梁思成为终身伴侣的原因是什么？\",......,\"NBA 球员因什么打篮球？他们理由各不同，邓肯因怕鲨鱼而打球\",\"......\"]","METRIC":"[24.0,19.0,6.0,6.0,6.0]"}
String：SSK	{"ID":"[\"林徽因什么理由拒绝了徐志摩而选择梁思成为终身伴侣？\",\"林徽因拒绝徐志摩而选择梁思成为终身伴侣的原因是什么？\",......]","METRIC":"[0.9786452262557865,0.7201100809120543,0.17344669301981408,0.17344669301981408,0.17344669301981408]"}
String：COSINE	{"ID":"[\"林徽因什么理由拒绝了徐志摩而选择梁思成为终身伴侣？\",\"林徽因拒绝徐志摩而选择梁思成为终身伴侣的原因是什么？\",......]","METRIC":"[0.9789450103725609,0.7089490077940542,0.18860838403857944,0.16718346377260584,0.16718346377260584]"}

可以观察到如下信息。

- 原数据中包含这个新闻标题，每个算法都找到了完全一致的那个文本，并且排在了第一位。
- "林徽因拒绝徐志摩而选择梁思成为终身伴侣的原因是什么？"是目标文本的一个改写，意思相同，大部分文字也相同，各个算法都把它找出来了，而且都排在了第二位。这也说明了各种算法的有效性。
- 对于第三个及后面的选项，各算法的差异很大。

我们再以单词为单位，将文本数据进行分词，再使用 TextNearestNeighbor 组件进行计算。与 StringNearestNeighbor 组件类似，需要输入两个数据，一个是文本数据所在的列，另一个是 IdCol 列名称。

```
for (String metric : new String[] {"LEVENSHTEIN", "LCS", "SSK", "COSINE"}) {
  new Pipeline()
    .add(
      new Segment()
        .setSelectedCol(TXT_COL_NAME)
        .setOutputCol("segmented_title")
    )
    .add(
      new TextNearestNeighbor()
        .setMetric(metric)
        .setSelectedCol("segmented_title")
        .setIdCol(TXT_COL_NAME)
        .setTopN(5)
```

```
        .setOutputCol("similar_titles")
    )
    .fit(source)
    .transform(target)
    .lazyPrint(-1, "TextNearestNeighbor + " + metric.toString());
  BatchOperator.execute();
}
```

运行结果如表 21-23 所示,左边一列为 TextNearestNeighbor 组件的度量参数 metric 值。

表 21-23 文本最近邻计算结果

metric 值	林徽因什么理由拒绝了徐志摩而选择梁思成为终身伴侣?
Text : LEVENSHTEIN	{"ID":"[\"林徽因什么理由拒绝了徐志摩而选择梁思成为终身伴侣?\",\"林徽因拒绝徐志摩而选择梁思成为终身伴侣的原因是什么?\",\"什么造就了今年的季后赛隆多\",\"是什么因素阻止了伊涅斯塔加盟中超?\",\"是什么让你选择留在鄂州?\"]","METRIC":"[1.0,8.0,9.0,9.0,9.0]"}
Text : LCS	{"ID":"[\"林徽因什么理由拒绝了徐志摩而选择梁思成为终身伴侣?\",\"林徽因拒绝徐志摩而选择梁思成为终身伴侣的原因是什么?\",......,\"小米就要港股上市了,那么为什么选择香港而没有选择上海?\"]","METRIC":"[11.0,8.0,3.0,3.0,3.0]"}
Text : SSK	{"ID":"[\"林徽因什么理由拒绝了徐志摩而选择梁思成为终身伴侣?\",\"林徽因拒绝徐志摩而选择梁思成为终身伴侣的原因是什么?\",......]","METRIC":"[0.9519716503395643,0.5072036680937279,0.1021051630028303,0.08756718614037064,0.081940159723 62995]"}
Text : COSINE	{"ID":"[\"林徽因什么理由拒绝了徐志摩而选择梁思成为终身伴侣?\",\"林徽因拒绝徐志摩而选择梁思成为终身伴侣的原因是什么?\",......]","METRIC":"[0.9534625892455924,0.45643546458763845,0.13187609467 91574,0.12909944487358055,0.10540925533894598]"}

可以观察到如下信息。

- 排在第一、二位的结果都一致,与前面基于字符计算得到的前两名结果也一样,说明了各种算法的有效性。
- 在第三位及后面的选项中,各算法差异明显,而且与基于字符计算的结果差异更大。"林徽因"作为一个单词或者三个字符,在相似度计算中所占的比重有很大差别,所以在以单词为单位的 TopN 计算结果中,和"林徽因"相关的其他标题没有排到前五位。

我们再将"京城最值得你来场文化之旅的博物馆"相关的结果汇总起来,如表 21-24 所示,左边一列为所用组件(StringNearestNeighbor 或者 TextNearestNeighbor)的度量参数 metric 值。

表 21-24 最近邻计算结果

metric 值	京城最值得你来场文化之旅的博物馆
String : LEVENSHTEIN	{"ID":"[\"京城最值得你来场文化之旅的博物馆\",\"全国教师文化之旅 第一季\",\"《刮痧》文化之间的差异\",\"青岛长沙路小学学生走进海关博物馆\",\"中国传统文化之楼阁\"]","METRIC":"[0.0,12.0,12.0,13.0,13.0]"}
String : LCS	{"ID":"[\"京城最值得你来场文化之旅的博物馆\",\"第十四届文博会书画艺术文化之旅活动在深圳合正艺术博物馆开幕\",\"除了故宫,北京还有哪些值得去的博物馆?\",\""金城文化名家"王作宝丝绸之路风情画 被敦煌市博物馆收藏\",\"彰显着青海深厚的历史文化的博物馆都在这了!\"]","METRIC":"[16.0,7.0,7.0,7.0,6.0]"}
String : SSK	{"ID":"[\"京城最值得你来场文化之旅的博物馆\",\"杭州有哪些值得参观的博物馆?\",\"吐鲁番博物馆,最值得看的文物有哪些?\",\"青州 5A 级景区值得你来\",\"彰显着青海深厚的历史文化的博物馆都在这了!\"]","METRIC":"[1.0,0.25123248562505274,0.23162579414326565,0.22439441671450996,0.22091900671835765]"}
String : COSINE	{"ID":"[\"京城最值得你来场文化之旅的博物馆\",\"杭州有哪些值得参观的博物馆?\",\"吐鲁番博物馆,最值得看的文物有哪些?\",\"青州 5A 级景区值得你来\",\"超级链接的博物馆:93 项活动亮相北京 518 博物馆日\"]","METRIC":"[1.0,0.2864459496157732,0.2504897164340598,0.24494897427831783,0.24397501823713333]"}
Text : LEVENSHTEIN	{"ID":"[\"京城最值得你来场文化之旅的博物馆\",\"教你判断 P2P 平台的背景\",\"庄浪,你是人间五月的天堂\",\"诗词与企业文化的关系\",\"圣迭戈有哪些地方你不容错过的? \"]","METRIC":"[0.0,7.0,7.0,7.0,7.0]"}
Text : LCS	{"ID":"[\"京城最值得你来场文化之旅的博物馆\",\"6 款最值得买的合资品牌 SUV 全在这儿,有没有你等的?\",\"美媒报道中国 40 个最美丽最值得游览的景点,这里有你喜欢的吗?\",\"初夏,最值得你去的旅行地——安康,有 4 处神秘之地,你去过几个\",\"世界上最奇葩的水果,你未必都见过,吃过其中的一种我就服\"]","METRIC":"[9.0,4.0,4.0,4.0,3.0]"}
Text : SSK	{"ID":"[\"京城最值得你来场文化之旅的博物馆\",\"初夏,最值得你去的旅行地——安康,有 4 处神秘之地,你去过几个\",\"彰显着青海深厚的历史文化的博物馆都在这了!\",\"最值得购买的 5 款中级轿车\",\"最值得购买的 5 款 MPV\"]","METRIC":"[1.0,0.13138047152000644,0.13063966852069117,0.11823132268957977,0.11823132268957977]"}

续表

metric 值	京城最值得你来场文化之旅的博物馆
Text : COSINE	{"ID":"[\"京城最值得你来场文化之旅的博物馆\",\"初夏,最值得你去的旅行地——安康,有4处神秘之地,你去过几个\",\"广西什么地方最值得去旅游\",\"兴义哪个景色最值得去?\",\"最值得购买的5款中级轿车\"]","METRIC":"[1.0,0.15430334996920919,0.14433756729740646,0.14433756729740646,0.14433756729740646]"}

可以观察到如下信息。

- 针对原数据中包含的这个新闻标题，每个算法都找到了与其完全一致的那个文本，并且排在了第一位。
- 在以字符为单位的 Top5 结果中，以"文化之旅"和"博物馆"为主；在以单位为单位的 Top5 结果中，以"最值得"为主。

最后看一下标题"什么是超写实绘画？"相关的 Top5 结果汇总情况，如表 21-25 所示，StringNearestNeighbor 或者 TextNearestNeighbor 的结果差异小了一些。

表 21-25　最近邻计算结果

metric 值	什么是超写实绘画?
String : LEVENSHTEIN	{"ID":"[\"什么是超写实绘画? \",\"什么是文人画? \",\"什么是区块链? \",\"什么是区块链? \",\"什么是爱情? \"]","METRIC":"[0.0,4.0,5.0,5.0,5.0]"}
String : LCS	{"ID":"[\"什么是超写实绘画? \",\"超写实绘画只是在描绘现实吗? \",\"QQ飞车手游中什么段位是实力的分界线? \",\"你怎么看吕建军的写实油画? \",\"QQ飞车手游中什么段位是实力的分界线? \"]","METRIC":"[9.0,6.0,5.0,5.0,5.0]"}
String : SSK	{"ID":"[\"什么是超写实绘画? \",\"超写实绘画只是在描绘现实吗? \",\"什么是文人画? \",\"解答：什么是超载? \",\"什么是人情,什么是世故? \"]","METRIC":"[1.0,0.3911627417302326,0.37272313358225584,0.3525653656235633,0.3268835910175176]"}
String : COSINE	{"ID":"[\"什么是超写实绘画? \",\"什么是文人画? \",\"超写实绘画只是在描绘现实吗? \",\"解答：什么是超载? \",\"什么是人情,什么是世故? \"]","METRIC":"[1.0,0.433012701892219535,0.3922322702763681,0.375,0.3651483716701107]"}
Text : LEVENSHTEIN	{"ID":"[\"什么是超写实绘画? \",\"什么是区块链? \",\"什么是区块链? \",\"什么是爱情? \",\"什么是伊朗核协议? \"]","METRIC":"[0.0,3.0,3.0,3.0,3.0]"}

续表

metric 值	什么是超写实绘画？
Text：LCS	{"ID":"[\"什么是超写实绘画？\",\"超写实绘画只是在描绘现实吗？\",\"什么是方法论？怎么掌握方法论？\",\"什么是以租代购？以租代购过户吗？车子什么有保障吗\",\"什么是混改，为什么要联通混改，而不是电信或者移动呢？\"]","METRIC":"[6.0,4.0,3.0,3.0,3.0]"}
Text：SSK	{"ID":"[\"什么是超写实绘画？\",\"超写实绘画只是在描绘现实吗？\",\"什么是四维空间\",\"什么是教育\",\"什么是化性\"]","METRIC":"[1.0,0.2872548320403265,0.2672519210676725,0.2672519210676725,0.2672519210676725]"}
Text：COSINE	{"ID":"[\"什么是超写实绘画？\",\"超写实绘画只是在描绘现实吗？\",\"什么是四维空间\",\"什么是教育\",\"什么是化性\"]","METRIC":"[1.0,0.31622776601683794,0.31622776601683794,0.31622776601683794,0.31622776601683794]"}

2. 近似算法

文本相似度算法虽然能获得较好的分类效果，但是计算时间较长。在实际的大规模文本比较的场景中，我们常常希望在效率和效果中有所折中，那么近似相似度就派上了用场。

Alink 提供了 StringApproxNearestNeighbor 组件，以字符为单位，可以选择多种近似计算相似 TopN 的方式。与 StringNearestNeighbor 组件类似，需要输入两个数据，一个是文本数据所在的列，另一个是 IdCol 列名称。如下代码所示，循环尝试各种近似相似度计算。

```
for (String metric : new String[] {"JACCARD_SIM", "MINHASH_JACCARD_SIM", "SIMHASH_HAMMING_SIM",}) {
    new StringApproxNearestNeighbor()
        .setMetric(metric)
        .setSelectedCol(TXT_COL_NAME)
        .setIdCol(TXT_COL_NAME)
        .setTopN(5)
        .setOutputCol("similar_titles")
        .fit(source)
        .transform(target)
        .lazyPrint(-1, "StringApproxNearestNeighbor + " + metric.toString());
    BatchOperator.execute();
}
```

Alink 提供了 TextApproxNearestNeighbor 组件，以单词为单位，可以选择多种近似计算相似 TopN 的方式。与 TextNearestNeighbor 组件类似，需要先将文本数据进行分词，该组件需要输入两个数据，一个是文本数据所在的列，另一个是 IdCol 列名称。具体代码如下，尝试各种近似度量，将输出样例结果进行比较。

```
for (String metric : new String[] {"JACCARD_SIM", "MINHASH_JACCARD_SIM", "SIMHASH_HAMMING_SIM"}) {
    new Pipeline()
        .add(
            new Segment()
                .setSelectedCol(TXT_COL_NAME)
                .setOutputCol("segmented_title")
        )
        .add(
            new TextApproxNearestNeighbor()
                .setMetric(metric)
                .setSelectedCol("segmented_title")
                .setIdCol(TXT_COL_NAME)
                .setTopN(5)
                .setOutputCol("similar_titles")
        )
        .fit(source)
        .transform(target)
        .lazyPrint(-1, "TextApproxNearestNeighbor + " + metric.toString());
    BatchOperator.execute();
}
```

我们先将新闻标题"林徽因什么理由拒绝了徐志摩而选择梁思成为终身伴侣？"在上面两个组件和各种度量情况下的计算结果汇总如表 21-26 所示。

表 21-26　近似最近邻计算结果

metric 值	林徽因什么理由拒绝了徐志摩而选择梁思成为终身伴侣？
String：JACCARD_SIM	{"ID":"[\"林徽因什么理由拒绝了徐志摩而选择梁思成为终身伴侣？\",\"林徽因拒绝徐志摩而选择梁思成为终身伴侣的原因是什么？\",......,\"林黛玉为什么忽然成了肺结核\"]","METRIC":"[0.96,0.7241379310344828,0.20588235294117646,0.2,0.1935483870967742]"}
String：MINHASH_JACCARD_SIM	{"ID":"[\"林徽因什么理由拒绝了徐志摩而选择梁思成为终身伴侣？\",\"林徽因拒绝徐志摩而选择梁思成为终身伴侣的原因是什么？\",\"绝地求生：蓝战非终于完成了摩托车特技？摩托车脾气上来直接自爆\",\"中国股市终于有人把布林线指标浓缩成精髓了，少走二十年弯路\",......]","METRIC":"[1.0,0.9,0.4,0.4,0.3]"}
String：SIMHASH_HAMMING_SIM	{"ID":"[\"林徽因什么理由拒绝了徐志摩而选择梁思成为终身伴侣？\",\"迷你世界愤怒的小鸟\",\"《极限挑战》除了张艺兴还缺席了一个人，但观众却都很高兴\",\"扬州一中学探索因材施教 老师分层次布置作业\",\"忽而今夏：白宇为她设计了一款游戏，女生最羡慕的异地恋游戏！\"]","METRIC":"[0.953125,0.921875,0.90625,0.90625,0.890625]"}

续表

metric 值	林徽因什么理由拒绝了徐志摩而选择梁思成为终身伴侣？
Text：JACCARD_SIM	{"ID":"[\"林徽因什么理由拒绝了徐志摩而选择梁思成为终身伴侣？\",\"林徽因拒绝徐志摩而选择梁思成为终身伴侣的原因是什么？\",\"小米香港上市将为小米带来了什么？\",\"公务员考试失利后你选择了做什么？\",\"OPPO为迎接5G时代做了些什么？\"]","METRIC":"[0.9166666666666666,0.6,0.16666666666666666,0.16666666666666666,0.16666666666666666]"}
Text：MINHASH_JACCARD_SIM	{"ID":"[\"林徽因什么理由拒绝了徐志摩而选择梁思成为终身伴侣？\",\"林徽因拒绝徐志摩而选择梁思成为终身伴侣的原因是什么？\",......,\"孝顺儿子接我进城养老，准备走时接了个电话，我找个理由不去了\",\"汽车停产理由多，而这几款车，只因长得太丑就被停产了\"]","METRIC":"[1.0,0.7,0.3,0.3,0.3]"}
Text：SIMHASH_HAMMING_SIM	{"ID":"[\"林徽因什么理由拒绝了徐志摩而选择梁思成为终身伴侣？\",\"日本为什么不敢研发干线客机？俄军专家：发动机不仅要买且被锁死\",\"果农注意了，近期柑橘溃疡病大爆发！\",......,\"重庆市民投资对象变化：理财品成首选 房地产退居第四\"]","METRIC":"[0.9375,0.78125,0.765625,0.75,0.734375]"}

对于原始数据包含的同样的标题，在各种度量下都能被找出，并排在第一位。而对与之相近的标题"林徽因拒绝徐志摩而选择梁思成为终身伴侣的原因是什么？"，在多数度量中都排进了前五位。

再看另一个标题"什么是超写实绘画？"，汇总结果见表21-27。

表21-27 近似最近邻计算结果

metric 值	什么是超写实绘画？
String：JACCARD_SIM	{"ID":"[\"什么是超写实绘画？\",\"超写实主义画作存在的意义是什么？\",\"什么是文人画？\",\"超写实绘画只是在描绘现实吗？\",\"鲁本斯绘画的长处是什么？\"]","METRIC":"[1.0,0.47058823529411764,0.45454545454545453,0.4375,0.4]"}
String：MINHASH_JACCARD_SIM	{"ID":"[\"什么是超写实绘画？\",\"为什么子弹超音速时没有音爆？\",\"什么是妖股？\",\"什么是财务自由？\",\"合肥的市花是什么？\"]","METRIC":"[1.0,0.6,0.6,0.6,0.6]"}
String：SIMHASH_HAMMING_SIM	{"ID":"[\"什么是超写实绘画？\",\"叙利亚的军事实力如何？\",\"安徽省临近江浙沪，为什么发展却跟不上？\",\"电子维修有前途吗？\",\"郴州有哪些比较好的技校？\"]","METRIC":"[1.0,0.9375,0.90625,0.90625,0.90625]"}

续表

metric 值	什么是超写实绘画？
Text：JACCARD_SIM	{"ID":"[\"什么是超写实绘画？\",\"鲁本斯绘画的长处是什么？\",\"什么是咖喱？\",\"什么是都江堰？\",\"什么是妖股？\"]","METRIC":"[1.0,0.4444444444444444,0.42857142857142855,0.428571 42857142855,0.42857142857142855]"}
Text：MINHASH_JACCARD_SIM	{"ID":"[\"什么是超写实绘画？\",\"什么是妖股？\",\"web 服务器是什么？\",\"什么是都江堰？\",\"什么是子网？什么是子网掩码？\"]","METRIC":"[1.0,0.7,0.7,0.7,0.7]"}
Text：SIMHASH_HAMMING_SIM	{"ID":"[\"什么是超写实绘画？\",\"不同国家的军人在一起，军衔还论高低吗？\",\"现代海战中，能否俘获敌方完整战舰？已经绝无可能？\",\"共享单车好骑吗？\",\"刀郎的徒弟是谁？\"]","METRIC":"[1.0,0.84375,0.84375,0.84375,0.84375]"}

3. 计算性能

本节将分别选取计算近似与非近似相似度算法来测试模型训练时间，并针对不同数据量的测试预测时间，以便读者对算法性能有一个初步的印象，便于在实际使用中选择合适的方法。

非近似算法组件 StringNearestNeighbor 选择 LEVENSHTEIN（编辑距离）度量，近似算法组件 StringApproxNearestNeighbor 选择 JACCARD_SIM 度量。构建 Pipeline 的代码如下。

```
Pipeline snn = new Pipeline()
    .add(
        new StringNearestNeighbor()
            .setMetric("LEVENSHTEIN")
            .setSelectedCol(TXT_COL_NAME)
            .setIdCol(TXT_COL_NAME)
            .setTopN(5)
            .setOutputCol("similar_titles")
    );

Pipeline approx_snn = new Pipeline()
    .add(
        new StringApproxNearestNeighbor()
            .setMetric("JACCARD_SIM")
            .setSelectedCol(TXT_COL_NAME)
            .setIdCol(TXT_COL_NAME)
            .setTopN(5)
            .setOutputCol("similar_titles")
    );
```

计算非近似算法模型训练的时间,代码如下,将模型保存到文件路径。

```
static final String SNN_MODEL_FILE = "snn_model.ak";
... ...
snn.fit(source)
   .save(DATA_DIR + SNN_MODEL_FILE);
BatchOperator.execute();
```

得到任务执行时间为 12.587 秒,模型文件大小为 52.7MB。

计算近似算法模型训练的时间,代码如下,将模型保存到文件路径。

```
static final String APPROX_SNN_MODEL_FILE = "approx_snn_model.ak";
... ...
approx_snn
   .fit(source)
   .save(DATA_DIR + APPROX_SNN_MODEL_FILE);
BatchOperator.execute();
```

任务执行时间为 21.218 秒,模型文件大小为 356.7MB。

下面我们看预测时间。首先构造两个预测集,希望数据量适中,既可以演示差异,也不会运行时间过长。在 21.1 节中,我们知道各个新闻类别的数据条数,其中新闻类"stock"有 340 条,新闻类"news_story"有 6273 条。我们可以分别将其过滤出来作为测试集,具体代码如下。

```
BatchOperator <?> target_stock = source.filter("category_name = 'stock'");
BatchOperator <?> target_news_story = source.filter("category_name = 'news_story'");
```

测试非近似算法模型对 340 条 stock 数据的预测时间,代码如下。

```
PipelineModel
   .load(DATA_DIR + SNN_MODEL_FILE)
   .transform(target_stock)
   .lazyPrint(10, "StringNearestNeighbor + LEVENSHTEIN");
BatchOperator.execute();
```

运行时间为 4 分 35 秒。

测试近似算法模型对 340 条 stock 数据的预测时间,代码如下。

```
PipelineModel
   .load(DATA_DIR + APPROX_SNN_MODEL_FILE)
   .transform(target_stock)
   .lazyPrint(10, "MINHASH_SIM + stock");
BatchOperator.execute();
```

运行时间为 33 秒,去除模型加载的时间,实际预测的时间会更短。我们可以尝试更大的数据集。

测试非近似算法模型对 6273 条 news_story 数据的预测时间，代码如下。

```
PipelineModel
    .load(DATA_DIR + APPROX_SNN_MODEL_FILE)
    .transform(target_news_story)
    .lazyPrint(10, "MINHASH_SIM + news_story");
BatchOperator.execute();
```

运行时间为 2 分 59 秒。

综上，使用近似文本相似算法可以在预测阶段大幅提升性能。

4. 流式预测

我们将做两个流式预测实验，一个是使用非近似算法模型预测 4 条数据，与批式结果相比较，看看结果是否相同；另一个是使用近似算法模型预测 stock 数据，并将流式预测的速度与批式预测相比较。

沿用前面实验中的 4 条新闻标题，使用组件 MemSourceStreamOp 构造流式数据源，通过 PipeModel 加载非近似算法模型文件，然后对流式数据源进行预测，输出预测结果，便于和前面批式预测的结果进行对比。具体代码如下。

```
Row[] rows = new Row[] {
    Row.of("林徽因什么理由拒绝了徐志摩而选择梁思成为终身伴侣？"),
    Row.of("发酵床的垫料种类有哪些？哪种更好？"),
    Row.of("京城最值得你来场文化之旅的博物馆"),
    Row.of("什么是超写实绘画？")
};
StreamOperator <?> stream_target
    = new MemSourceStreamOp(rows, new String[] {TXT_COL_NAME});

PipelineModel
    .load(DATA_DIR + SNN_MODEL_FILE)
    .transform(stream_target)
    .print();
StreamOperator.execute();
```

运行结果如下，与前面计算的批式预测结果完全一致。

```
news_title|similar_titles
---------|---------------
林徽因什么理由拒绝了徐志摩而选择梁思成为终身伴侣|{"ID":"[\"林徽因什么理由拒绝了徐志摩而选择梁思成为终身伴侣？\",\"林徽因拒绝徐志摩而选择梁思成为终身伴侣的原因是什么？\",……]","METRIC":"[1.0,12.0,19.0,20.0,20.0]"}
发酵床的垫料种类有哪些？哪种更好？|{"ID":"[\"发酵床的垫料种类有哪些？哪种更好？\",\"发酵饲料有哪些好处\",\"教育的分类有哪些？\",\"烤猪蹄的佐料有哪些？\",\"普洱茶的原料有哪些讲究？\"]","METRIC":"[0.0,10.0,11.0,11.0,11.0]"}
京城最值得你来场文化之旅的博物馆|{"ID":"[\"京城最值得你来场文化之旅的博物馆\",\"全国教师文化之旅 第一季\",\"《刮痧》文化之间的差异\",\"青岛长沙路小学学生走进海关博物馆\",\"中国传统文化之楼阁
```

\"]","METRIC":"[0.0,12.0,12.0,13.0,13.0]"}
什么是超写实绘画？|{"ID":"[\"什么是超写实绘画？\",\"什么是文人画？\",\"什么是区块链？\",\"什么是区块链？\",\"什么是爱情？\"]","METRIC":"[0.0,4.0,5.0,5.0,5.0]"}

我们再用流的方式读取原始数据文件，并对流式数据进行过滤，保留类别为"stock"的数据作为流式预测的数据。利用 PipeModel 加载近似算法模型文件，然后对流式数据源进行预测，由于预测条数较多，因此进行了流式采样操作（采样率为 0.02），随后输出采样的预测结果。具体代码如下。

```
StreamOperator.setParallelism(1);
... ...
StreamOperator <?> stream_target_stock
   = getStreamSource().filter("category_name = 'stock'");

sw.reset();
sw.start();
PipelineModel
   .load(DATA_DIR + APPROX_SNN_MODEL_FILE)
   .transform(stream_target_stock)
   .sample(0.02)
   .print();
StreamOperator.execute();
sw.stop();
System.out.println(sw.getElapsedTimeSpan());
```

注意，由于近似算法模型很大，而每个流式预测 worker 都会保留一份完整模型，这里设置流式任务的并发度为 1，即只有一个 worker 进行预测；由于构造的流式数据量有限，预测完成后会自动终止流式任务，在流式任务的执行前后设置了计时，统计任务运行时间。

这里选取一条输出的新闻标题"三大股指暴跌，后市该如何走？"，预测的相似标题结果如下。

{"ID":"[\"三大股指暴跌，后市该如何走？\",\"周三（3月28日）股市大跌，明天如何走？\",\"大盘筑底，个股冷不丁暴跌，该如何选股？\",\"股票开盘跌停，该如何卖出？\",\"央行加息，股市应该如何投资？\"]","METRIC":"[1.0,0.4166666666666667,0.375,0.35,0.3333333333333333]"}

流式任务的总运行时间为 1 分 7 秒，考虑到其流式任务的并发度为 1，而对应的批式预测任务的并发度为 2，运行时间为 35 秒。可以看出，批式和流式方式对于单条数据的预测速度差不多是一样的。

21.7 主题模型

本章所用的数据集是新闻文档的集合，虽然每篇新闻的内容不同，但是我们看过若干篇后，会发现有些新闻讲的是同一方面的事情,譬如都是关于电影的或者足球意大利甲级联赛的,我们称之为"主题"。本节要讲的内容,是让计算机直接从文本内容上计算其中有多少主题。为了控制主题的粒度,我们使用一个参数,即指定主题的总个数。

计算主题模型(Topic Model)有几个经典的方法:潜在语义分析(Latent Semantic Analysis,LSA)、概率潜在语义分析(Probabilistic Latent Semantic Analysis,PLSA),以及本节重点介绍的潜在狄里克莱分布(Latent Dirichlet Allocation,LDA)。这些方法的共同点是,我们会忽略每个文档中单词出现的顺序,文档是由出现了哪些单词、每个单词出现的个数决定的,即词袋(bag of words)。

21.7.1 LDA模型

LDA 模型将贝叶斯估计的思想引入话题模型,使得参数具备了概率分布。LDA 是关于多个参数的分布,为了使读者容易理解此模型,我们先介绍简单的情形,即两个参数的情形,与此相关的两个分布分别是二项分布和 Beta 分布。

1. 二项分布(Binomial Distribution)

二项分布即重复 n 次独立的伯努利试验。在每次试验中只有两种可能的结果,而且两种结果发生与否是互相对立的,并且相互独立,与其他各次试验结果无关,事件发生与否的概率在每一次独立试验中都保持不变,则这一系列试验总称为 n 重伯努利实验。譬如投硬币的实验,出现正面的概率为 x,则出现负面的概率为 $1-x$,那么,出现 α_1 次正面,α_2 次负面的概率函数为

$$P(\alpha_1, \alpha_2; x) = \binom{\alpha_1 + \alpha_2}{\alpha_1} x^{\alpha_1}(1-x)^{\alpha_2}$$

2. Beta分布

给定参数 $\alpha_1, \alpha_2 > 0$,取值范围为 $[0,1]$ 的随机变量 x 的概率密度函数为

$$\begin{aligned} P(x; \alpha_1, \alpha_2) &= \frac{1}{B(\alpha_1, \alpha_2)} x^{\alpha_1-1}(1-x)^{\alpha_2-1} \\ &= \frac{\Gamma(\alpha_1+\alpha_2)}{\Gamma(\alpha_1)\Gamma(\alpha_2)} x^{\alpha_1-1}(1-x)^{\alpha_2-1} \end{aligned}$$

其中,$B(\alpha_1, \alpha_2)$ 为 Beta 函数,可以用 Gamma 函数 $\Gamma(\cdot)$ 表示。如图 21-3 所示,列出 3 个 Beta 分布的概率密度分布图,参数的数值越大,在局部的概率密度会越大。

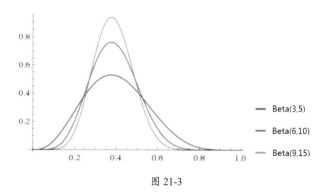

图 21-3

还是以投硬币的场景为例,我们来看这两种分布。二项分布是把概率x作为参数,即在概率x的值确定的情况下,不同的正反面次数α_1和α_2对应的概率分布;Beta 分布是把α_1和α_2作为参数,即在正反面出现次数确定的情况下,给出概率x对应的概率密度分布。

由贝叶斯理论,Beta 分布为二项分布的共轭先验分布。"轭"为驾车时搁在牛马颈上的曲木,可以起到束缚、控制的作用,使牛马行走同步。"共轭"可以理解为按照一定规律匹配的一对。

下面我们将"二项"推广为"多项",将投硬币的场景扩展到投骰子的场景。在"二项"的时候,只给出一种结果的概率x,譬如硬币为正面的概率,另一种结果的概率可以用$1-x$表示。但在"多项"的情况下,K种结果的概率要分别以$x_1,x_2,\cdots,x_K$来表示,还要满足一个限制条件——所有结果的概率和为 1,即

$$x_1+x_2+\cdots+x_K=1$$

3. 多项分布(Multinomial Distribution)

某随机实验如果有K个可能结局$A_1,A_2,\cdots,A_K$,譬如骰子有 6 个面,各种结果出现的概率分别为 $x_1,x_2,\cdots,x_K$,那么在$\alpha_1+\alpha_2+\ldots+\alpha_K$次试验结果中,$A_1$出现$\alpha_1$次、$A_2$出现$\alpha_2$次、……、$A_K$出现$\alpha_K$次的概率,即多项分布的概率密度函数为

$$P\left(\alpha_1,\alpha_2,\cdots,\alpha_K\,;\,x_1,x_2,\cdots,x_K\right)=\frac{(\sum_{k=1}^{K}\alpha_k)!}{\prod_{k=1}^{K}\alpha_k!}\cdot\prod_{k=1}^{K}x_k^{\alpha_k}$$

其中

$$\sum_{k=1}^{K}x_k=1,\ x_k\geq 0,k=1,2,\cdots,K$$

4. Dirichlet分布

参数为K个正数$\alpha_1, \alpha_2, \cdots, \alpha_K$，其概率密度函数为

$$P(x_1, x_2, \cdots, x_K; \alpha_1, \alpha_2, \cdots, \alpha_K) = \frac{\Gamma(\sum_{k=1}^{K} \alpha_k)}{\prod_{k=1}^{K} \Gamma(\alpha_k)} \cdot \prod_{k=1}^{K} x_k^{\alpha_k - 1}$$

其中，变量$x_1, x_2, \cdots, x_K$的定义域为

$$\sum_{k=1}^{K} x_k = 1, \quad x_k \geq 0, k = 1, 2, \cdots, K$$

以投骰子的场景为例，多项分布是把各面出现的概率$x_1, x_2, \cdots, x_6$作为参数，即在概率确定的情况下，给出各面出现次数$\alpha_1, \alpha_2, \cdots, \alpha_6$对应的概率分布。Dirichlet 分布是把各面出现的次数$\alpha_1, \alpha_2, \cdots, \alpha_6$作为参数，即在各面出现次数确定的情况下，给出各面出现概率$x_1, x_2, \cdots, x_6$对应的概率密度分布。

LDA 主题模型如图 21-4 所示，详细内容请参考 David M. Blei、Andrew Y. Ng 和 Michael I. Jordan 在 2003 年发表的论文（参见链接 21-2）。其中，带阴影的圆圈表示可以观测到的变量，即文档内容；不带阴影的圆圈都是潜在变量。每篇文章的主题分布是按照参数为α的 Dirichlet 分布采样得到的多项分布，每个主题的词分布是按照参数为β的 Dirichlet 分布采样得到的多项分布。

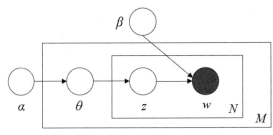

图 21-4

21.7.2 新闻的主题模型

下面以头条新闻数据集为例，根据新闻标题内容计算主题模型。首先，在原始数据中选出本节需要的两列数据，新闻标题列用来计算主题模型，在 21.8 节将通过标签列来验证主题模型的效果。因为 LDA 模型是基于分词结果计算的，所以还需使用分词和停用词组件，具体代码如下。

```
BatchOperator <?> docs = getSource()
    .select(LABEL_COL_NAME + ", " + TXT_COL_NAME)
    .link(new SegmentBatchOp().setSelectedCol(TXT_COL_NAME))
    .link(new StopWordsRemoverBatchOp().setSelectedCol(TXT_COL_NAME));

docs.lazyPrint(10);
```

处理后的结果数据如表 21-28 所示。

表 21-28　新闻标题进行数据处理后的结果

category_name	news_title
news_sports	詹姆斯 吃饭 睡觉 猛龙 猛龙 很萌
news_sports	上港 球迷 远征 亚冠 客场 扬 国旗 举 围巾 霸气 呐喊助威
news_finance	沪 指 重新 站上 3100 点 市场 氛围 转好 优质 票 可选
news_finance	需 小心 地雷 炸伤 大盘 积极 搭台 创业板 唱戏 反弹 围绕 一点 位 展开
news_finance	人民日报 欧盟 预算 因何 众口难调
news_finance	天津 七 一二 通信 广播 股份 有限公司 召开 2017 年度 业绩 现金 分红 说明 会 公告
news_finance	股票 不能 碰
news_finance	券商 四大 传统 业务 承压 期货 海外 业务 加速 成长
news_finance	文安 产业 新城 签约 追梦 教育 集团 助力 文安 教育 发展
news_finance	泰达 股份 北方 信托 混改 有利于 优化 公司 竞争力

下面讲解本节的重点——LDA 模型训练。使用 LDA 模型训练组件，设置 Topic 个数为 10，设置迭代次数 NumIter=200，设置词汇表的容量为 20000（会根据各单词在整个语料中出现的频率进行排序，取前 20000 个单词）；选择输出 LDA 模型的信息，并将 LDA 模型保存到文件中。另外，训练组件的 0 号侧输出（SideOutput[0]）还会输出单词主题概率表（单词属于各 Topic 的概率），也将其保存到文件，便于后面分析使用。相关代码如下。

```
static final String LDA_MODEL_FILE = "lda_model.ak";
static final String LDA_PWZ_FILE = "lda_pwz.ak";
... ...
LdaTrainBatchOp lda = new LdaTrainBatchOp()
    .setTopicNum(10)
    .setNumIter(200)
    .setVocabSize(20000)
    .setSelectedCol(TXT_COL_NAME)
    .setRandomSeed(123);

docs.link(lda);
```

```
lda.lazyPrintModelInfo();
lda.link(
    new AkSinkBatchOp().setFilePath(DATA_DIR + LDA_MODEL_FILE)
);
lda.getSideOutput(0)
    .link(
        new AkSinkBatchOp().setFilePath(DATA_DIR + LDA_PWZ_FILE)
    );
BatchOperator.execute();
```

显示 LDA 模型信息如下，可以看到 Topic 的个数、词汇表的个数，以及两个重要指标 logPerplexity 和 logLikelihood。

```
---------------------------- LdaModelInfo ----------------------------
|topic number|vocabulary size|logPerplexity|logLikelihood|
|------------|---------------|-------------|-------------|
|        10.0|          20000|    1247.7366|-24954732.9439|
```

21.7.3 主题与原始分类的对比

我们首先使用 LDA 模型预测各新闻标题所属的主题（Topic），具体代码如下。

```
new LdaPredictBatchOp()
    .setSelectedCol(TXT_COL_NAME)
    .setPredictionCol(PREDICTION_COL_NAME)
    .setPredictionDetailCol("predinfo")
    .linkFrom(
        new AkSourceBatchOp().setFilePath(DATA_DIR + LDA_MODEL_FILE),
        docs
    )
    .lazyPrint(5)
    .link(
        new EvalClusterBatchOp()
            .setLabelCol(LABEL_COL_NAME)
            .setPredictionCol(PREDICTION_COL_NAME)
            .lazyPrintMetrics()
    );
```

使用 LdaPredictBatchOp 组件，设置参数 SelectedCol 为标题文本数据，设置预测结果列 PredictionCol，并设置预测详细信息列 PredictionDetailCol（其输出为一个向量，各分量对应着属于各 Topic 的概率）。该预测组件需要输入两个数据，一个为 LDA 模型，另一个为所要预测的数据。预测结果中会输出 5 条数据，由于主题模型的预测结果与聚类预测结果相似，可以使用聚类评估组件，看看预测结果与原始标签的联系。

预测结果如表 21-29 所示，右边两列分别为主题索引值和该新闻属于各主题的概率。

表 21-29 LDA 计算结果

category_name	news_title	pred	predinfo
news_house	山东 最 近几个月 好 地段 楼盘 几乎 翻 一倍 看	9	0.08846173326746984 0.08790198371425065 0.08695652173913526 0.11132784130175362 0.08704271106524446 0.09637847076941840 0.10143139656072783 0.087337078787098 0.10146142566592847 0.15170083712897348
news_house	中国 两百万 以上 资产 家庭 大概	9	0.09230769231252751 0.10769197296805640 0.09230769783775432 0.09230769273934723 0.09230770566005040 0.10769311957764260 0.09279286831550560 0.09230769230822215 0.10769728241983208 0.12258708348094007
news_house	买 二手房 要交 税 最全 计算 方式 一览	9	0.08955223880597014 0.11107779409223460 0.09083815954230492 0.09070069094553912 0.11264249230503576 0.09498002328302255 0.10447764165673845 0.08955223910814286 0.10294956612073876 0.11322900731016139
news_house	广东 惠州 买房 升值 空间	9	0.09230769230769281 0.09230769235793640 0.09231051293661477 0.09230769245814259 0.09230769232219097 0.09230769239314511 0.10770124162569400 0.09230805945158871 0.10769998096127562 0.13844174318571903
news_car	a2 驾照 能开 23 座 客车	6	0.09096819616171696 0.09096908235666334 0.10775203096378642 0.10633489746936109 0.09595360701940348 0.09443902433286673 0.13459232009550584 0.09593454722532924 0.09099265512255543 0.09206363925281145

输出评估结果如下，我们主要关注标签值与主题索引值间的交叉矩阵。

```
------------------------- Metrics: -------------------------
k:10
ARI:0.2466 NMI:0.2958 Purity:0.4515

|    Label\Cluster|     9|     8|     7|     6|     5|     4|     3|     2|     1|     0|
|-----------------|------|------|------|------|------|------|------|------|------|------|
|            stock|    83|   151|     8|    16|    23|    12|    11|    29|     3|     4|
|       news_world|   911|   773|   529|   541|  3547|   554|  4236| 12875|   889|  2054|
|      news_travel|  2121|  2876|   486|  1250|  4895|   648|   753|   974|  5829|  1590|
|        news_tech|  4248|  6215|   720|  2643|  2158| 18977|  2986|  2093|   759|   744|
|       news_story|   400|    61|    58|   100|   645|   147|   118|   123|   963|  3658|
|      news_sports|   782|   750| 28915|   565|  1417|   616|   707|  1095|   761|  1960|
|    news_military|   592|   778|   497|   756|  2487|   470| 13919|  3949|   716|   820|
|       news_house| 11619|  1528|   197|   980|   903|   639|   241|   832|   535|   198|
|        news_game|  1133|  6492|  2220|   953|  2313|  1745|  7274|  1497|  1220|  4453|
|      news_finance|  5814| 10018|   609|  1546|  2176|  1319|  1398|  3149|   665|   391|
|news_entertainment|   855|   611|  1360|   621|  2421|   711|  1090|   891|  3590| 27246|
|         news_edu|  1497| 12657|   551|  1336|  1752|  4846|   761|   989|  1761|   908|
|     news_culture|   697|  1577|   586|   522|  3392|   784|  1059|  1102| 17085|  1227|
|         news_car|  1313|  1105|   309| 23783|  3049|  2418|   716|  1182|  1028|   882|
| news_agriculture|  3134|  3961|   279|   827|  5706|  1043|   818|  1055|  1902|   597|
```

接着导入单词主题概率表，先输出几条数据，了解数据的格式，然后找出主题中概率最高的 20 个单词，这些单词上可以帮助我们理解各主题的内容。具体代码如下。

```
AkSourceBatchOp pwz = new AkSourceBatchOp().setFilePath(DATA_DIR + LDA_PWZ_FILE);

pwz.sample(0.001).lazyPrint(10);

for (int t = 0; t < 10; t++) {
    pwz.select("word, topic_" + t)
        .orderBy("topic_" + t, 20, false)
        .lazyPrint(-1, "topic" + t);
}
```

单词主题概率表中的数据如表 21-30 所示，第一列为单词，后面是属于各主题的概率。可以看到，横向的概率和为 1，单词"顶配"在 topic_6 上的概率最大，为 0.7334；再从前面的交叉矩阵中可以看到，topic_6 中 news_car 的新闻出现最多。

表 21-30　单词主题概率表

word	topic_0	topic_1	topic_2	topic_3	topic_4	topic_5	topic_6	topic_7	topic_8	topic_9
顶配	0.0214	0.0214	0.0214	0.0532	0.0426	0.0214	0.7334	0.0214	0.0426	0.0214
问询	0.0140	0.0140	0.1943	0.0695	0.0556	0.0417	0.0695	0.0556	0.4717	0.0140
鉴别	0.2118	0.2118	0.0607	0.0305	0.2420	0.0305	0.0607	0.0607	0.0305	0.0607
这款	0.0033	0.0044	0.0011	0.0111	0.0066	0.0033	0.9557	0.0011	0.0078	0.0055

续表

word	topic_0	topic_1	topic_2	topic_3	topic_4	topic_5	topic_6	topic_7	topic_8	topic_9
花卉	0.0862	0.0346	0.0690	0.2239	0.0690	0.1723	0.0518	0.0862	0.1207	0.0862
犯下	0.0288	0.0288	0.2282	0.1427	0.1142	0.0573	0.0573	0.0858	0.1427	0.1142
沪市	0.0646	0.0325	0.0968	0.1611	0.0325	0.0646	0.0968	0.0646	0.2897	0.0968
宿迁	0.0572	0.1000	0.0144	0.1428	0.1000	0.1285	0.0144	0.0715	0.2712	0.1000
姐	0.1311	0.1475	0.0656	0.0329	0.1147	0.1147	0.0493	0.0656	0.2293	0.0493
同一个	0.1939	0.0449	0.1939	0.1045	0.0896	0.0896	0.0896	0.0300	0.0896	0.0747

各主题中概率最高的 20 个单词，计算汇总如表 21-31 所示。

表 21-31　各主题高频单词

topic_0	topic_1	topic_2	topic_3	topic_4
word|topic_0	word|topic_1	word|topic_2	word|topic_3	word|topic_4
----|-------	----|-------	----|-------	----|-------	----|-------
网友|0.9967	上联|0.9973	美国|0.9959	以色列|0.9892	手机|0.9976
岁|0.9904	下联|0.9969	世界|0.9926	求生|0.9880	联想|0.9946
活动|0.9871	黄山|0.9774	特朗普|0.9917	俄罗斯|0.9868	老师|0.9933
穿|0.9849	泰山|0.9729	退出|0.9904	战机|0.9866	学生|0.9899
现身|0.9837	赵本山|0.9729	伊核|0.9900	俄|0.9843	看待|0.9886
出席|0.9834	对联|0.9611	协议|0.9854	击落|0.9842	华为|0.9885
戛纳|0.9781	春风|0.9601	伊朗|0.9832	绝地|0.9824	5g|0.9879
rng|0.9763	接|0.9565	国家|0.9821	称|0.9813	高通|0.9876
kz|0.9745	写|0.9535	黄金|0.9811	警告|0.9811	响个|0.9868
范冰冰|0.9744	公鸡|0.9526	宣布|0.9768	普京|0.9808	家长|0.9866
机场|0.9706	故事|0.9503	伊朗核|0.9736	将会|0.9782	事|0.9864
儿子|0.9687	打架|0.9466	会|0.9693	军事基地|0.9765	不停|0.9850
女儿|0.9625	人生|0.9408	成为|0.9677	链|0.9743	苹果|0.9844
粉丝|0.9620	甄|0.9315	制裁|0.9639	轰炸|0.9736	赔|0.9823
腿|0.9580	求下联|0.9301	影响|0.9616	叙利亚|0.9711	发票|0.9815
鹿晗|0.9567	千金|0.9257	总统|0.9591	航母|0.9694	小米|0.9808
赵丽颖|0.9550	头|0.9253	事件|0.9585	美军|0.9667	上课时|0.9795
谢娜|0.9544	媾|0.9250	走|0.9581	战场|0.9658	票|0.9790
明星|0.9519	经典|0.9208	今后|0.9522	歼|0.9606	摔|0.9781
红毯|0.9517	一夜|0.9203	要求|0.9493	币|0.9589	投给|0.9776

续表

topic_5	topic_6	topic_7	topic_8	topic_9
word|topic_5	word|topic_6	word|topic_7	word|topic_8	word|topic_9
----|-------	----|-------	----|-------	----|-------	----|-------
说|0.9930	万|0.9941	火箭|0.9918	荣耀|0.9923	月|0.9940
人|0.9873	suv|0.9895	詹姆斯|0.9890	王者|0.9899	日|0.9906
没有|0.9801	车|0.9877	勇士|0.9850	万元|0.9765	房子|0.9905
中国|0.9795	奥迪|0.9856	nba|0.9849	2017|0.9740	马云|0.9903
日本|0.9792	买|0.9846	球迷|0.9842	股份|0.9727	5|0.9902
象棋|0.9759	上市|0.9789	球员|0.9839	教育|0.9717	房价|0.9878
印度|0.9738	奔驰|0.9771	骑士|0.9836	有限公司|0.9674	言论|0.9821
古代|0.9719	款|0.9768	保罗|0.9819	营收|0.9666	惊人|0.9808
大象|0.9680	大众|0.9767	决赛|0.9810	亿元|0.9658	8|0.9806
源自|0.9646	宝马|0.9737	1|0.9810	专业|0.9625	买房|0.9798
农村|0.9613	座|0.9702	西决|0.9809	招生|0.9618	后|0.9797
说法|0.9602	国产|0.9698	恒大|0.9806	净赚|0.9591	可信|0.9786
方舟子|0.9588	新车|0.9690	比赛|0.9800	大学|0.9587	不值钱|0.9752
很多|0.9551	仅|0.9685	猛龙|0.9783	公告|0.9587	年|0.9722
吃|0.9508	车型|0.9664	0|0.9779	建设|0.9574	银行|0.9644
俗语|0.9388	万起|0.9664	中超|0.9749	产业|0.9554	上涨|0.9634
股市|0.9170	丰田|0.9664	主场|0.9743	项目|0.9548	9|0.9614
有人|0.9166	1|0.9656	联赛|0.9737	公司|0.9511	支付宝|0.9570
认为|0.9076	全新|0.9639	哈登|0.9730	教师|0.9468	钱|0.9558
这句|0.8928	性价比|0.9621	分|0.9713	新|0.9440	出|0.9552

有了这个主题单词表，再加上前面的标签值与主题索引值间的交叉矩阵，我们就能更容易地了解各主题的内容。这里抛砖引玉，分析以下两个主题。

- 前面的讨论涉及了 topic_6，其中 news_car 的新闻占了绝大部分，而且其中概率较高的单词有"SUV、车、奥迪、奔驰、大众、宝马、新车"等，显然是关于汽车的主题。
- topic_0 中 news_entertainment 的新闻占了绝大部分，高概率出现的单词有"戛纳、范冰冰、鹿晗、赵丽颖、谢娜、粉丝、明星、红毯"等，非常明显，这是娱乐新闻。

21.8　组件使用小结

本节用思维导图描绘前面介绍的各种文本组件的调用关系及所需的数据依赖，如图 21-5 所示。

第 21 章 文本分析

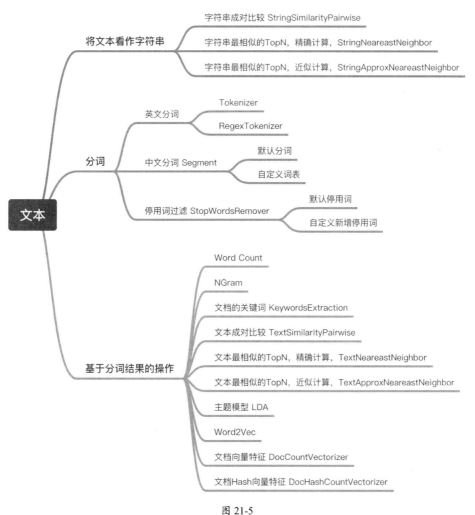

图 21-5

22 单词向量化

单词是文本的基本单位,单词向量化是文本向量化的基础。通过向量化将文本转化为机器学习所需的向量特征,也可以通过计算向量间的距离判断单词或文本间的相似度。

把一个单词表示成一个向量,主要有两种方法。

- 独热(One-Hot)编码表示:这是最简单的表示方式,仅与词典相关。为词典中的每个单词赋予一个序号,对于每个单词,向量维度是词典中单词的个数,将该单词序号的位置设置为 1,其余位置都设置为 0。整个向量中只有一个位置不为 0,只有一个热点,即 One-Hot。
- 嵌入式(Embedding)表示:这种表示方式不仅需要词典,还需要训练语料,根据每个单词出现的上下文(context),需要指定向量的维度(一般选 50、100),显然,对比独热编码表示,维度数显著减少,每个维度不再对应一个单词,可以看作某个隐含变量。每个单词会通过各个隐含变量上的权重值表示出来,也称为"Word Embedding"。对于该种表示方法得出的结果,可以通过向量间的距离判断单词间的相似度。

独热编码的优势是简单直观,各个单词对应一个向量分量的位置,在文本中各个单词所在向量分量的位置设置数值,就得到了文本的向量表示。向量分量值可以设置为 1,表示文档中是否出现了该单词,也可以设置为单词出现的个数,或者更能表达单词区分度的指标(譬如 TF、IDF、TF-IDF 等)。在实际应用中,还可以使用 hash 的方式,简化单词到向量位置的映射操作。

单词表示为嵌入式向量,需要对语料库数据进行训练才能得到。而在不同的语料库数据和不同的训练参数情况下,每个词得到的单词向量表示可能是不一样的。嵌入式单词向量将单词映射到高维空间上的点,点之间距离可以看作对应的两个词之间的"距离",即两个词在语法、语义上的相似性。

22.1 单词向量预训练模型

这里我们选择了两个容易查到的英文的单词向量预训练模型。
- 一个模型是文件尺寸小一些的，便于下载，参见链接 22-1，由维基百科相关（Wikipedia Dependency）的一些语料训练而成。
- 另一个模型来自 GloVe，也是选择其中较小的一个下载，参见链接 22-2，其中包含了多个维度的向量模型。

22.1.1 加载模型

将 Wiki 单词向量预训练模型文件（参见链接 22-1）下载到本地，并解压为 deps.words，使用文本编辑器打开，如图 22-1 所示。

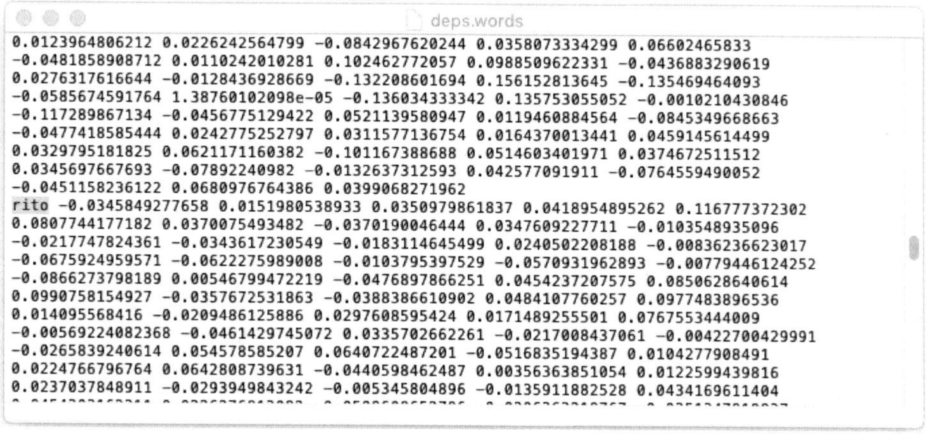

图 22-1

数据格式比较简单，单词与向量之间用空格隔开，向量的各分量间使用空格分隔（这与 Alink 稠密向量的序列化为字符串的格式一样）。可以通过组件 TextSourceBatchOp 将一行看作一条文本数据，然后以第一个出现的空格为界，划分为 word 和 vec 两部分，如下代码所示。

```
static BatchOperator <?> getWikiDependency() {
  return new TextSourceBatchOp().setFilePath(DATA_DIR + WIKI_DEPENDENCY)
      .setTextCol("txt")
```

```
    .select("SUBSTRING(txt FROM 1 FOR POSITION(' ' IN txt)-1 ) AS word, "
        + "SUBSTRING(txt FROM POSITION(' ' IN txt) + 1 ) AS vec");
}
```

将 glove.6B 单词向量预训练模型文件（参见链接 22-2）下载到本地，解压后会出现 4 个文本文件，名称分别为 glove.6B.50d.txt、glove.6B.100d.txt、glove.6B.200d.txt 和 glove.6B.300d.txt。从名称上就可以看出，各文件的差别在于对应向量的维度。我们选择 100 维的情况作为试验的目标。使用文本编辑器打开，如图 22-2 所示。

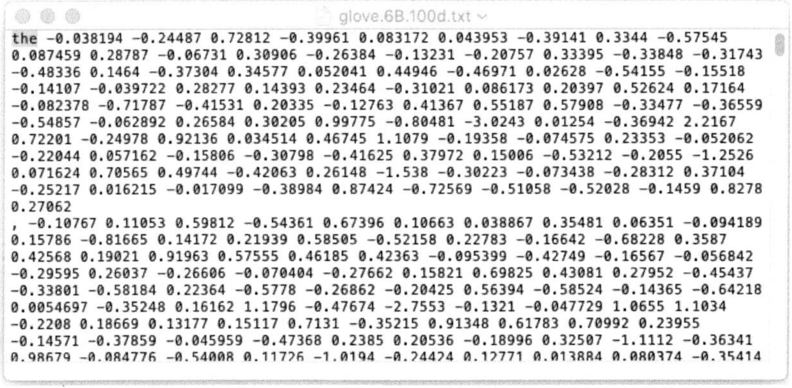

图 22-2

数据格式与 deps.words 一样，单词与向量之间用空格隔开，向量的各分量间使用空格隔开（这与 Alink 稠密向量的序列化为字符串的格式一样）。仍然通过 TextSourceBatchOp 组件将一行看作一条文本数据，然后以第一个出现的空格为界，划分为 word 和 vec 两部分。

```
static BatchOperator <?> getGlove6B100d() {
    return new TextSourceBatchOp()
        .setFilePath(DATA_DIR + GLOVE_6B_100D)
        .setTextCol("txt")
        .select("SUBSTRING(txt FROM 1 FOR POSITION(' ' IN txt)-1 ) AS word, "
            + "SUBSTRING(txt FROM POSITION(' ' IN txt) + 1 ) AS vec");
}
```

经过上述处理得到的数据集包含两列，分别为 word 和 vec，与 Alink 提供的单词向量模型的数据格式完全一样。

22.1.2　查找相似的单词

可以使用向量最近邻组件 VectorNearestNeighbor，其使用方式与第 21 章介绍的 StringNearestNeighbor 组件类似，只是把用来计算和比较的内容换成了向量。具体代码如下：

第 22 章 单词向量化

```
BatchOperator.setParallelism(1);
for (BatchOperator <?> word2vec : new BatchOperator <?>[] {getWikiDependency(), getGlove6B100d()}) {
    for (String metric : new String[] {"EUCLIDEAN", "COSINE"}) {
        new VectorNearestNeighbor()
            .setIdCol("word")
            .setSelectedCol("vec")
            .setMetric(metric)
            .setOutputCol("similar_words")
            .setTopN(7)
            .fit(word2vec)
            .transform(word2vec.filter("word='king'"))
            .select("word, similar_words")
            .lazyPrint(-1, metric);
        BatchOperator.execute();
    }
}
```

外层循环选择不同的单词向量模型，内层循环使用不同的距离度量（EUCLIDEAN 与 COSINE），每次计算的目标都是单词"king"对应的向量。

先看 WikiDependency 向量模型计算出来的结果，如下所示。

EUCLIDEAN	{"ID":"[\"king\",\"norodom\",\"songtsän\",\"queen\",\"bhumibol\",\"monarch\",\"archduke\"]","METRIC":"[2.9802322387695312E-8, 0.8055086190497484, 0.8063912079032907, 0.8214690626774828, 0.8229473168473302, 0.8253910038487656, 0.8260947822801127]"}
COSINE	{"ID":"[\"king\",\"norodom\",\"songtsän\",\"queen\",\"bhumibol\",\"monarch\",\"archduke\"]","METRIC":"[-7.771561172376096E-16, 0.32442206768187054, 0.3251333900919796, 0.3374057104681471, 0.338621143153173, 0.340635154617282, 0.3412162946552998]"}

可以看到，使用两种度量方式计算出来的前 7 个词都是一样的，甚至顺序都一样，但两种方式下的度量值不同。被搜索的词库中含有该单词，所以该单词被排在了第一位，单词 queen （皇后）、monarch（帝王）和 archduke（大公）显然和 king 有紧密的联系；对于其他 3 个单词，通过搜索引擎获取其含义如表 22-1 所示，都是著名的"国王"。

表 22-1　3 个单词的含义

norodom	songtsän	bhumibol
诺罗敦	Songtsen Gampo（松赞干布）	Bhumibol Adulyadej（普密蓬·阿杜德）
诺罗敦，柬埔寨国王，1860 年至 1904 年在位。本名安瓦戴。他是柬埔寨诺罗敦王室的始祖	松赞干布，悉补野氏，是吐蕃雅鲁王朝第 33 任赞普，也是吐蕃帝国的建立者，约 629 年至 650 年在位。他是前任赞普囊日论赞的儿子。原名赤松赞，"松赞干布"是其尊号，意思是"心胸深邃的松赞"	普密蓬·阿杜德，亦称普密蓬大帝，泰国却克里王朝第九代国王，亦称拉玛九世。1946 年 6 月 9 日即位，2016 年 10 月 13 日驾崩，过世时为世界史上在任时间第二长的国家元首，亦为泰国历史上统治时间最长的国君，总共在位 70 年 126 天

401

再看 GLOVE_6B_100D 向量模型计算出来的结果，如下所示。

EUCLIDEAN	{"ID":"[\"king\",\"prince\",\"queen\",\"monarch\",\"brother\",\"uncle\",\"nephew\"]","METRIC":"[1.6858739404357614E-7,4.092165813215177,4.281252149113389,4.474171532331215,4.536668410815254,4.668967636875079,4.695680189674975]"}
COSINE	{"ID":"[\"king\",\"prince\",\"queen\",\"son\",\"brother\",\"monarch\",\"throne\"]","METRIC":"[2.220446049250313E-16,0.23176711289070218,0.2492309206376152,0.29791115594092876,0.30142244166461385,0.3022109473347163,0.3080009470368401]"}

可以看到，使用两种度量方式计算出来的前 3 个词都是一样的，都为 king、prince 和 queen。后面的词主要有两类：monarch（帝王）和 throne（君主）；brother（兄弟）、uncle（叔伯）、nephew（侄子）和 son（儿子）。

综合上面的测试结果可知，不同语料来源的单词向量模型是有区别的；单词的相似度可以通过向量距离来衡量，但"相似"的含义受具体单词向量模型的影响。

22.1.3 单词向量

本节我们会聚焦 4 个单词：man、woman、king、queen。

首先，我们提取并输出这 4 个单词所对应的向量，具体代码如下。

```
getWikiDependency()
    .filter("word IN ('man', 'woman', 'king', 'queen')")
    .lazyPrint(-1);

getGlove6B100d()
    .filter("word IN ('man', 'woman', 'king', 'queen')")
    .lazyPrint(-1);

BatchOperator.execute();
```

输出 WikiDependency 中这 4 个单词的向量值，见表 22-2。

表 22-2　WikiDependency 中 4 个单词的向量值

word	vec
man	-0.00220404170083 0.0678135463787 …… -0.0164630158472 -0.0441532936764
king	0.0330381329194 0.0665698704227 …… -0.013237880883 -0.0142381059934
woman	-0.0923953735088 0.062354665309 …… -0.0518072294457 0.0179050510523
queen	0.0199859507483 0.1526205478 …… 0.0267236279156 0.0352932794572

输出 Glove6B100d 中这 4 个单词的向量值，见表 22-3。

第 22 章 单词向量化

表 22-3　Glove6B100d 中 4 个单词的向量值

word	vec
man	0.37293 0.38503 …… 0.039138 -0.53911
king	-0.32307 -0.87616 …… 0.16483 -0.98878
woman	0.59368 0.44825 …… 0.15162 -0.30754
queen	-0.50045 -0.70826 …… 0.13347 -0.56075

Alink 定义了稠密向量（DenseVector）类型，可以处理向量计算。下面将由提取出来的向量字符串构造 DenseVector，随后从向量角度进行分析。

先分析 WikiDependency 中这 4 个单词的向量值，具体代码如下。

```
DenseVector vec_man = VectorUtil.parseDense(
    "-0.00220404170083 0.0678135463787 ... ... -0.0164630158472 -0.0441532936764");

DenseVector vec_woman = VectorUtil.parseDense(
    "-0.0923953735088 0.062354665309 ... ... -0.0518072294457 0.0179050510523");

DenseVector vec_king = VectorUtil.parseDense(
    "0.0330381329194 0.0665698704227 ... ... -0.013237880883 -0.0142381059934");

DenseVector vec_queen = VectorUtil.parseDense(
    "0.0199859507483 0.1526205478 ... ... 0.0267236279156 0.0352932794572");

System.out.println("'man' vector normL2 : " + vec_man.normL2());
System.out.println("'woman' vector normL2 : " + vec_woman.normL2());
System.out.println("'king' vector normL2 : " + vec_king.normL2());
System.out.println("'queen' vector normL2 : " + vec_queen.normL2());
System.out.println("'man - woman' normL2 : " + vec_man.minus(vec_woman).normL2());
System.out.println("'king - queen' normL2 : " + vec_king.minus(vec_queen).normL2());
System.out.println("(man - woman) - (king - queen) normL2 : " +
    vec_man.minus(vec_woman).minus(vec_king.minus(vec_queen)).normL2());
```

运行结果如下。

```
'man' vector normL2 : 0.999999999999892
'woman' vector normL2 : 0.9999999999997525
'king' vector normL2 : 0.9999999999997967
'queen' vector normL2 : 1.0000000000000957
'man - woman' normL2 : 0.6768403103559452
'king - queen' normL2 : 0.8214690626774831
(man - woman) - (king - queen) normL2 : 0.9766210180698275
```

显然，这 4 个单词向量的 2-范数都为 1；"man" 与 "woman" 的向量差 "man - woman" 的 2-范数为 0.6768403103559452；"king" 与 "queen" 的向量差 "king - queen" 的 2-范数为 0.8214690626774831。

由于 2-范数不同，向量"man-woman"与"king-queen"应该不同，但差异有多大呢？我们计算这两个向量的差，即"(man-woman)-(king-queen)"，其 2-范数为 0.9766210180698275，说明差异不小。所以从向量值的角度，有

$$\text{man-woman} \neq \text{king-queen}$$

但是向量"man-woman"所指向的方向是否能给我们带来一些有意思的结论呢？在下面的实验中，我们会构造两个新的向量"king-man+woman"与"queen-woman+man"，看看有哪些与之相似的单词？具体代码如下。

```
new VectorNearestNeighbor()
    .setIdCol("word")
    .setSelectedCol("vec")
    .setMetric(Metric.EUCLIDEAN)
    .setOutputCol("similar_words")
    .setTopN(5)
    .fit(
        getWikiDependency()
    )
    .transform(
        new MemSourceBatchOp(
            new Row[] {
                Row.of("king", vec_king),
                Row.of("king-man+woman", vec_king.minus(vec_man).plus(vec_woman)),
                Row.of("queen", vec_queen),
                Row.of("queen-woman+man", vec_queen.minus(vec_woman).plus(vec_man)),
            },
            new String[] {"word", "vec"}
        )
    )
    .select("word, similar_words")
    .print();
```

运行结果如表 22-4 所示。

表 22-4　相似单词

word	similar_words
king	{"ID":"["king","norodom","songtsän","queen","bhumibol"]", "METRIC":"[2.9802322387695312E-8,0.8055086190497484,0.8063912079032907, 0.8214690626774828,0.8229473168473302]"}
king-man+woman	{"ID":"["king","queen","monarch","princess","norodom"]", "METRIC":"[0.6768403103559449,0.976621101980627,0.9920983527502455, 1.023152990110213,1.0314874003333323]"}

续表

word	similar_words
queen	{"ID":"["queen","tsarina","princess","king","tsaritsa"]", "METRIC":"[2.9802322387695312E-8,0.8076690691199051,0.8195939124642033, 0.8214690626774828,0.8629289238663901]"}
queen-woman+man	{"ID":"["queen","king","tsarina","knave","overlord"]", "METRIC":"[0.6768403103559447,0.9766210180698268,1.0186863394683976, 1.0294727515416788,1.0405324064187942]"}

对比"king-man+woman"与"king",最相邻的有皇后"queen"和公主"princess";而和"queen-woman+man"最相邻的是皇后"queen"和国王"king"。

我们继续做这个实验,分析Glove6B100d中这4个单词的向量值,具体代码如下。

```
DenseVector vec_man = VectorUtil.parseDense(
    "0.37293 0.38503 0.71086 ... ... -0.93711 0.039138 -0.53911");

DenseVector vec_woman = VectorUtil.parseDense(
    "0.59368 0.44825 0.5932 ... ... -0.54648 0.15162 -0.30754");

DenseVector vec_king = VectorUtil.parseDense(
    "-0.32307 -0.87616 0.21977 ... ... -0.52881 0.16483 -0.98878");

DenseVector vec_queen = VectorUtil.parseDense(
    "-0.50045 -0.70826 0.55388 ... ... -0.36032 0.13347 -0.56075");

System.out.println("'man' vector normL2 : " + vec_man.normL2());
System.out.println("'woman' vector normL2 : " + vec_woman.normL2());
System.out.println("'king' vector normL2 : " + vec_king.normL2());
System.out.println("'queen' vector normL2 : " + vec_queen.normL2());
System.out.println("'man - woman' normL2 : " + vec_man.minus(vec_woman).normL2());
System.out.println("'king - queen' normL2 : " + vec_king.minus(vec_queen).normL2());
System.out.println("(man - woman) - (king - queen) normL2 : " +
    vec_man.minus(vec_woman).minus(vec_king.minus(vec_queen)).normL2());
```

运行结果如下。

```
'man' vector normL2 : 5.593625640618958
'woman' vector normL2 : 5.961704845063365
'king' vector normL2 : 6.1176004287966546
'queen' vector normL2 : 6.006716940168814
'man - woman' normL2 : 3.364067897373334
'king - queen' normL2 : 4.281252149113388
(man - woman) - (king - queen) normL2 : 4.081078579600476
```

显然,这4个单词向量的2-范数都在6左右;向量"(man-woman) - (king-queen)"的2-范

数为 4.081078579600476，说明差异不小。所以从向量值的角度，有

$$man-woman \neq king-queen$$

与 WikiDependency 单词向量模型计算两个新的向量 "king-man+woman" 和 "queen-woman+man" 的代码类似，我们同样可计算基于 Glove6B100d 单词向量模型的情形，这里略去具体代码，只显示运行结果，如表 22-5 所示。

表 22-5　基于 Glove6B100d 单词向量模型的相似单词

word	similar_words
king	{"ID":"["king","prince","queen","monarch","brother"]", "METRIC":"[1.6858739404357614E-7,4.092165813215177, 4.281252149113389,4.474171532331215,4.536668410815254]"}
king-man+woman	{"ID":"["king","queen","monarch","throne","elizabeth"]", "METRIC":"[3.364067897373334,4.081078579600477, 4.642907390892788,4.905500707490234,4.921558914642319]"}
queen	{"ID":"["queen","princess","elizabeth","king","lady"]", "METRIC":"[0.0,3.8532465383426704,4.159615241047596, 4.281252149113389,4.467098262191517]"}
queen-woman+man	{"ID":"["queen","king","prince","royal","majesty"]", "METRIC":"[3.364067897373337,4.0810785796004785, 5.034758702597461,5.069448877059333,5.189214516972033]"}

对比 "king-man+woman" 与 "king" 的结果，最相邻的有皇后 "queen" 和 "elizabeth"；对比 "queen-woman+man" 与 "queen" 的结果，最相邻的是国王 "king" 和王子 "prince"。

22.2　单词映射为向量

本节以四大名著之一的《三国演义》为例，将整本书的内容作为语料，使用 Word2Vec 算法将单词映射为向量，并使用向量距离分析各人物。

《三国演义》文本文件的下载地址为链接 22-3，里面不仅有《三国演义》，还有《水浒传》、《红楼梦》和《西游记》。感兴趣的读者可以仿照本节的方法，使用机器学习 "读" 一下其他的名著。

下载到本地后，使用文本编辑器打开，如图 22-3 所示。

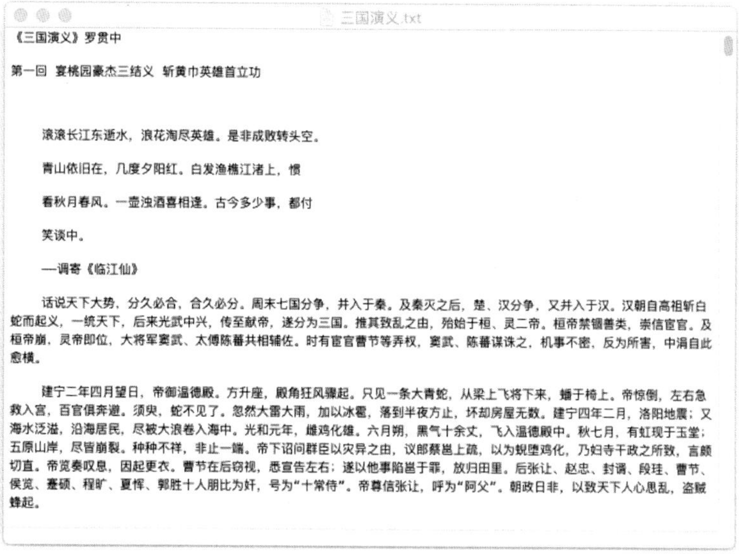
三国演义.txt

《三国演义》罗贯中

第一回 宴桃园豪杰三结义 斩黄巾英雄首立功

滚滚长江东逝水，浪花淘尽英雄。是非成败转头空。

青山依旧在，几度夕阳红。白发渔樵江渚上，惯

看秋月春风。一壶浊酒喜相逢。古今多少事，都付

笑谈中。

—调寄《临江仙》

话说天下大势，分久必合，合久必分。周末七国分争，并入于秦。及秦灭之后，楚、汉分争，又并入于汉。汉朝自高祖斩白蛇而起义，一统天下，后来光武中兴，传至献帝，遂分为三国。推其致乱之由，殆始于桓、灵二帝。桓帝禁锢善类，崇信宦官。及桓帝崩，灵帝即位，大将军窦武、太傅陈蕃共相辅佐。时有宦官曹节等弄权，窦武、陈蕃谋诛之，机事不密，反为所害，中涓自此愈横。

建宁二年四月望日，帝御温德殿。方升座，殿角狂风骤起。只见一条大青蛇，从梁上飞将下来，蟠于椅上。帝惊倒，左右急救入宫，百官俱奔避。须臾，蛇不见了。忽然大雷大雨，加以冰雹，落到半夜方止，坏却房屋无数。建宁四年二月，洛阳地震；又海水泛溢，沿海居民，尽被大浪卷入海中。光和元年，雌鸡化雄。六月朔，黑气十余丈，飞入温德殿中。秋七月，有虹现于玉堂；五原山岸，尽皆崩裂。种种不祥，非止一端。帝下诏问群臣以灾异之由，议郎蔡邕上疏，以为蜺堕鸡化，乃妇寺干政之所致，言颇切直。帝览奏叹息，因起更衣。曹节在后窃视，悉宣告左右；遂以他事陷邕于罪，放归田里。后张让、赵忠、封谞、段珪、曹节、侯览、蹇硕、程旷、夏恽、郭胜十人朋比为奸，号为"十常侍"。帝尊信张让，呼为"阿父"。朝政日非，以致天下人心思乱，盗贼蜂起。

图 22-3

有了原始数据文件，如何读取它，使之成为数据源呢？我们可以将文档中的每个段落作为一条记录，即每个换行符分隔一条记录。Alink 提供了一个文本数据源组件，可以直接读取文本文件。

```
static final String ORIGIN_FILE = "三国演义.txt";
… … …
TextSourceBatchOp source = new TextSourceBatchOp().setFilePath(DATA_DIR + ORIGIN_FILE);

source.lazyPrint(8);
```

输出前 8 条数据，显示如下。

```
《三国演义》罗贯中
第一回 宴桃园豪杰三结义 斩黄巾英雄首立功

    滚滚长江东逝水，浪花淘尽英雄。是非成败转头空。
    青山依旧在，几度夕阳红。白发渔樵江渚上，惯
    看秋月春风。一壶浊酒喜相逢。古今多少事，都付
    笑谈中。
    ——调寄《临江仙》
```

接下来做一下分析前的预处理工作，先使用分词组件将句子拆分为单词，然后使用停用词组件过滤标点符号。我们先使用 WordCountBatchOp 组件统计单词出现的频率，具体代码如下。

列出一些常用的人名，存放在字符串数组 CHARACTER_DICT 中，用于为分词的自定义词典参数 UserDefinedDict 的赋值；计算完词频后，按词频降序选择前 100 个并输出。

```
final String[] CHARACTER_DICT = new String[]{
  "曹操", "孔明", "玄德", "刘玄德", "刘备", "关羽", "张飞",
  "赵云", "曹孟德", "诸葛亮", "张郃", "孙权", "张辽", "鲁肃"
};
source
  .link(
    new SegmentBatchOp()
      .setSelectedCol("text")
      .setUserDefinedDict(CHARACTER_DICT)
  )
  .link(
    new StopWordsRemoverBatchOp()
      .setSelectedCol("text")
  )
  .link(
    new WordCountBatchOp()
      .setSelectedCol("text")
  )
  .orderBy("cnt", 100, false)
  .print();
```

运行结果如表 22-6 所示，"曰"出现的次数最多，有 7558 次；在各人物中，"曹操"出现的次数最多，有 908 次，而"操"也常用来指代"曹操"，出现了 665 次，可见曹操是三国演义中出场次数最多的人物；紧随其后的是"玄德"和"孔明"。通过词频统计结果可以看出，我们使用的停用词过滤主要针对现代文，对于古文中的一些词，例如"曰""吾""皆"等，还需要扩充停用词词典再进行处理。

表 22-6 单词出现的频率

C cword	cnt
曰	7558
吾	1916
皆	1132
去	1100
曹操	908
不	890
见	826
遂	818

续表

C cword	cnt
人	796
将军	759
玄德	752
孔明	731
操	665
今	664
欲	651

下面是本节的核心内容，使用 Word2VecTrainBatchOp 组件训练单词向量模型。扩充了停用词表，设置参数 MinCount=10，即只计算词频数在 10 以上的单词，选择训练迭代次数为 50，并将结果保存为 AK 格式数据文件。具体代码如下。

```
source
  .link(
    new SegmentBatchOp()
      .setSelectedCol("text")
      .setUserDefinedDict(CHARACTER_DICT)
  )
  .link(
    new StopWordsRemoverBatchOp()
      .setSelectedCol("text")
      .setStopWords(
        "亦", "曰", "遂", "吾", "己", "去", "二人", "令", "使", "中", "知",
        "不", "见", "都", "令", "却", "欲", "请", "人", "谓", "不可", "闻",
        "前", "后", "皆", "便", "问", "日", "时", "耳", "不敢", "问", "回", "才",
        "之事", "之人", "之时", "料", "今日", "令人", "受", "说", "出", "已毕",
        "不得", "使人", "众", "何不", "不知", "再", "处", "无", "即日", "诸", "此时",
        "只", "下", "还", "上", "杀", "将军", "却说", "兵", "汝", "走", "言", "寨",
        "不能", "斩", "死", "商议", "听", "军士", "军", "左右", "军马", "引兵", "次日",
        "二", "看", "耶", "退", "更", "毕", "正", "一人", "原来", "大笑", "车胄", "口",
        "引", "大喜", "其事", "助", "事", "未", "大", "至此", "讫", "心中", "敢"
      )
  )
  .link(
    new Word2VecTrainBatchOp()
      .setSelectedCol("text")
      .setMinCount(10)
      .setNumIter(50)
```

```
)
.link(
   new AkSinkBatchOp()
      .setFilePath(DATA_DIR + W2V_MODEL_FILE)
);
```

训练完成后，我们再导入模型数据，使用向量最近邻组件，看一下这些人物的关系。具体代码如下。

```
AkSourceBatchOp word2vec
   = new AkSourceBatchOp().setFilePath(DATA_DIR + W2V_MODEL_FILE);
new VectorNearestNeighbor()
   .setIdCol("word")
   .setSelectedCol("vec")
   .setTopN(20)
   .setOutputCol("similar_words")
   .fit(word2vec)
   .transform(
      word2vec.filter("word IN ('曹操', '操', '玄德', '刘备', '孔明', "
         + "'亮', '卧龙', '周瑜', '吕布', '貂蝉', '云长', '孙权')")
   )
   .select("word, similar_words")
   .print();
```

整理输出内容如表 22-7 所示，为了节省篇幅，这里省略了每个单词对应的距离指标，并将"曹操"和"操"、"玄德"和"刘备"的结果放在相邻位置，便于比对。

表 22-7　相似单词的计算结果

word	similar_words
曹操	{"ID":"["曹操","玄德","玄德曰","徐州","刘备","怒","其言","郡","云长","吕布","以为","进兵","守","大军","朝廷","东吴","此事","许都","操","之兵"]"}
操	{"ID":"["操","玄德曰","天子","曹操","玄德","许都","何如","朝廷","何故","主公","公","徐州","刘备","大事","破","投","忽","吕布","英雄","坐"]"}
玄德	{"ID":"["玄德","玄德曰","曹操","徐州","云长","吕布","许都","投","操","坐","刘备","教","丞相","只得","主公","大事","相见","关公","命","引军"]"}
刘备	{"ID":"["刘备","徐州","曹操","主公","玄德曰","将士","书","急","袁术","孙策","起兵","之计","攻","朝廷","郡","东吴","玄德","其言","进兵","袁绍"]"}
孔明	{"ID":"["孔明","魏延","先主","孔明曰","曹真","东吴","司马懿","魏兵","蜀","军中","马岱","祁山","奏","关兴","姜维","懿","魏","以为","成都","中原"]"}

续表

word	similar_words
亮	{"ID":"["亮","权","东吴","甚","都督","鲁肃","恪","孙权","诸葛","内","孙","周瑜","先主","瑜","道","不见","何故","先生","年","罢"]"}
卧龙	{"ID":"["卧龙","叹","先生","年","汉室","道","军中","张","瑜","罢","后人","鲁肃","玄德","诗","周瑜","真","入城","城中","马","必"]"}
吕布	{"ID":"["吕布","布","军中","徐州","玄德","陈宫","玄德曰","张辽","曹操","曹兵","城","主公","投","相见","云长","只得","城中","袁术","城下","三军"]"}
云长	{"ID":"["云长","张飞","玄德曰","玄德","何故","张","急","关公","徐州","曹操","飞","吕布","只见","救","主公","厮杀","关","之计","贼","大叫"]"}
孙权	{"ID":"["孙权","江东","东吴","魏兵","鲁肃","三军","提兵","江南","玄德曰","权","汉中","救","无不","陆逊","主公","奏","周瑜","蜀","之计","先主"]"}
周瑜	{"ID":"["周瑜","鲁肃","瑜","之计","军中","曹操","吴侯","孙权","东吴","孙策","刘备","救","江东","将士","命","汉中","云长","厮杀","提兵","道"]"}
貂蝉	{"ID":"["貂蝉","卓","妾","董卓","入","罢","允","玄德曰","布","而来","立于","忽见","军中","吕布","间","叱","岂","天下","肃","否"]"}

可以观察到如下信息。
- "曹操"和"操"的距离比较近，相近的单词内容也比较相似，都出现了"玄德""玄德曰""许都""朝廷""徐州""吕布"。
- "玄德"和"刘备"的距离较近，都关联着"曹操"和"徐州"；但有趣的是，关联的单词中没有"诸葛亮"。
- 我们看一下诸葛亮的3个称呼："孔明""亮"和"卧龙"。"孔明"出现的频率最高，与其最相似的单词为"魏延""司马懿""曹真""东吴"，还有一个词比较特殊——"先主"，应该是指刘备。"亮"最相近的单词有"权""东吴""鲁肃""周瑜"，可能"亮"这个称呼在赤壁大战前后使用的比较多。与"卧龙"相近的词中出现了"玄德"，它和"先生""汉室"的距离也较近。
- 与"云长"最近的是"张飞""玄德曰""玄德"，也有"关公"，当然也少不了"曹操"。

23 情感分析

本章主要讲解一个文本分析领域非常有用的应用——情感分析。我们在网上购买商品（或浏览电影信息、寻找饭馆信息）时，经常看到一个区域显示着用户的评论。我们会习惯性地参考其他人的评论，如果大家都说好，就更坚定了我们买下此商品的信心；如果有人给了差评，那么我们会很自然地想到，我要是购买了此商品是否也会遇到同样的问题。如何通过机器学习算法高效地判断评论的情感倾向呢？

我们选择一个在情感分析方面常用的公开数据集（参见链接23-1），是用户对电影的评论，并有标签标明用户评论的核心意思是喜欢还是不喜欢。每条评论都比较长，这里选择了一条完整内容，显示如下。虽然有 "good" 和 "laughed loud" 等好评的词，但整篇读下来，可以知道用户并不看好这部电影，由此可见情感分析的难度。

> Oh dear. good cast, but to write and direct is an art and to write wit and direct wit is a bit of a task. Even doing good comedy you have to get the timing and moment right. Im not putting it all down there were parts where i laughed loud but that was at very few times. The main focus to me was on the fast free flowing dialogue, that made some people in the film annoying. It may sound great while reading the script in your head but getting that out and to the camera is a different task. And the hand held camera work does give energy to few parts of the film. Overall direction was good but the script was not all that to me, but I'm sure you was reading the script in your head it would sound good. Sorry.

将数据文件下载并解压，目录结构如图23-1所示。

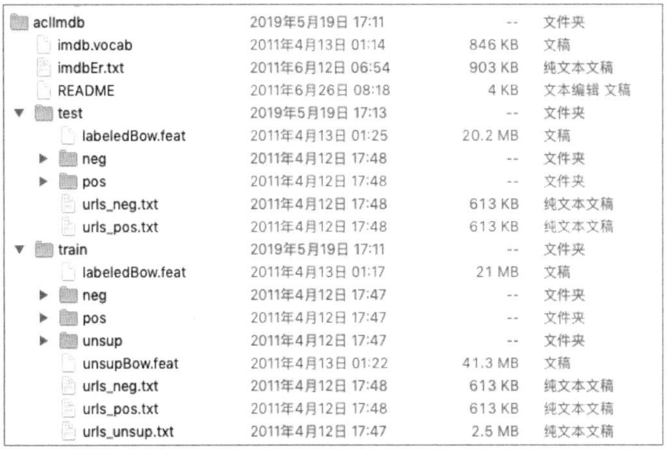

第 23 章　情感分析

图 23-1

其中，README 文件对数据集进行了详细介绍。数据集中包括原始文本数据，也包括从原始数据中使用词袋（bag of words）方式提取的特征数据。本章会先使用已提取特征的数据快速进行实验，得到模型效果；然后，对原始的英文数据进行预处理，展示完整的情感分析流程。

23.1　使用提供的特征

下面了解一下数据，使用词袋的方式将每个词作为一个特征，而该词在样本中出现的个数作为特征值。在具体的实现方式上，需要将所有出现的单词作为词汇表，放在文件 imdb.vocab 中，内容如图 23-2 所示。每行为一个单词，将词汇表的单词总数作为稀疏特征向量的维度，每个单词在词汇表的排序位置（从 0 开始）即为其在特征向量的索引位置。

图 23-2

我们再用文本编辑器打开 train/labeledBow.feat 文件，如图 23-3 所示。

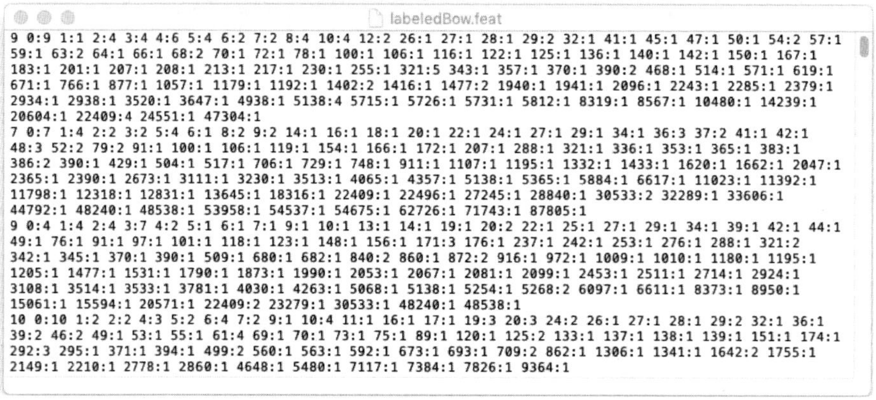

```
labeledBow.feat
9 0:9 1:1 2:4 3:4 4:6 5:4 6:2 7:2 8:4 10:4 12:2 26:1 27:1 28:1 29:2 32:1 41:1 45:1 47:1 50:1 54:2 57:1
59:1 63:2 64:1 66:1 68:2 70:1 72:1 78:1 100:1 106:1 116:1 122:1 125:1 136:1 140:1 142:1 150:1 167:1
183:1 201:1 207:1 208:1 213:1 217:1 230:1 255:1 321:5 343:1 357:1 370:1 390:2 468:1 514:1 571:1 619:1
671:1 766:1 877:1 1057:1 1179:1 1192:1 1402:2 1416:1 1477:2 1940:1 1941:1 2096:1 2243:1 2285:1 2379:1
2934:1 2938:1 3520:1 3647:1 4938:1 5138:4 5715:1 5726:1 5731:1 5812:1 8319:1 8567:1 10480:1 14239:1
20604:1 22409:4 24551:1 47304:1
7 0:7 1:4 2:2 3:2 5:4 6:1 8:2 9:2 14:1 16:1 18:1 20:1 22:1 24:1 27:1 29:1 34:1 36:3 37:2 41:1 42:1
48:3 52:2 79:2 91:1 100:1 106:1 119:1 154:1 166:1 172:1 207:1 288:1 321:1 336:1 353:1 365:1 383:1
386:2 390:1 429:1 504:1 517:1 706:1 729:1 748:1 911:1 1107:1 1195:1 1332:1 1433:1 1620:1 1662:1 2047:1
2365:1 2390:1 2673:1 3111:1 3230:1 3513:1 4065:1 4357:1 5138:1 5365:1 5884:1 6617:1 11023:1 11392:1
11798:1 12318:1 12831:1 13645:1 18316:1 22409:1 22496:1 27245:1 28840:1 30533:2 32289:1 33606:1
44792:1 48240:1 48538:1 53958:1 54537:1 54675:1 62726:1 71743:1 87805:1
9 0:4 1:4 2:4 3:7 4:2 5:1 6:1 7:1 9:1 10:1 13:1 14:1 19:1 20:2 22:1 25:1 27:1 29:1 34:1 39:1 42:1 44:1
49:1 76:1 91:1 97:1 101:1 118:1 123:1 148:1 156:1 171:3 176:1 237:1 242:1 253:1 276:1 288:1 321:2
342:1 345:1 370:1 390:1 509:1 680:1 682:1 840:2 860:1 872:2 916:1 972:1 1009:1 1010:1 1180:1 1195:1
1205:1 1477:1 1531:1 1790:1 1873:1 1990:1 2053:1 2067:1 2081:1 2099:1 2453:1 2511:1 2714:1 2924:1
3108:1 3514:1 3533:1 3781:1 4030:1 4263:1 5068:1 5138:1 5254:1 5268:2 6097:1 6611:1 8373:1 8950:1
15061:1 15594:1 20571:1 22409:2 23279:1 30533:1 48240:1 48538:1
10 0:10 1:2 2:2 4:3 5:2 6:4 7:2 9:1 10:4 11:1 16:1 17:1 19:3 20:3 24:2 26:1 27:1 28:1 29:2 32:1 36:1
39:2 46:2 49:1 53:1 55:1 61:4 69:1 70:1 73:1 75:1 89:1 120:1 125:2 133:1 137:1 138:1 139:1 151:1 174:1
292:3 295:1 371:1 394:1 499:2 560:1 563:1 592:1 673:1 693:1 709:2 862:1 1306:1 1341:1 1642:2 1755:1
2149:1 2210:1 2778:1 2860:1 4648:1 5480:1 7117:1 7384:1 7826:1 9364:1
```

图 23-3

这是 LIBSVM 格式的文件，关于该格式的详细介绍请参考 3.2.2 节。

可以看到，train/labeledBow.feat 文件中的单词索引是从 0 开始的，在词汇表文件 imdb.vocab 中可以找到与索引相对应的单词。譬如，文件中第一行出现的 0:9，表示词汇表 imdb.vocab 中的第一个单词（the）在该评论中出现了 9 次。

我们可以方便地使用 Alink 组件读取数据，需要设置文件的路径，因为 LIBSVM 文件通常以 1 作为起始索引，但是该数据以 0 作为起始索引，所以还需设置参数 StartIndex 为 0。具体代码如下所示。

```
static String ORIGIN_DATA_DIR = DATA_DIR + "aclImdb" + File.separator;
... ...
BatchOperator <?> train_set = new LibSvmSourceBatchOp()
   .setFilePath(ORIGIN_DATA_DIR + "train" + File.separator + "labeledBow.feat")
   .setStartIndex(0);

train_set.lazyPrint(10, "train_set");
```

最后一行代码用于输出前 10 行数据，图 23-4 显示了部分结果。

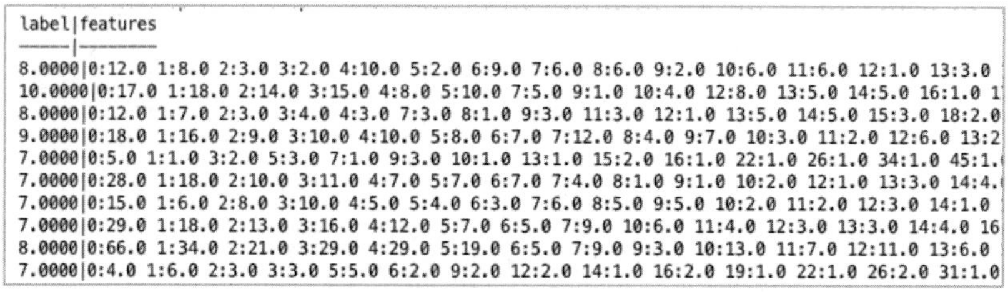

label	features
8.0000	0:12.0 1:8.0 2:3.0 3:2.0 4:10.0 5:2.0 6:9.0 7:6.0 8:6.0 9:2.0 10:6.0 11:6.0 12:1.0 13:3.0
10.0000	0:17.0 1:18.0 2:14.0 3:15.0 4:8.0 5:10.0 7:5.0 9:1.0 10:4.0 12:8.0 13:5.0 14:5.0 16:1.0 1
8.0000	0:12.0 1:7.0 2:3.0 3:4.0 4:3.0 7:3.0 8:1.0 9:3.0 11:3.0 12:1.0 13:5.0 14:5.0 15:3.0 18:2.0
9.0000	0:18.0 1:16.0 2:9.0 3:10.0 4:10.0 5:8.0 6:7.0 7:12.0 8:4.0 9:7.0 10:3.0 11:2.0 12:6.0 13:2
7.0000	0:5.0 1:1.0 3:2.0 5:3.0 7:1.0 9:3.0 10:1.0 13:1.0 15:2.0 16:1.0 22:1.0 26:1.0 34:1.0 45:1.
7.0000	0:28.0 1:18.0 2:10.0 3:11.0 4:7.0 5:7.0 6:7.0 7:4.0 8:1.0 9:1.0 10:2.0 12:1.0 13:3.0 14:4.
7.0000	0:15.0 1:6.0 2:8.0 3:10.0 4:5.0 5:4.0 6:3.0 7:6.0 8:5.0 9:5.0 10:2.0 11:2.0 12:3.0 14:1.0
7.0000	0:29.0 1:18.0 2:13.0 3:16.0 4:12.0 5:7.0 6:5.0 7:9.0 10:6.0 11:4.0 12:3.0 13:3.0 14:4.0 16
8.0000	0:66.0 1:34.0 2:21.0 3:29.0 4:29.0 5:19.0 6:5.0 7:9.0 9:3.0 10:13.0 11:7.0 12:11.0 13:6.0
7.0000	0:4.0 1:6.0 2:3.0 3:3.0 5:5.0 6:2.0 9:2.0 12:2.0 14:1.0 16:2.0 19:1.0 22:1.0 26:2.0 31:1.0

图 23-4

显然，数据分为两列，features 列为稀疏特征向量，label 列记录的是评分值。我们再看一下 label 列数据的分布情况，代码如下。

```
train_set
    .groupBy("label", "label, COUNT(label) AS cnt")
    .orderBy("label", 100)
    .lazyPrint(-1, "labels of train_set");
```

显示结果如下。

```
label|cnt
-----|---
1.0000|5100
2.0000|2284
3.0000|2420
4.0000|2696
7.0000|2496
8.0000|3009
9.0000|2263
10.0000|4732
```

可以看到，最低分 1.0 出现的次数最多，为 5100；最高分 10.0 出现的次数也很多，为 4732；其他分值出现的次数比较平均；无法明显体现情感倾向的中间评分 5.0 和 6.0 均未出现。

同样，我们看一下预测集的情况。

```
BatchOperator <?> test_set = new LibSvmSourceBatchOp()
    .setFilePath(ORIGIN_DATA_DIR + "test" + File.separator + "labeledBow.feat")
    .setStartIndex(0);
test_set
    .lazyPrint(10, "test_set")
    .groupBy("label", "label, COUNT(label) AS cnt")
    .orderBy("label", 100)
    .lazyPrint(-1, "labels of test_set");
```

预测集 test_set 的评分分布如下。

```
label|cnt
-----|---
1.0000|5022
2.0000|2302
3.0000|2541
4.0000|2635
7.0000|2307
8.0000|2850
9.0000|2344
10.0000|4999
```

其评分分布情况与训练集大致相同，以 1.0 和 10.0 居多，其他分值出现的次数比较平均，无法明显体现情感倾向的中间评分 5.0 和 6.0 均未出现。

我们的目标是二分类问题，需要按评分多少转换为"pos"（正向评论、好评）或"neg"（负向评论、差评）标签值。使用字符串常量 VECTOR_COL_NAME 来统一向量特征列的命名。

```
private static final String LABEL_COL_NAME = "label";
private static final String VECTOR_COL_NAME = "vec";
... ...

train_set = train_set.select("CASE WHEN label>5 THEN 'pos' ELSE 'neg' END AS label, "
    + "features AS " + VECTOR_COL_NAME);
test_set = test_set.select("CASE WHEN label>5 THEN 'pos' ELSE 'neg' END AS label, "
    + "features AS " + VECTOR_COL_NAME);

train_set.lazyPrint(10, "train_set");
```

看一下变换后的训练数据集，向量数据较长，图 23-5 显示了部分数据。

```
label|vec
-----|---
neg|0:7.0 1:2.0 3:3.0 4:1.0 5:2.0 6:5.0 9:1.0 10:1.0 15:1.0 22:1.0 24:1.0 27:4.0 28:1.0 33:2.0 34:1.0 35:1.0 37:1.
neg|0:6.0 1:4.0 2:2.0 3:1.0 6:2.0 8:2.0 9:5.0 10:3.0 11:3.0 12:1.0 14:1.0 15:3.0 16:2.0 17:1.0 18:1.0 19:1.0
neg|0:1.0 1:1.0 3:2.0 4:1.0 5:2.0 6:1.0 7:1.0 9:3.0 10:1.0 13:2.0 15:1.0 16:1.0 22:1.0 24:1.0 25:1.0 31:3.0 33:1.0
neg|0:3.0 1:3.0 2:3.0 3:2.0 4:2.0 6:1.0 7:1.0 8:2.0 9:4.0 10:1.0 11:3.0 12:1.0 13:3.0 15:2.0 16:3.0 17:2.0 18:1.0
neg|0:2.0 1:6.0 2:4.0 3:2.0 4:1.0 5:1.0 6:2.0 8:5.0 9:2.0 10:3.0 14:1.0 15:3.0 16:1.0 18:1.0 19:1.0 24:1.0
neg|0:12.0 1:8.0 2:11.0 3:7.0 4:6.0 5:12.0 6:5.0 7:6.0 8:2.0 9:6.0 10:3.0 11:2.0 12:2.0 13:3.0 14:2.0 15:8.0 16:2.
neg|0:6.0 2:8.0 4:5.0 5:2.0 6:3.0 7:2.0 8:2.0 11:4.0 14:2.0 15:1.0 16:5.0 18:1.0 19:1.0 20:8.0 21:1.0 22:1.0 25:1.
neg|0:12.0 1:7.0 2:5.0 3:3.0 4:8.0 5:2.0 6:7.0 7:8.0 9:1.0 10:1.0 12:1.0 13:1.0 14:1.0 15:2.0 16:2.0 17:1.0 18:1.0
neg|0:24.0 1:12.0 2:11.0 3:10.0 4:8.0 5:5.0 6:5.0 7:1.0 9:4.0 10:6.0 11:1.0 12:3.0 14:4.0 15:1.0 16:5.0 17:4.0 18:
neg|0:10.0 1:5.0 2:8.0 3:6.0 4:5.0 5:7.0 6:4.0 7:2.0 8:1.0 9:3.0 10:4.0 12:5.0 13:2.0 14:4.0 15:5.0 16:4.0 19:2.0
```

图 23-5

23.1.1 使用朴素贝叶斯方法

针对文本的朴素贝叶斯分类器与通用的朴素贝叶斯分类器是有区别的，一个明显的不同之处是输入的数据格式不同，通用的朴素贝叶斯分类器支持数据列的输入，每个数据列都是数值型或枚举型数据；针对文本的朴素贝叶斯分类器输入的数据为稀疏向量形式，向量的维度较大，一般为语料中不同单词的个数。

针对文本的朴素贝叶斯算法有如下两种。
- 多项式朴素贝叶斯（Multinomial Naive Bayes）：特征向量 $x = (x_1, x_2, \cdots, x_m)$，其中各分量为对应单词在该文档中出现的个数；设 p_{ki} 为第 i 个单词出现在第 k 个类别的概率，则 x 属于第 k 个类别的可能性为

$$P(x|C_k) = \frac{(\sum_i x_i)!}{\prod_i x_i!} \prod_i p_{ki}^{x_i}$$

- 伯努利朴素贝叶斯（Bernoulli Naive Bayes）：特征向量 $\boldsymbol{x} = (x_1, x_2, \cdots, x_m)$，其中各分量的取值为 0 或 1，代表对应的单词是否在该文档中出现；设 p_{ki} 为第 i 个单词出现在第 k 个类别的概率，则 $\boldsymbol{x}$ 属于第 k 个类别的可能性为

$$P(\boldsymbol{x}|C_k) = \prod_i p_{ki}^{x_i}(1 - p_{ki})^{(1-x_i)}$$

可以看到它们的差异，在输入数据的形式方面，多项式朴素贝叶斯以单词出现的次数为特征值，为非负整数；伯努利朴素贝叶斯以 0 或 1 为特征值，表示单词是否出现。从类别可能性的公式上看，在多项式朴素贝叶斯中，单词出现的次数对分类可能性有影响；而在伯努利朴素贝叶斯中，单词出现 1 次和多次对分类可能性的影响是一样的。

当前训练、预测数据集的特征向量的各个分量值正是对应单词在当前文档中出现的个数，所以我们可以直接使用 NaiveBayesTextClassifier 组件，设置 ModelType 参数为"Multinomial"，尝试多项式朴素贝叶斯算法。

```
new NaiveBayesTextClassifier()
  .setModelType("Multinomial")
  .setVectorCol(VECTOR_COL_NAME)
  .setLabelCol(LABEL_COL_NAME)
  .setPredictionCol(PREDICTION_COL_NAME)
  .setPredictionDetailCol(PRED_DETAIL_COL_NAME)
  .enableLazyPrintModelInfo()
  .fit(train_set)
  .transform(test_set)
  .link(
    new EvalBinaryClassBatchOp()
      .setPositiveLabelValueString("pos")
      .setLabelCol(LABEL_COL_NAME)
      .setPredictionDetailCol(PRED_DETAIL_COL_NAME)
      .lazyPrintMetrics("NaiveBayesTextClassifier + Multinomial")
  );
BatchOperator.execute();
```

贝叶斯模型信息显示如下。

```
-------------------------- NaiveBayesTextModelInfo --------------------------

=========================== model meta info ===========================
{model type: Multinomial, vector size: 89527, vector col name: vec}
===================== label proportion information =====================

|neg|pos|
|---|---|
|0.5|0.5|
```

```
==================== feature probability information ====================
|vector index|   neg|   pos|
|------------|------|------|
|           0|0.0552|0.0573|
|           1|0.0251|0.0296|
|           2|0.0268|0.0276|
|         ...|   ...|   ...|
|       89524|     0|     0|
|       89525|     0|     0|
|       89526|     0|     0|
```

计算结果如下。

```
----------------------------- Metrics: -----------------------------
AUC:0.8905  Accuracy:0.8136  Precision:0.8591   Recall:0.7504   F1:0.8011   LogLoss:1.6084
|Pred\Real| pos|  neg|
|---------|----|-----|
|      pos|9380| 1539|
|      neg|3120|10961|
```

再尝试使用伯努利朴素贝叶斯算法，仍然使用 NaiveBayesTextClassifier 组件，需要设置 ModelType 参数为"Bernoulli"。因为伯努利朴素贝叶斯算法要求特征向量的分量值只能为 0 或 1，所以我们需要加上预处理组件 Binarizer，以某个阈值为界，将数据转化为 0 和 1（小于等于阈值会被转化为 0，大于阈值则为 1），该组件默认的阈值为 0。另外，为了查看该组件处理结果，可以使用 enableLazyPrintTransformData 方法。

```
new Pipeline()
  .add(
    new Binarizer()
        .setSelectedCol(VECTOR_COL_NAME)
        .enableLazyPrintTransformData(5, "After Binarizer")
  )
  .add(
    new NaiveBayesTextClassifier()
        .setModelType("Bernoulli")
        .setVectorCol(VECTOR_COL_NAME)
        .setLabelCol(LABEL_COL_NAME)
        .setPredictionCol(PREDICTION_COL_NAME)
        .setPredictionDetailCol(PRED_DETAIL_COL_NAME)
        .enableLazyPrintModelInfo()
  )
  .fit(train_set)
  .transform(test_set)
  .link(
```

```
new EvalBinaryClassBatchOp()
    .setPositiveLabelValueString("pos")
    .setLabelCol(LABEL_COL_NAME)
    .setPredictionDetailCol(PRED_DETAIL_COL_NAME)
    .lazyPrintMetrics("Binarizer + NaiveBayesTextClassifier + Bernoulli")
);
```

模型信息显示如下。

```
------------------------ NaiveBayesTextModelInfo ------------------------
========================= model meta info =========================
{model type: Bernoulli, vector size: 89527, vector col name: vec}
===================== label proportion information =====================

|pos|neg|
|---|---|
|0.5|0.5|


===================== feature probability information =====================

|vector index|   pos|   neg|
|------------|------|------|
|           0|0.0068|0.0069|
|           1|0.0067|0.0066|
|           2|0.0066|0.0067|
|         ...|   ...|   ...|
|       89524|     0|     0|
|       89525|     0|     0|
|       89526|     0|     0|
```

计算结果如下,各项指标要优于使用多项式朴素贝叶斯算法。

```
----------------------------- Metrics: -----------------------------
AUC:0.908  Accuracy:0.8274  Precision:0.8674    Recall:0.7729    F1:0.8174  LogLoss:1.0788
|Pred\Real| pos|  neg|
|---------|----|-----|
|      pos|9661| 1477|
|      neg|2839|11023|
```

23.1.2 使用逻辑回归算法

使用朴素贝叶斯文本分类器可以快速获取一个不错的模型,我们也可以尝试使用 LogisticRegression 分类器,具体代码如下。

```
new LogisticRegression()
    .setVectorCol(VECTOR_COL_NAME)
```

```
    .setLabelCol(LABEL_COL_NAME)
    .setPredictionCol(PREDICTION_COL_NAME)
    .setPredictionDetailCol(PRED_DETAIL_COL_NAME)
    .enableLazyPrintTrainInfo("< LR train info >")
    .enableLazyPrintModelInfo("< LR model info >")
    .fit(train_set)
    .transform(test_set)
    .link(
        new EvalBinaryClassBatchOp()
            .setPositiveLabelValueString("pos")
            .setLabelCol(LABEL_COL_NAME)
            .setPredictionDetailCol(PRED_DETAIL_COL_NAME)
            .lazyPrintMetrics("LogisticRegression")
    );
```

计算结果如下，相比于文本朴素贝叶斯算法，各项指标又有所提升。

```
-------------------------------- Metrics: --------------------------------
AUC:0.94  Accuracy:0.873  Precision:0.8649  Recall:0.8842  F1:0.8744  LogLoss:0.8491
|Pred\Real|  pos  | neg  |
|---------|------|------|
|   pos   |11052 | 1727 |
|   neg   | 1448 |10773 |
```

逻辑回归训练过程的信息如下。

```
-------------------------- train meta info --------------------------
{model name: Logistic Regression, num feature: 89527}
-------------------------- train importance info --------------------------
|colName|importnaceValue|colName| weightValue |
|-------|---------------|-------|-------------|
|   427 |   1.82374809  | 6126  | 11.24306434 |
|    77 |   1.78636621  | 9808  | 10.48445417 |
|   240 |   1.63787453  | 8898  |  9.63473157 |
|  ... ...              |  ... ...            |
| 21750 |   0.00000059  | 8157  |-10.27395642 |
| 44833 |   0.00000050  | 7339  |-11.29044369 |
| 46409 |   0.00000036  | 4111  |-13.70695642 |
-------------------------- train convergence info --------------------------
step:0  loss:0.69288129  gradNorm:0.05158780  learnRate:0.40000000
step:1  loss:0.58087868  gradNorm:0.05154487  learnRate:1.60000000
step:2  loss:0.42480889  gradNorm:0.05234910  learnRate:4.00000000
... ... ...
step:97  loss:0.01554132  gradNorm:0.00131275  learnRate:4.00000000
step:98  loss:0.01439589  gradNorm:0.00141001  learnRate:4.00000000
step:99  loss:0.01365725  gradNorm:0.00162553  learnRate:4.00000000
```

在第三部分中，显示了每次迭代中 3 个重要的指标值（loss、gradNorm、learnRate）。此次训练是到了最大迭代次数（默认值 100）而终止的，而在终止的时候，loss 的值还在不断下降，

第 23 章 情感分析

gradNorm 和 learnRate 的值也不小。这时可能需要调高迭代次数，使 loss 值收敛；可能训练过程有些"过拟合"，过多次的迭代可以获得更低的 loss 值，但我们关心的 AUC、Accuracy 等指标不一定是最优的。此时最好的办法是进行超参数搜索，确定较优的参数。

使用交叉验证方式网格搜索参数，具体代码如下。

```
AlinkGlobalConfiguration.setPrintProcessInfo(true);

LogisticRegression lr = new LogisticRegression()
    .setVectorCol(VECTOR_COL_NAME)
    .setLabelCol(LABEL_COL_NAME)
    .setPredictionCol(PREDICTION_COL_NAME)
    .setPredictionDetailCol(PRED_DETAIL_COL_NAME);

GridSearchCV gridSearch = new GridSearchCV()
    .setEstimator(
        new Pipeline().add(lr)
    )
    .setParamGrid(
        new ParamGrid()
            .addGrid(lr, LogisticRegression.MAX_ITER,
                new Integer[] {10, 20, 30, 40, 50, 60, 80, 100})
    )
    .setTuningEvaluator(
        new BinaryClassificationTuningEvaluator()
            .setLabelCol(LABEL_COL_NAME)
            .setPredictionDetailCol(PRED_DETAIL_COL_NAME)
            .setTuningBinaryClassMetric(TuningBinaryClassMetric.AUC)
    )
    .setNumFolds(6)
    .enableLazyPrintTrainInfo();

GridSearchCVModel bestModel = gridSearch.fit(train_set);

bestModel
    .transform(test_set)
    .link(
        new EvalBinaryClassBatchOp()
            .setPositiveLabelValueString("pos")
            .setLabelCol(LABEL_COL_NAME)
            .setPredictionDetailCol(PRED_DETAIL_COL_NAME)
            .lazyPrintMetrics("LogisticRegression")
    );
BatchOperator.execute();
```

单独定义 LogisticRegression 实例 lr，便于后面指定调参对象。通过组件 GridSearchCV 进行网格搜索，并设置交叉验证的折数（NumFolds）为 6，设置搜索参数的列表，最大迭代次数

（MAX_ITER）会尝试{10, 20, 30, 40, 50, 60, 80, 100}。还有一个重要参数就是调参的目标，在 TuningEvaluator 中设置，这里我们选择了 AUC。最后，使用搜索出来的最优参数对整个训练数据集进行训练，得到 bestModel，再用测试集进行评估。

需要注意的是，网格搜索所需时间较长，我们一般会允许组件输出中间结果信息，便于我们看到搜索的进展，运行 AlinkGlobalConfiguration 的 setPrintProcessInfo(true) 方法即可。需要屏幕显示网格搜索结果时可以使用 enableLazyPrintTrainInfo 方法。

网格搜索的结果如下，最大迭代次数为 30 时，AUC 的指标更高。

```
Metric information:
  Metric name: AUC
  Larger is better: true
Tuning information:
|                AUC|             stage| param|value|
|-------------------|------------------|------|-----|
|0.9495710102806689 |LogisticRegression|maxIter|  30|
|0.9480044722419051 |LogisticRegression|maxIter|  40|
|0.9478236485040631 |LogisticRegression|maxIter|  20|
|0.9474024334078258 |LogisticRegression|maxIter|  60|
|0.9471031004488054 |LogisticRegression|maxIter|  50|
| 0.943939344834118 |LogisticRegression|maxIter|  80|
|0.9426986353478771 |LogisticRegression|maxIter| 100|
|0.9293614922795674 |LogisticRegression|maxIter|  10|
```

对于使用最大迭代次数 30 训练出来的 bestModel，其评估结果如下，明显优于前面使用默认最大迭代次数 100 得到的结果。

```
---------------------------- Metrics: ----------------------------
AUC:0.9472  Accuracy:0.8814  Precision:0.8868    Recall:0.8744    F1:0.8806  LogLoss:0.3727
|Pred\Real|  pos|  neg|
|---------|-----|-----|
|      pos|10930| 1395|
|      neg| 1570|11105|
```

另外，如果将调参目标变为 Accuracy，可以得到如下的搜索结果，仍然是当最大迭代次数为 30 时结果最优。

```
Metric information:
  Metric name: Accuracy
  Larger is better: true
Tuning information:
|           Accuracy|             stage| param|value|
|-------------------|------------------|------|-----|
|0.8880398353537382 |LogisticRegression|maxIter|  30|
|0.8859196401109802 |LogisticRegression|maxIter|  40|
```

```
|0.8833198384935733|LogisticRegression|maxIter|  60|
|0.8821201265012576|LogisticRegression|maxIter|  50|
|0.8815198768736043|LogisticRegression|maxIter|  80|
|0.8809601136853843|LogisticRegression|maxIter| 100|
|0.8781601840551746|LogisticRegression|maxIter|  20|
|0.8568795725338613|LogisticRegression|maxIter|  10|
```

23.2 如何提取特征

在 23.1 节中，我们对如何基于已经抽取的特征进行了建模和预测。本节我们会进一步讲解如何从原始文本数据中提取特征。原始评论数据的每条样本都是一个文本文件，正负样本分别位于 pos 文件夹和 neg 文件夹中。提取文本内容和标签后，将数据保存为 AK 格式文件，便于后续多次调用该训练和测试数据集。具体代码如下。

```
ArrayList <Row> trainRows = new ArrayList <>();
ArrayList <Row> testRows = new ArrayList <>();

for (String label : new String[] {"pos", "neg"}) {
   File subfolder = new File(ORIGIN_DATA_DIR + "train" + File.separator + label);
   for (File f : subfolder.listFiles()) {
      trainRows.add(Row.of(label, readFileContent(f)));
   }
}
for (String label : new String[] {"pos", "neg"}) {
   File subfolder = new File(ORIGIN_DATA_DIR + "test" + File.separator + label);
   for (File f : subfolder.listFiles()) {
      testRows.add(Row.of(label, readFileContent(f)));
   }
}

new MemSourceBatchOp(trainRows, COL_NAMES)
   .link(
      new AkSinkBatchOp()
         .setFilePath(DATA_DIR + TRAIN_FILE)
   );
new MemSourceBatchOp(testRows, COL_NAMES)
   .link(
      new AkSinkBatchOp()
         .setFilePath(DATA_DIR + TEST_FILE)
   );
BatchOperator.execute();
```

使用前面保存的 AK 格式文件构建训练和测试数据源，并输出 5 条数据，查看数据内容。具体代码如下。

```
AkSourceBatchOp train_set = new AkSourceBatchOp().setFilePath(DATA_DIR + TRAIN_FILE);
AkSourceBatchOp test_set = new AkSourceBatchOp().setFilePath(DATA_DIR + TEST_FILE);

train_set.lazyPrint(5);
```

由于每条样本的文本内容较多，这里只显示其中的两条，如表 23-1 所示。

表 23-1 两条样本数据

label	review
pos	This was a good film with a powerful message of love and redemption. I loved the transformation of the brother and the repercussions of the horrible disease on the family. Well-acted and well-directed. If there were any flaws, I'd have to say that the story showed the typical suburban family and their difficulties again. What about all people of all cultural backgrounds? I would love to see a movie where all of these cultures are shown - like in real life. Nevertheless, the film soared in terms of its values and its understanding of the how a disease can bring someone closer to his or her maker. Loved the film and it brought tears to my eyes
pos	Hello. This movie is.......well.......okay. Just kidding! ITS AWESOME! It's NOT a Block Buster smash hit. It's not meant to be. But its a big hit in my world. And my sisters. We are rockin' Rollers. GO RAMONES!!!! This is a great movie............. For ME!

有了英文文本数据，构造 23.1 节中所用的特征向量就很简单了。首先使用组件 RegexTokenizer，得到空格分隔的英文单词序列；然后使用组件 DocCountVectorizer，计算各文档中单词出现的数目，并根据指定的类型产生特征向量，这里选择的是"WORD_COUNT"，即计算文档内各单词出现的次数；最后，直接对生成的特征向量进行建模，通过模型的评估指标验证我们所生成的特征向量的有效性。具体代码如下。

```
new Pipeline()
    .add(
        new RegexTokenizer()
            .setPattern("\\W+")
            .setSelectedCol(TXT_COL_NAME)
    )
    .add(
        new DocCountVectorizer()
            .setFeatureType("WORD_COUNT")
            .setSelectedCol(TXT_COL_NAME)
            .setOutputCol(VECTOR_COL_NAME)
            .enableLazyPrintTransformData(5)
    )
    .add(
```

```
      new LogisticRegression()
        .setMaxIter(30)
        .setVectorCol(VECTOR_COL_NAME)
        .setLabelCol(LABEL_COL_NAME)
        .setPredictionCol(PREDICTION_COL_NAME)
        .setPredictionDetailCol(PRED_DETAIL_COL_NAME)
    )
    .fit(train_set)
    .transform(test_set)
    .link(
      new EvalBinaryClassBatchOp()
        .setPositiveLabelValueString("pos")
        .setLabelCol(LABEL_COL_NAME)
        .setPredictionDetailCol(PRED_DETAIL_COL_NAME)
        .lazyPrintMetrics("DocCountVectorizer")
    );
```

计算评估结果如下,与直接使用已提供的特征向量训练得到的模型的效果相当,从而验证了我们构造的特征向量的有效性。

```
-------------------------------- Metrics: --------------------------------
AUC:0.9479   Accuracy:0.882   Precision:0.8754   Recall:0.8909   F1:0.8831   LogLoss:0.3467
|Pred\Real|  pos |  neg |
|---------|-----|------|
|     pos |11136| 1585 |
|     neg | 1364|10915 |
```

另外,我们还可以通过 hash 方式同样实现单词到向量索引的映射。使用 DocHashCountVectorizer 组件,同样是用模型的评估指标验证我们所生成的特征向量的有效性,具体代码如下。

```
new Pipeline()
  .add(
    new RegexTokenizer()
      .setPattern("\\W+")
      .setSelectedCol(TXT_COL_NAME)
  )
  .add(
    new DocHashCountVectorizer()
      .setFeatureType("WORD_COUNT")
      .setSelectedCol(TXT_COL_NAME)
      .setOutputCol(VECTOR_COL_NAME)
      .enableLazyPrintTransformData(5)
  )
  .add(
    new LogisticRegression()
      .setMaxIter(30)
      .setVectorCol(VECTOR_COL_NAME)
      .setLabelCol(LABEL_COL_NAME)
```

```
            .setPredictionCol(PREDICTION_COL_NAME)
            .setPredictionDetailCol(PRED_DETAIL_COL_NAME)
    )
    .fit(train_set)
    .transform(test_set)
    .link(
        new EvalBinaryClassBatchOp()
            .setPositiveLabelValueString("pos")
            .setLabelCol(LABEL_COL_NAME)
            .setPredictionDetailCol(PRED_DETAIL_COL_NAME)
            .lazyPrintMetrics("DocHashCountVectorizer")
    );
BatchOperator.execute();
```

使用 hash 方式计算得到的评估结果如下,在评估指标上,与精确匹配方式得到的结果相当,这就验证了使用 hash 方式生成特征向量的有效性。

```
-------------------------- Metrics: --------------------------------
AUC:0.9477  Accuracy:0.883  Precision:0.8816  Recall:0.8849  F1:0.8833  LogLoss:0.351
|Pred\Real|  pos  |  neg  |
|---------|------|-------|
|    pos  |11061 | 1485  |
|    neg  | 1439 |11015  |
```

23.3 构造更多特征

特征生成方法主要利用了单词是否在该文档中出现的性质,并没有考虑各单词之间的先后关系,本节将利用 NGram 构造新的特征。

将文档看作单词的序列,NGram 是将序列中相邻的 n 个单词作为一个片段,从片段的出现次数上可以看出哪些单词组合在语料中最为常用。就像每个单词可以作为特征向量的一个分量,每个片段也可以作为特征向量的一个分量。先进行分词,按单词的出现次数生成特征向量;然后使用 NGram 组件,设置参数 $N=2$,即相邻两个单词作为一个片段;再将片段当作单词看待,为了对产生的数据有一个直观印象,这里激活了该组件处理结果的输出功能。同样可以使用 DocCountVectorizer 组件生成特征向量,最后使用 VectorAssembler 组件,将两个生成的特征向量组装为一个;接着和前面介绍的流程一样,使用逻辑回归算法进行训练、预测和评估模型,具体代码如下。为了节省篇幅,这里略去一些前面章节出现过的代码。

```
New Pipeline()
    .add(
```

```
    new RegexTokenizer()
        .setPattern("\\W+")
        .setSelectedCol(TXT_COL_NAME)
)
.add(
    new DocCountVectorizer()
        .setFeatureType("WORD_COUNT")
        .setSelectedCol(TXT_COL_NAME)
        .setOutputCol(VECTOR_COL_NAME)
)
.add(
    new NGram()
        .setN(2)
        .setSelectedCol(TXT_COL_NAME)
        .setOutputCol("v_2")
        .enableLazyPrintTransformData(1, "2-gram")
)
.add(
    new DocCountVectorizer()
        .setFeatureType("WORD_COUNT")
        .setSelectedCol("v_2")
        .setOutputCol("v_2")
)
.add(
    new VectorAssembler()
        .setSelectedCols(VECTOR_COL_NAME, "v_2")
        .setOutputCol(VECTOR_COL_NAME)
)
.add(new LogisticRegression()... ...)
.fit(train_set)
.transform(test_set)
.link(new EvalBinaryClassBatchOp()... ...);
BatchOperator.execute();
```

2-gram 操作后,输出了一条数据,整理如表 23-2 所示,左边是其所在的数据列名称,右边是数据值。

表 23-2 一条数据的各列取值

label	pos
review	i really like this show it has drama romance and comedy all rolled into one i am 28 and i am a married mother so i can identify both with lorelei s and rory s experiences in the show i have been watching mostly the repeats on the family channel lately so i am not up to date on what is going on now i think females would like this show more than males but i know some men out there would enjoy it i really like that is an hour long and not a half hour as th hour seems to fly by when i am watching it give it a chance if you have never seen the show i think lorelei and luke are my favorite characters on the show though mainly because of the way they are with one another how could you not see something was there or take that long to see it i guess i should say br br happy viewing

续表

label	pos
vec	$74629$0:6.0 1:5.0 2:3.0 3:1.0 4:3.0 5:2.0 6:2.0 7:5.0 8:1.0 9:13.0 10:2.0 11:2.0 12:2.0 13:1.0 14:1.0 16:2.0 18:1.0 21:2.0 22:4.0 23:3.0 25:2.0 27:2.0 29:2.0 30:1.0 32:1.0 33:1.0 34:1.0 36:2.0 38:3.0 39:2.0 41:1.0 44:1.0 45:1.0 46:1.0 47:1.0 48:1.0 50:1.0 51:1.0 53:1.0 55:1.0 59:1.0 60:2.0 64:2.0 65:2.0 72:1.0 76:1.0 82:1.0 85:1.0 86:1.0 93:1.0 96:1.0 101:2.0 102:1.0 107:1.0 113:1.0 120:5.0 121:1.0 132:1.0 141:1.0 142:1.0 147:2.0 148:1.0 150:1.0 160:1.0 169:1.0 185:1.0 191:1.0 194:2.0 197:1.0 201:1.0 208:1.0 213:1.0 242:4.0 319:1.0 336:1.0 356:1.0 420:1.0 451:1.0 480:1.0 508:1.0 528:3.0 569:1.0 636:1.0 660:1.0 822:1.0 872:1.0 1006:1.0 1273:1.0 1289:1.0 1423:1.0 2031:1.0 2211:1.0 2486:1.0 3481:1.0 4466:1.0 5027:1.0 5232:1.0 5721:1.0 6263:1.0 7568:1.0 10706:1.0 10978:1.0
v_2	i_really really_like like_this this_show show_it it_has has_drama drama_romance romance_and and_comedy comedy_all all_rolled rolled_into into_one one_i i_am am_28 28_and and_i i_am am_a a_married married_mother mother_so so_i i_can can_identify identify_both both_with with_lorelei lorelei_s s_and and_rory rory_s s_experiences experiences_in in_the the_show show_i i_have have_been been_watching watching_mostly mostly_the the_repeats repeats_on on_the the_family family_channel channel_lately lately_so so_i i_am am_not not_up up_to to_date date_on on_what what_is is_going going_on on_now now_i i_think think_females females_would would_like like_this this_show show_more more_than than_males males_but but_i i_know know_some some_men men_out out_there there_would would_enjoy enjoy_it it_i i_really really_like like_that that_is is_an an_hour hour_long long_and and_not not_a a_half half_hour hour_as as_th th_hour hour_seems seems_to to_fly fly_by by_when when_i i_am am_watching watching_it it_give give_it it_a a_chance chance_if if_you you_have have_never never_seen seen_the the_show show_i i_think think_lorelei lorelei_and and_luke luke_are are_my my_favorite favorite_characters characters_on on_the the_show show_though though_mainly mainly_because because_of of_the the_way way_they they_are are_with with_one one_another another_how how_could could_you you_not not_see see_something something_was was_there there_or or_take take_that that_long long_to to_see see_it it_i i_guess guess_i i_should should_say say_br br_br br_happy happy_viewing

其中 v_2 列就是对原始文本数据进行 2-gram 的结果，相邻的两个单词用下画线"_"连接起来，成为一个片段，片段之间使用空格分隔。这里将片段看作"单词"，还可以使用 DocCountVectorizer 组件生成特征向量。

最后的评估结果如下，AUC 和 Accuracy 指标有明显提升，说明了新特征的有效性。

```
------------------------------- Metrics: -------------------------------
AUC:0.9603  Accuracy:0.893  Precision:0.9407    Recall:0.839    F1:0.8869   LogLoss:0.6826
|Pred\Real|   pos |  neg |
|---------|------|------|
|     pos |10487 |  661 |
|     neg | 2013 |11839 |
```

既然 2-gram 产生了作用，那么再增加片段长度，譬如 3-gram，是否能获得更好的效果呢？

在 2-gram 代码的基础上，增加 3-gram 特征向量的生成，并和其他两个特征向量一起组装成一个向量，具体代码如下。

```
new Pipeline()
  .add(new RegexTokenizer()... ...)
  .add(new DocCountVectorizer()... ...)
  .add(new NGram()... ...)
  .add(new DocCountVectorizer()... ...)
  .add(
    new NGram()
       .setN(3)
       .setSelectedCol(TXT_COL_NAME)
       .setOutputCol("v_3")
  )
  .add(
    new DocCountVectorizer()
       .setFeatureType("WORD_COUNT")
       .setVocabSize(10000)
       .setSelectedCol("v_3")
       .setOutputCol("v_3")
  )
  .add(
    new VectorAssembler()
       .setSelectedCols(VECTOR_COL_NAME, "v_2", "v_3")
       .setOutputCol(VECTOR_COL_NAME)
  )
  .add(new LogisticRegression()... ...)
  .fit(train_set)
  .transform(test_set)
  .link(new EvalBinaryClassBatchOp()... ...);
BatchOperator.execute();
```

注意，由于不同的 3-gram 片段的数量较大，我们只使用出现频次较高的那些片段，在 DocCountVectorizer 组件设置了 VocabSize 为 10000。

运行得到的评估结果如下，AUC 和 Accuracy 指标再一次被提升。

```
------------------------------- Metrics: -------------------------------
AUC:0.9641  Accuracy:0.9057  Precision:0.9039    Recall:0.9078   F1:0.9059   LogLoss:0.4472
|Pred\Real|   pos |  neg |
|---------|------|------|
|     pos |11348 | 1206 |
|     neg | 1152 |11294 |
```

23.4 模型保存与预测

前面的实验流程都是基于 Pipeline 的,而 Pipeline 执行 fit 方法就得到了 PipelineModel,我们可以使用 save 方法将模型保存到文件系统。这里省略了 Pipeline 构建的代码,save 方法不会触发任务执行,所以在其后显式调用 execute 函数,具体代码如下。

```
static String PIPELINE_MODEL = "pipeline_model.ak";
... ...
new Pipeline()
   .add(new RegexTokenizer()... ...)
   ... ...
   .add(new LogisticRegression()... ...)
   .fit(train_set)
   .save(DATA_DIR + PIPELINE_MODEL);
BatchOperator.execute();
```

运行成功后,我们可以在相应路径中看到新生成的模型文件。

23.4.1 批式/流式预测任务

对于批式/流式预测任务,我们需要从模型文件中导入 PipelineModel,可以使用如下代码。

```
PipelineModel pipeline_model = PipelineModel.load(DATA_DIR + PIPELINE_MODEL);
```

对于批式数据,可以直接使用 PipelineModel 的 transform 方法进行预测,后接预测评估组件,检验模型效果,具体代码如下。

```
AkSourceBatchOp test_set = new AkSourceBatchOp().setFilePath(DATA_DIR + TEST_FILE);
pipeline_model
   .transform(test_set)
   .link(
      new EvalBinaryClassBatchOp()
         .setPositiveLabelValueString("pos")
         .setLabelCol(LABEL_COL_NAME)
         .setPredictionDetailCol(PRED_DETAIL_COL_NAME)
         .lazyPrintMetrics("NGram 2 and 3")
   );
BatchOperator.execute();
```

其评估结果如下所示,与前面批式实验的结果基本一致。

```
-------------------------------- Metrics: --------------------------------
AUC:0.9644 Accuracy:0.9056 Precision:0.9185    Recall:0.8901    F1:0.9041 LogLoss:0.4387
```

```
|Pred\Real|  pos|  neg|
|---------|-----|-----|
|      pos|11126|  987|
|      neg| 1374|11513|
```

对于流式数据，我们也可以直接使用 PipelineModel 的 transform 方法，输入为流式数据，输出的预测结果也为流式数据。由于数据较多，我们进行了采样（采样率为 0.001），并选择输出最主要的 3 列，具体代码如下。

```
AkSourceStreamOp test_stream = new AkSourceStreamOp().setFilePath(DATA_DIR + TEST_FILE);
pipeline_model
    .transform(test_stream)
    .sample(0.001)
    .select(PREDICTION_COL_NAME + ", " + LABEL_COL_NAME + ", " + TXT_COL_NAME)
    .print();
StreamOperator.execute();
```

可以看到，随着流式任务的运行，会不断输出预测结果。

23.4.2 嵌入式预测

在实际应用中，我们经常需要将机器学习模型预测嵌入应用中，通过 SDK 方式直接调用。Alink 提供的 LocalPredictor 就是为了解决此问题的，LocalPredictor 的输入和输出都为 Flink Row 类型。

PipelineModel 提供了 collectLocalPredictor 方法，可以直接得到 LocalPredictor 的实例。在 Pipeline 中，通过数据列名称来指定所要处理的数据，我们输入的 Row 类型数据没有列名信息，所以需要在构建 LocalPredictor 时加入当前输入数据的列名和类型信息，一般使用 Schema String 方式描述（详见 2.8 节）。具体代码如下。

```
String str
    = "Oh dear. good cast, but to write and direct is an art and to write wit and direct wit is a bit of a "
    + "task. Even doing good comedy you have to get the timing and moment right. Im not putting it all down "
    + "there were parts where i laughed loud but that was at very few times. The main focus to me was on the "
    + "fast free flowing dialogue, that made some people in the film annoying. It may sound great while "
    + "reading the script in your head but getting that out and to the camera is a different task. And the "
    + "hand held camera work does give energy to few parts of the film. Overall direction was good but the "
    + "script was not all that to me, but I'm sure you was reading the script in your head it would sound good"
    + ". Sorry.";

Row pred_row;

LocalPredictor local_predictor = pipeline_model.collectLocalPredictor("review string");

System.out.println(local_predictor.getOutputSchema());
```

```
pred_row = local_predictor.map(Row.of(str));

System.out.println(pred_row.getField(4));
```

其中，使用 getOutputSchema 方法获得当前预测输出结果的 Schema，内容如下。

```
root
 |-- review: STRING
 |-- vec: LEGACY(GenericType<com.alibaba.alink.common.linalg.Vector>)
 |-- v_2: LEGACY(GenericType<com.alibaba.alink.common.linalg.SparseVector>)
 |-- v_3: LEGACY(GenericType<com.alibaba.alink.common.linalg.SparseVector>)
 |-- pred: STRING
 |-- predinfo: STRING
```

"pred"列是预测标签列，其前面有 4 列，可以按索引值 4 从预测输出 pred_row 中获取预测标签值，即 pred_row.getField(4)。输出结果如下。

```
neg
```

输入的电影评论表达的是用户不喜欢该影片。

从 PipelineModel 获取 LocalPredictor 更符合我们的直观理解，但这需要启动 Flink 任务才能完成。在实际使用时，我们可以用另一种更轻量的方法——直接使用 LocalPredictor 的构造方法，第一个参数为模型文件地址，第二个参数为 SchemaStr，具体代码如下。

```
LocalPredictor local_predictor_2
    = new LocalPredictor(DATA_DIR + PIPELINE_MODEL, "review string");

System.out.println(local_predictor_2.getOutputSchema());

pred_row = local_predictor_2.map(Row.of(str));

System.out.println(pred_row.getField(4));
```

运行结果如下。

```
root
 |-- review: STRING
 |-- vec: LEGACY(GenericType<com.alibaba.alink.common.linalg.Vector>)
 |-- v_2: LEGACY(GenericType<com.alibaba.alink.common.linalg.SparseVector>)
 |-- v_3: LEGACY(GenericType<com.alibaba.alink.common.linalg.SparseVector>)
 |-- pred: STRING
 |-- predinfo: STRING

neg
```

构建推荐系统

本章讨论的主题是"推荐",即通过参考已知的数据,对每个用户所关注的商品或信息给出个性化、精准的推荐。本章会围绕一个电影推荐的具体场景展开,需要用到两个数据集——电影信息和用户对电影的评分记录,二者均来自 MovieLens 网站,下载地址为链接 24-1。

电影信息数据集的数据表名为 movielens_movies,包含 1700 部电影的信息,内容如图 24-1 所示。

movieid	title	genres
1	Toy Story (1995)	Adventure|Animation|Children|Comedy|Fantasy
2	Jumanji (1995)	Adventure|Children|Fantasy
3	Grumpier Old Men (1995)	Comedy|Romance
4	Waiting to Exhale (1995)	Comedy|Drama|Romance
5	Father of the Bride Part II (1995)	Comedy
6	Heat (1995)	Action|Crime|Thriller
7	Sabrina (1995)	Comedy|Romance
8	Tom and Huck (1995)	Adventure|Children
9	Sudden Death (1995)	Action
10	GoldenEye (1995)	Action|Adventure|Thriller
11	American President, The (1995)	Comedy|Drama|Romance
12	Dracula: Dead and Loving It (1995)	Comedy|Horror
13	Balto (1995)	Animation|Children
14	Nixon (1995)	Drama
15	Cutthroat Island (1995)	Action|Adventure|Romance
16	Casino (1995)	Crime|Drama

图 24-1

其中共有 3 个数据列,第一列名称为 movieid,即电影 ID,在下面的数据集中,通过电影

ID 关联评论信息；第二列名称为 title，是电影名称，电影名称后括号内为电影发行的年份；第三列名称为 genres，是电影类别（喜剧、动作片、爱情喜剧等）。

用户对电影的评分记录集，其数据表名为 movielens_ratings，包含 100000 条用户对电影的打分记录，内容如图 24-2 所示。

userid	movieid	rating
1	122	5
1	185	5
1	231	5
1	292	5
1	316	5
1	329	5
1	355	5
1	356	5
1	362	5
1	364	5

图 24-2

同前面的数据表一样，评分记录表也只有 3 列，第一列名称为 userid，用一个数字 ID 代表某个用户；第二列名称为 movieid，是电影 ID，在前面介绍数据集时已经提过，可以使用此 ID 与具体的电影信息对应；第三列名称为 rating，是用户为该影片的打分，范围为 0.5、1、1.5、……、4.5、5。

24.1 与推荐相关的组件介绍

Alink 提供了一系列与推荐相关的组件，从组件使用的角度来看，需要重点关注如下 3 个方面。

1. 算法选择

推荐领域有很多算法，常用的有基于物品/用户的协同过滤、ALS、FM 算法等。对于不同的数据场景，算法也会在计算方式上有很大的变化，譬如 ALS 算法针对数据是显式评分还是隐式反馈的情况，会采用不用的目标函数进行模型求解，得出截然不同的模型。

2. 推荐方式

输入信息可以有多种选择，输入结果也有多种情况。

（1）同时输入一个用户的信息和一个物品的信息，计算用户对此物品的评分。

（2）输入用户的信息，可以推荐适合此用户的相关物品，也可以计算与其相似的用户。

（3）输入物品的信息，推荐给可能喜欢该物品的用户，也可以计算与其相似的物品。

3. 使用方法

在应用推荐引擎时，可能是在离线任务中进行批量推荐，也可能是在实时任务中对流式数据进行推荐，还可以通过使用 Alink Java SDK 将推荐引擎嵌入用户的应用系统中。

算法选择和推荐方式是通过"推荐模型"这个桥梁连接起来的，如图 24-3 所示。选择适合的推荐算法，基于训练数据集得到推荐模型，基于训练出来的模型可以执行多种推荐方式。

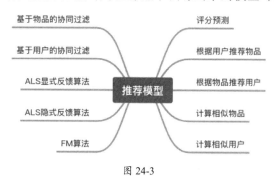

图 24-3

下面以基于物品的协同过滤（Item-based Collaborative Filtering）算法为例，看一下 Alink 相关的组件。模型训练为离线批式训练，对应组件为 ItemCfTrainBatchOp，得到 ItemCf 模型。基于此模型可以进行多种推荐，但不是每种推荐方式使用该 ItemCf 算法都能得到较好的效果。我们只提供了适合该算法的推荐方式——评分预测（ItemCfRate）、根据用户推荐物品（ItemCfItemsPerUser）、计算相似物品（ItemCfSimilarItems），没有提供"根据物品推荐用户"和"计算相似用户"组件。考虑到每种推荐需要支持多种使用方式，每种方法都提供了 3 种组件——批式推荐（RecommBatchOp）、流式推荐（RecommStreamOp）和 Pipeline 节点（后缀为 Recommender，并可由此获得 LocalPredictor，嵌入用户的应用中推荐）。如图 24-4 所示。

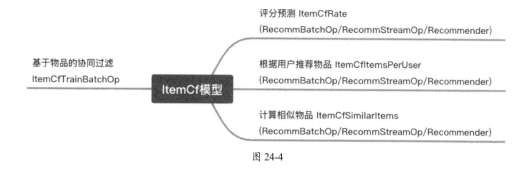

图 24-4

再看另一个有代表性的推荐算法——交替最小二乘法，其基本思路为交替固定用户特征向量和物品特征向量的值，每次求解一个最小二乘问题，直到满足求解条件。根据用户—物品矩阵中值的含义是评分值，还是行为次数、观看/收听时长，分别选用显式反馈算法（AlsTrainBatchOp）与隐式反馈算法（AlsImplicitTrainBatchOp）。两种计算方式得到 ALS 模型格式是一样的，后面可以选用 5 种推荐方式，而且每种方法都提供了 3 种组件——批式推荐（RecommBatchOp）、流式推荐（RecommStreamOp）和 Pipeline 节点（后缀为 Recommender，并可由此获得 LocalPredictor，进行嵌入用户的应用中推荐）。如图 24-5 所示。

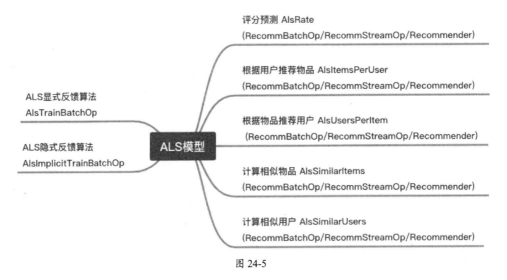

图 24-5

Alink 在推荐方面提供的组件较多，但是规律性很强。
- 模型训练组件一律是"算法名+TrainBatchOp"。
- 推荐组件的名称为"算法名+推荐方式+使用方法"。
- 现在支持的算法名如下。
    - 基于物品的协同过滤（ItemCf）
    - 基于用户的协同过滤（UserCf）
    - ALS 显式反馈算法（Als）
    - ALS 隐式反馈算法（AlsImplicit）
    - FM 算法（Fm）
- 推荐方法如下。
    - 评分预测（Rate）

- 根据用户推荐物品（ItemsPerUser）
- 根据物品推荐用户（UsersPerItem）
- 计算相似物品（SimilarItems）
- 计算相似用户（SimilarUsers）
- 使用方法如下。
  - 批式推荐（RecommBatchOp）
  - 流式推荐（RecommStreamOp）
  - Pipeline 节点（后缀为 Recommender，并可由此获得 LocalPredictor，进行嵌入用户的应用中推荐）

24.2 常用推荐算法

本节介绍两种常用的推荐算法，分别是协同过滤和交替最小二乘法。

24.2.1 协同过滤

协同过滤（Collaborative Filtering）是一种利用集体智慧的方法，从众多用户的历史行为中收集目标用户的相似信息，从而发现用户潜在的兴趣偏好。

我们的日常生活中就在使用协同过滤的思想，比如你想看电影，可能会询问与自己品味相似的朋友，他们最近看了什么；你喜欢的电影如果出了续集，也可能吸引你走进电影院。这就引出了两种常用的协同过滤方法。

（1）基于用户（User-Based）的协同过滤。即找到与你相似的用户，他们喜欢的很有可能也是你喜欢的，例如，可以通过在已经看过的影片历史记录中寻找和你的记录相似的用户，作为"相似用户"。

（2）基于物品（Item-Based）的协同过滤。由历史记录出发，同时被更多人观看过的影片，相似度更高。也就是根据你的观影记录，将关联相似度更高的影片推荐给你。

在实际处理中，用户与物品的关系用矩阵来表示，如图 24-6 所示。用户有 M 个，记为 $u_1, u_2, \cdots, u_M$，物品有 N 个，记为 $i_1, i_2, \cdots, i_N$。如果用户 u_m 购买过物品 i_n，则可在矩阵的第 m 行 n 列记录购买次数，没有购买行为的位置可以标记为 0，这样我们就得到了用户与物品关系的矩阵。

	i_1	i_2	...	i_n	...	i_N
u_1						
u_2						
⋮						
u_m				1		
⋮						
u_M						

图 24-6

由用户与物品关系的矩阵，我们可以衡量用户间的相似程度、物品间的相似程度。具体的方法是：对于每个用户，将其所在矩阵的行看作其向量值，用户间的距离或相似程度可以由其对应的向量进行计算；对于每个物品，将其所在的矩阵的列看作其向量值，物品间的相似程度可以由其对应的向量进行计算。

向量间比较常用的是余弦相似度，公式如下。

$$\text{CosSim}(\boldsymbol{a},\boldsymbol{b}) = \frac{\boldsymbol{a} \cdot \boldsymbol{b}}{\|\boldsymbol{a}\|_2 \|\boldsymbol{b}\|_2} = \frac{\sum a_k b_k}{\sqrt{(\sum a_k^2)(\sum b_k^2)}}$$

如果我们忽略向量的具体数值，而在意每个位置是否为 0，则可以使用集合的 Jaccard 相似度。

$$\text{JaccardSim}(A,B) = \frac{|A \cap B|}{|A \cup B|}$$

24.2.2 交替最小二乘法

本小节简单介绍交替最小二乘法（Alternating Least Square，ALS）。设用户的打分矩阵为 $R_{M \times N}$，每个用户和物品都可以用 K 维向量表示，用户矩阵为 $U_{M \times K}$，物品矩阵为 $I_{N \times K}$。评分 R_{mn} 可以表示为用户向量 U_m 和物品向量 I_n 的内积，即

$$R_{mn} \approx U_m \cdot I_n^{\mathrm{T}}$$

我们希望计算用户矩阵 $U_{M \times K}$ 和物品矩阵 $I_{N \times K}$，使其乘积逼近，采用最小平方误差定义损失函数。

$$L(\boldsymbol{U},\boldsymbol{I}) = \sum_{R_{mn}\ is\ known} (R_{mn} - \boldsymbol{U}_m \cdot \boldsymbol{I}_n^{\mathrm{T}})^2$$

为了防止过拟合,加入正则项系数λ。

$$L(\boldsymbol{U},\boldsymbol{I}) = \sum_{R_{mn}\ is\ known} (R_{mn} - \boldsymbol{U}_m \cdot \boldsymbol{I}_n^{\mathrm{T}})^2 + \lambda \left(\sum_m |\boldsymbol{U}_m|^2 + \sum_n |\boldsymbol{I}_n|^2 \right)$$

固定$\boldsymbol{I}_{N\times K}$,对$\boldsymbol{U}_m$求导,得

$$\frac{\partial L}{\partial \boldsymbol{U}_m} = -2 \sum_{R_{mn}\ is\ known} \boldsymbol{I}_n (R_{mn} - \boldsymbol{U}_m \cdot \boldsymbol{I}_n^{\mathrm{T}}) + 2\lambda \boldsymbol{U}_m$$

令$\frac{\partial L}{\partial \boldsymbol{U}_m} = 0$,可以得到相应的$\boldsymbol{U}_m$,进而得到新的$\boldsymbol{U}_{M\times K}$。

同理,固定$\boldsymbol{U}_{M\times K}$,可以得到新的$\boldsymbol{I}_{N\times K}$。如此交替执行,直到误差满足阈值条件或者达到最大迭代次数上限。

上面处理的是显式反馈的情况,对于隐式反馈,R_{mn}不再是评分值,而是表示动作次数,譬如购买、加购物车、收藏、点击的次数,或者观看/收听的时长等,取值范围是$[0, +\infty)$。

定义损失函数如下:

$$L(\boldsymbol{U},\boldsymbol{I}) = \sum_{m,n} C_{mn} (P_{mn} - \boldsymbol{U}_m \cdot \boldsymbol{I}_n^{\mathrm{T}})^2 + \lambda \left(\sum_m |\boldsymbol{U}_m|^2 + \sum_n |\boldsymbol{I}_n|^2 \right)$$

其中

$$P_{mn} = \begin{cases} 1 & if\ R_{mn} > 0 \\ 0 & if\ R_{mn} = 0 \end{cases}$$

$$C_{mn} = 1 + \alpha\, R_{mn}$$

求解方式与显式反馈一样,对$\boldsymbol{U}_{M\times K}$和$\boldsymbol{I}_{N\times K}$交替固定求解,直到误差满足阈值条件或者达到最大迭代次数上限。

24.3 数据探索

本章会围绕一个电影推荐的具体场景,需要用到两个数据集——电影信息和用户对电影的

评分记录，二者均来自 MovieLens 网站，下载地址为链接 24-2。

根据评论数的多少，MovieLens 提供了多个数据集。

- MovieLens 100K Dataset
- MovieLens 1M Dataset
- MovieLens 10M Dataset
- MovieLens 20M Dataset
- MovieLens 25M Dataset
- MovieLens 1B Synthetic Dataset

最小的数据集中有 10 万条评论，最大的数据集中有 2500 万条评论，利用向量生成方式得到的更大数据集中有 10 亿条评论。

除了浏览其主页上的介绍，还可以从链接 24-3 中看到可以访问的全部数据，如图 24-7 所示。

Index of /datasets/movielens

Name	Last modified	Size	Description
Parent Directory		-	
ml-1m-README.txt	2019-12-03 11:14	5.4K	
ml-1m.zip	2019-12-03 11:14	5.6M	
ml-1m.zip.md5	2019-12-03 11:14	51	
ml-10m-README.html	2019-12-03 11:14	11K	
ml-10m.zip	2019-12-03 11:14	63M	
ml-10m.zip.md5	2019-12-03 11:14	52	
ml-20m-README.html	2019-12-03 11:14	12K	
ml-20m-youtube-READM..>	2019-12-03 11:14	2.5K	
ml-20m-youtube.zip	2019-12-03 11:14	639K	
ml-20m-youtube.zip.md5	2019-12-03 11:14	56	
ml-20m.zip	2019-12-03 11:15	189M	
ml-20m.zip.md5	2019-12-03 11:14	52	
ml-20mx16x32-README.txt	2019-12-03 11:14	5.0K	
ml-20mx16x32.tar	2019-12-03 11:15	3.1G	
ml-20mx16x32.tar.md5	2019-12-03 11:14	51	
ml-25m-README.html	2019-12-03 11:14	12K	
ml-25m.zip	2019-12-03 11:14	250M	
ml-25m.zip.md5	2019-12-03 11:14	45	
ml-100k-README.txt	2019-12-03 11:14	6.6K	
ml-100k.zip	2019-12-03 11:14	4.7M	
ml-100k.zip.md5	2019-12-03 11:14	53	
ml-100k/	2019-05-08 11:18	-	
ml-latest-README.html	2019-12-03 11:14	11K	
ml-latest-small-READ..>	2019-12-03 11:14	9.7K	
ml-latest-small.zip	2019-12-03 11:14	955K	
ml-latest-small.zip.md5	2019-12-03 11:14	61	
ml-latest.zip	2019-12-03 11:15	264M	
ml-latest.zip.md5	2019-12-03 11:15	55	
old/	2019-05-08 11:18	-	

图 24-7

每个数据集包括多个数据文件，所有数据集都提供了压缩文件，可以下载到本地；100K 的小数据集还单独提供了文件夹 ml-100K，可以直接访问其数据文件。

在网页浏览器上访问链接 24-4，如图 24-8 所示。

Index of /datasets/movielens/ml-100k

Name	Last modified	Size	Description
Parent Directory		-	
allbut.pl	2019-12-03 11:14	716	
mku.sh	2019-12-03 11:14	643	
u.data	2019-12-03 11:14	1.9M	
u.genre	2019-12-03 11:14	202	
u.info	2019-12-03 11:14	36	
u.item	2019-12-03 11:14	231K	
u.occupation	2019-12-03 11:14	193	
u.user	2019-12-03 11:14	22K	
u1.base	2019-12-03 11:14	1.5M	
u1.test	2019-12-03 11:14	383K	
u2.base	2019-12-03 11:14	1.5M	
u2.test	2019-12-03 11:14	386K	
u3.base	2019-12-03 11:14	1.5M	
u3.test	2019-12-03 11:14	387K	
u4.base	2019-12-03 11:14	1.5M	
u4.test	2019-12-03 11:14	388K	
u5.base	2019-12-03 11:14	1.5M	
u5.test	2019-12-03 11:14	388K	
ua.base	2019-12-03 11:14	1.7M	
ua.test	2019-12-03 11:14	182K	
ub.base	2019-12-03 11:14	1.7M	
ub.test	2019-12-03 11:14	182K	

图 24-8

可以看到，该路径下有多个文件，下面我们选择 3 个常用的文件进行介绍。

1. 文件u.user

表示用户属性信息，图 24-9 是数据文件的部分内容。

u.user

```
1|24|M|technician|85711
2|53|F|other|94043
3|23|M|writer|32067
4|24|M|technician|43537
5|33|F|other|15213
6|42|M|executive|98101
7|57|M|administrator|91344
8|36|M|administrator|05201
9|29|M|student|01002
10|53|M|lawyer|90703
```

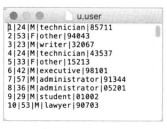

图 24-9

每个属性之间用“|”分隔，包含 5 个属性。

（1）user_id：用户 ID，对应 u.data 数据中的用户 ID 属性。

（2）age：年龄。

（3）gender：性别。

（4）occupation：职业。

（5）zip_code：邮政编码。

获取用户信息数据源的代码如下。

```
static final String USER_FILE = "u.user";

static final String USER_SCHEMA_STRING
    = "user_id long, age int, gender string, occupation string, zip_code string";
... ...

static CsvSourceBatchOp getSourceUsers() {
    return new CsvSourceBatchOp()
        .setFieldDelimiter("|")
        .setFilePath(DATA_DIR + USER_FILE)
        .setSchemaStr(USER_SCHEMA_STRING);
}
```

2. 文件u.item

表示物品（电影）属性信息，图 24-10 是数据文件的部分内容。

```
1|Toy Story (1995)|01-Jan-1995||http://us.imdb.com/M/title-exact?Toy%20Story%20(1995)|0|0|0|1|1|1|0|0|0|0|0|0|0|0|0|0|0|0|0
2|GoldenEye (1995)|01-Jan-1995||http://us.imdb.com/M/title-exact?GoldenEye%20(1995)|0|1|1|0|0|0|0|0|0|0|0|0|0|0|0|1|0|0
3|Four Rooms (1995)|01-Jan-1995||http://us.imdb.com/M/title-exact?Four%20Rooms%20(1995)|0|0|0|0|0|0|0|0|0|0|0|0|0|0|0|1|0|0
4|Get Shorty (1995)|01-Jan-1995||http://us.imdb.com/M/title-exact?Get%20Shorty%20(1995)|0|1|0|0|0|1|0|0|1|0|0|0|0|0|0|0|0|0
5|Copycat (1995)|01-Jan-1995||http://us.imdb.com/M/title-exact?Copycat%20(1995)|0|0|0|0|0|1|0|1|0|0|0|0|0|0|1|0|0
```

图 24-10

每个属性之间用"|"分隔，包含"5+19"个属性。

（1）item_id：物品（电影）ID，对应 u.data 数据中物品（电影）ID 属性。

（2）title：电影名称。

（3）release_date：电影上映日期。

（4）video_release_date：视频发布日期。

（5）imdb_url：IMDB 链接。

（6）最后 19 个字段是电影流派（未知、动作、冒险、动画、儿童、喜剧、犯罪、纪录片、戏剧、幻想、黑色电影、恐怖、音乐、神秘、浪漫、科幻、惊悚、战争、西部），1 表示电影是这种流派的，0 表示电影不是这种流派的；每部电影可以同时属于多种流派。

获取影片信息数据源的代码如下。

```
static final String ITEM_FILE = "u.item";
static final String ITEM_SCHEMA_STRING = "item_id long, title string, "
   + "release_date string, video_release_date string, imdb_url string, "
   + "unknown int, action int, adventure int, animation int, "
   + "children int, comedy int, crime int, documentary int, drama int, "
   + "fantasy int, film_noir int, horror int, musical int, mystery int, "
   + "romance int, sci_fi int, thriller int, war int, western int";
... ...

static CsvSourceBatchOp getSourceItems() {
   return new CsvSourceBatchOp()
      .setFieldDelimiter("|")
      .setFilePath(DATA_DIR + ITEM_FILE)
      .setSchemaStr(ITEM_SCHEMA_STRING);
}
```

3. 文件u.data

表示用户对电影的评级。其中包括 943 个用户对 1682 个物品（电影）进行的 10 万次评论；每个用户至少有 20 部电影；用户和物品（电影）都是从 1 开始编号的。图 24-11 是数据文件的部分内容。

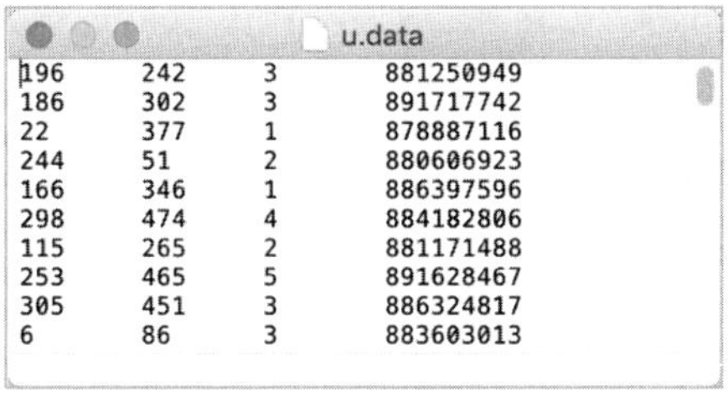

196	242	3	881250949
186	302	3	891717742
22	377	1	878887116
244	51	2	880606923
166	346	1	886397596
298	474	4	884182806
115	265	2	881171488
253	465	5	891628467
305	451	3	886324817
6	86	3	883603013

图 24-11

每个属性之间用制表符“\t”分隔，包含 4 个属性。

（1）user_id：用户 ID。

（2）item_id：物品（电影）ID。

（3）rating：评分。

（4）ts：Unix 时间戳，是自 1/1/1970 UTC 以来的 Unix 秒数。

获取评分信息数据源的代码如下。

```
static final String RATING_FILE = "u.data";
static final String RATING_SCHEMA_STRING
   = "user_id long, item_id long, rating float, ts long";
... ...

static TsvSourceBatchOp getSourceRatings() {
   return new TsvSourceBatchOp()
```

```
          .setFilePath(DATA_DIR + RATING_FILE)
          .setSchemaStr(RATING_SCHEMA_STRING);
}
```

目录下还有已拆分的训练与测试数据集。数据集 ua.base、ua.test、ub.base 和 ub.test 是由 u.data 拆分的训练集和测试集，测试集中每个用户正好有 10 个评级记录，且 ua.test 和 ub.test 不相交。数据集 u1.base 和 u1.test 到 u5.base 和 u5.test 是 5 折交叉验证数据集。

24.4　评分预测

本节的目标是基于已知的评分信息建立推荐模型，针对给定的用户 ID 和影片 ID 预测评分结果。

我们使用已拆分好的训练集 ua.base 和测试集 ua.test，使用评分数据的 SchemaStr，通过 TsvSourceBatchOp 组件获取数据源。

```
TsvSourceBatchOp train_set = new TsvSourceBatchOp()
   .setFilePath(DATA_DIR + RATING_TRAIN_FILE)
   .setSchemaStr(RATING_SCHEMA_STRING);

TsvSourceBatchOp test_set = new TsvSourceBatchOp()
   .setFilePath(DATA_DIR + RATING_TEST_FILE)
   .setSchemaStr(RATING_SCHEMA_STRING);
```

训练数据连接到 AlsTrainBatchOp 组件，需要设置用户列名、影片列名和评分值；获得 ALS 推荐模型，并将其保存为 AK 格式文件，具体代码如下。

```
train_set
   .link(
     new AlsTrainBatchOp()
        .setUserCol(USER_COL)
        .setItemCol(ITEM_COL)
        .setRateCol(RATING_COL)
        .setLambda(0.1)
        .setRank(10)
        .setNumIter(10)
   )
   .link(
     new AkSinkBatchOp()
        .setFilePath(DATA_DIR + ALS_MODEL_FILE)
   );
```

下面以 ID 为 1 的用户为例，演示评分预测过程。
- 评分预测需要使用 ALS 算法相应的评分推荐组件 AlsRateRecommender，该组件需要用到已经训练好的 ALS 模型，所以使用 setModelData 方法，从 AK 文件数据源获取模型。
- 为了便于理解评分结果，需要将影片 ID 转换为名称，我们调用组件 Lookup，其需要一个映射表，即原始影片信息数据表（前面已经介绍过，可通过封装的 getSourceItems 方法获取），设置映射的 Key 和 Value 所在列名。
- 对测试数据集 test_set 进行过滤，保留"user_id=1"的数据，然后选择 4 个主要数据列，并对评分值进行排序，便于我们查看和对比。

具体代码如下。

```
new PipelineModel
  (
    new AlsRateRecommender()
      .setUserCol(USER_COL)
      .setItemCol(ITEM_COL)
      .setRecommCol(RECOMM_COL)
      .setModelData(
        new AkSourceBatchOp()
          .setFilePath(DATA_DIR + ALS_MODEL_FILE)
      ),
    new Lookup()
      .setSelectedCols(ITEM_COL)
      .setOutputCols("item_name")
      .setModelData(getSourceItems())
      .setMapKeyCols("item_id")
      .setMapValueCols("title")
  )
  .transform(
    test_set.filter("user_id=1")
  )
  .select("user_id, rating, recomm, item_name")
  .orderBy("rating, recomm", 1000)
  .lazyPrint(-1);
```

运行结果如下。

```
user_id|rating|recomm|item_name
-------|------|------|---------
1|2.0000|2.5695|Dirty Dancing (1987)
1|3.0000|3.3675|Rock, The (1996)
1|3.0000|4.3740|Grand Day Out, A (1992)
1|4.0000|3.4306|Desperado (1995)
1|4.0000|3.6392|Angels and Insects (1995)
1|4.0000|3.7185|Glengarry Glen Ross (1992)
1|4.0000|3.7284|Hunt for Red October, The (1990)
1|4.0000|4.3246|Three Colors: White (1994)
```

```
1|5.0000|3.9633|Groundhog Day (1993)
1|5.0000|4.5148|Delicatessen (1991)
```

在对 10 部影片的预测结果中,最低分 2.5695 对应的影片为 Dirty Dancing (1987),原始的评分为 2.0 分;最高分 4.5148 对应的影片为 Delicatessen (1991),原始的用户评分为 5.0 分;其他预测结果与原始评分的差距似乎不大,要衡量其差距,可以使用回归评估组件。

我们对全部测试集 test_set 的数据进行推荐评分,然后使用回归评估组件 EvalRegressionBatchOp,详细代码如下。

```
new AlsRateRecommender()
   .setUserCol(USER_COL)
   .setItemCol(ITEM_COL)
   .setRecommCol(RECOMM_COL)
   .setModelData(
      new AkSourceBatchOp()
         .setFilePath(DATA_DIR + ALS_MODEL_FILE)
   )
   .transform(test_set)
   .link(
      new EvalRegressionBatchOp()
         .setLabelCol(RATING_COL)
         .setPredictionCol(RECOMM_COL)
         .lazyPrintMetrics()
   );
```

运行结果如下。

```
----------------------------- Metrics: -----------------------------
MSE:0.9092 RMSE:0.9535    MAE:0.7529 MAPE:28.0715    R2:0.2751
```

这里我们更关心每个预测值与原始用户评分的差距,即平均绝对误差(MAE)0.7529。

24.5 根据用户推荐影片

在本节和 24.6 节中,我们将使用基于物品的协同过滤算法建立推荐模型,并根据不同问题选择使用相应的推荐预测组件。

基于物品的协同过滤模型的训练与前面 ALS 算法的类似,需要设置用户列名、影片列名和评分值,随后将模型保存为 AK 格式文件,便于在后面的推荐过程中直接导入模型进行使用。具体代码如下。

```
getSourceRatings()
   .link(
      new ItemCfTrainBatchOp()
```

```
        .setUserCol(USER_COL)
        .setItemCol(ITEM_COL)
        .setRateCol(RATING_COL)
    )
    .link(
        new AkSinkBatchOp()
            .setFilePath(DATA_DIR + ITEMCF_MODEL_FILE)
    );
```

推荐任务如下代码所示。

```
MemSourceBatchOp test_data = new MemSourceBatchOp(new Long[] {1L}, "user_id");
new ItemCfItemsPerUserRecommender()
    .setUserCol(USER_COL)
    .setRecommCol(RECOMM_COL)
    .setModelData(
        new AkSourceBatchOp()
            .setFilePath(DATA_DIR + ITEMCF_MODEL_FILE)
    )
    .transform(test_data)
    .print();
```

先使用 MemSourceBatchOp 组件构造一个只有一列 "user_id" 的批式数据表，包含一条数据，即所要预测的用户 ID；接着使用根据用户推荐物品的组件，设置用户列名、推荐结果列名；还需要使用 setModelData 方法，从 AK 文件数据源获取推荐模型。

输出结果如表 24-1 所示，推荐结果中包括影片 ID 和相应的推荐分值。

表 24-1　推荐结果

user_id	recomm
1	{"item_id":"[174, 56, 176, 98, 50, 195, 172, 96, 181, 204]", "score":"[0.5446295202769522, 0.5421344455706749, 0.5310766611215199, 0.5306864538037679, 0.5261969546534264, 0.5179894090961199, 0.5168003816848815, 0.5124562838810977, 0.5076640796772237, 0.5065063397097225]"}

上面演示了批式推荐任务，同样可以使用相应的流式组件执行流式任务。接下来，演示如何使用 LocalPredictor 进行推荐，使推荐过程可以嵌入 Java 应用中。

对 ItemCfItemsPerUserRecommender 组件调用其 collectLocalPredictor 方法。注意：该方法必须输入预测数据的 SchemaStr。最后，我们再将 LocalPredictor 的输出 Schema 显示出来，具体代码如下。

```
LocalPredictor recomm_predictor = new ItemCfItemsPerUserRecommender()
    .setUserCol(USER_COL)
    .setRecommCol(RECOMM_COL)
    .setK(20)
```

```
    .setModelData(
        new AkSourceBatchOp()
            .setFilePath(DATA_DIR + ITEMCF_MODEL_FILE)
    )
    .collectLocalPredictor("user_id long");

System.out.println(recomm_predictor.getOutputSchema());
```

输出的 Schema 如下，共两列，一列为输入的用户 ID，另一列为推荐预测结果。

```
root
 |-- user_id: BIGINT
 |-- recomm: STRING
```

同样，定义从影片 ID 到名称映射的 kv_predictor，并将其输出的 Schema 显示出来，代码如下。

```
LocalPredictor kv_predictor = new Lookup()
    .setSelectedCols(ITEM_COL)
    .setOutputCols("item_name")
    .setModelData(getSourceItems())
    .setMapKeyCols("item_id")
    .setMapValueCols("title")
    .collectLocalPredictor("item_id long");

System.out.println(kv_predictor.getOutputSchema());
```

输出结果如下，一列为影片 ID，另一列为影片名称。

```
root
 |-- item_id: BIGINT
 |-- item_name: STRING
```

下面的代码演示了整个预测过程。

- 输入的是 Row 类型数据，只有一个 Long 型字段，可以写为 Row.*of*(1L)。
- 对 recomm_predictor 的 map 方法进行预测，结果为 2 个字段的 Row 类型数据，由前面输出的 Schema 信息可知，其索引号为 1 的字段为推荐结果（字符串类型）。
- 使用 JsonConverter 解析出推荐的影片 ID。
- 将每个推荐的影片 ID 映射到影片名称，并输出。

具体代码如下。

```
String recommResultStr = (String) recomm_predictor.map(Row.of(1L)).getField(1);

System.out.println(recommResultStr);

List <Long> recomm_ids = JsonConverter
    .fromJson(
        JsonConverter
            . <Map <String, String>>fromJson(
```

```
            recommResultStr,
            Map.class
        )
        .get("item_id"),
    new TypeReference <List <Long>>() {}.getType()
);
for (Long id : recomm_ids) {
    System.out.println(kv_predictor.map(Row.of(id)));
}
```

输出内容如下，前面为推荐结果，后面是各个推荐影片 ID 和名称。

```
{"item_id":"[174,56,176,98,50,195,172,96,181,204,183,89,168,173,79,228,210,234,423,144]","score":"[0.5
446295202769522,0.5421344455706749,0.5310766611215199,0.5306864538037679,0.5261969546534264,0.51798940
90961199,0.5168003816848815,0.5124562838810977,0.5076640796772237,0.5065063397097225,0.504903666688330
5,0.5037182261191273,0.503485951777993,0.4974758183778898,0.48814994943889534,0.4855370682711127,0.484
72553094018794,0.48321375585798926,0.48163621419180747,0.48119185189692276]"}
174,Raiders of the Lost Ark (1981)
56,Pulp Fiction (1994)
176,Aliens (1986)
98,Silence of the Lambs, The (1991)
50,Star Wars (1977)
195,Terminator, The (1984)
172,Empire Strikes Back, The (1980)
96,Terminator 2: Judgment Day (1991)
181,Return of the Jedi (1983)
204,Back to the Future (1985)
183,Alien (1979)
89,Blade Runner (1982)
168,Monty Python and the Holy Grail (1974)
173,Princess Bride, The (1987)
79,Fugitive, The (1993)
228,Star Trek: The Wrath of Khan (1982)
210,Indiana Jones and the Last Crusade (1989)
234,Jaws (1975)
423,E.T. the Extra-Terrestrial (1982)
144,Die Hard (1988)
```

在判断上面的推荐是否准确前，我们需要先了解 1 号用户（user_id=1）对影片的喜好。一个简单的办法是列出被 1 号用户评为高分的影片，就大概知道他喜欢什么类型影片，具体代码如下。

```
new Lookup()
    .setSelectedCols(ITEM_COL)
    .setOutputCols("item_name")
    .setModelData(getSourceItems())
    .setMapKeyCols("item_id")
    .setMapValueCols("title")
    .transform(
```

```
    getSourceRatings().filter("user_id=1 AND rating>4")
)
.select("item_name")
.orderBy("item_name", 1000)
.lazyPrint(-1);
```

从原始的评分数据中，过滤用户 ID 为 1 且评分大于 4 分的记录，并将其中的影片 ID 映射为影片名称。

输出结果如表 24-2 所示。

表24-2　1号用户喜欢的影片列表

12 Angry Men (1957)	Manon of the Spring (Manon des sources) (1986)
Alien (1979)	Mars Attacks! (1996)
Aliens (1986)	Maya Lin: A Strong Clear Vision (1994)
Amadeus (1984)	Mighty Aphrodite (1995)
Antonia's Line (1995)	Monty Python and the Holy Grail (1974)
Back to the Future (1985)	Monty Python's Life of Brian (1979)
Big Night (1996)	Mr. Holland's Opus (1995)
Blade Runner (1982)	Mystery Science Theater 3000: The Movie (1996)
Bound (1996)	Nightmare Before Christmas, The (1993)
Brazil (1985)	Nikita (La Femme Nikita) (1990)
Breaking the Waves (1996)	Pillow Book, The (1995)
Chasing Amy (1997)	Postino, Il (1994)
Chasing Amy (1997)	Priest (1994)
Cinema Paradiso (1988)	Princess Bride, The (1987)
Clerks (1994)	Professional, The (1994)
Contact (1997)	Raiders of the Lost Ark (1981)
Crumb (1994)	Remains of the Day, The (1993)
Cyrano de Bergerac (1990)	Return of the Jedi (1983)
Dead Man Walking (1995)	Ridicule (1996)
Dead Poets Society (1989)	Searching for Bobby Fischer (1993)
Delicatessen (1991)	Shanghai Triad (Yao a yao yao dao waipo qiao) (1995)
Dolores Claiborne (1994)	Shawshank Redemption, The (1994)
Eat Drink Man Woman (1994)	Sleeper (1973)
Empire Strikes Back, The (1980)	Sling Blade (1996)
Fargo (1996)	Star Trek: The Wrath of Khan (1982)
French Twist (Gazon maudit) (1995)	Star Wars (1977)
Full Monty, The (1997)	Swingers (1996)
Gattaca (1997)	Terminator 2: Judgment Day (1991)
Godfather, The (1972)	Terminator, The (1984)
Good, The Bad and The Ugly, The (1966)	Three Colors: Blue (1993)
Graduate, The (1967)	Three Colors: Red (1994)
Groundhog Day (1993)	Toy Story (1995)
Haunted World of Edward D. Wood Jr., The (1995)	Truth About Cats & Dogs, The (1996)
Henry V (1989)	Usual Suspects, The (1995)
Hoop Dreams (1994)	Wallace & Gromit: The Best of Aardman Animation (1996)
Horseman on the Roof, The (Hussard sur le toit, Le) (1995)	Welcome to the Dollhouse (1995)
	When Harry Met Sally... (1989)

	续表
Hudsucker Proxy, The (1994) Jean de Florette (1986) Jurassic Park (1993) Kids in the Hall: Brain Candy (1996) Kolya (1996) Lone Star (1996)	Wrong Trousers, The (1993) Young Frankenstein (1974)

我们对比推荐结果，发现很多是用户评价为 5 分的影片，推荐的影片的确都是用户喜欢的。但在实际使用中，用户更希望我们推荐新的影片，也就是用户没看过的影片。在 Alink 推荐组件中提供了一个方法 setExcludeKnown(true)，该方法的默认值为 false。构建一个新的 LocalPredictor，推荐用户没评价过的影片，如下代码所示。

```
... ...
LocalPredictor recomm_predictor_2 = new ItemCfItemsPerUserRecommender()
    .setUserCol(USER_COL)
    .setRecommCol(RECOMM_COL)
    .setK(20)
    .setExcludeKnown(true)
    .setModelData(
        new AkSourceBatchOp()
            .setFilePath(DATA_DIR + ITEMCF_MODEL_FILE)
    )
    .collectLocalPredictor("user_id long");
... ...
```

推荐结果如下，这次排在第 1 位的是 E.T. the Extra-Terrestrial (1982)，在上次的推荐结果中它排在十几位，前面都是用户评价过的影片。

```
{"item_id":"[423,318,357,655,568,385,684,496,403,566,732,433,651,527,432,474,405,550,588,435]","score":"[0.48163621419180747,0.4564074931610021,0.4425355598672598,0.4360613380052011,0.42484215121153635,0.41393159715117167,0.41252565376658934,0.4028040283264125,0.3992900166944061,0.39697125072097783,0.39490277720105643,0.3880766917892268,0.3843606351882687,0.3813688255025348,0.3749809650446883,0.37404643823803746,0.37202533566821233,0.369066737917194,0.36636674389371593,0.36300743588081913]"}
423,E.T. the Extra-Terrestrial (1982)
318,Schindler's List (1993)
357,One Flew Over the Cuckoo's Nest (1975)
655,Stand by Me (1986)
568,Speed (1994)
385,True Lies (1994)
684,In the Line of Fire (1993)
496,It's a Wonderful Life (1946)
403,Batman (1989)
566,Clear and Present Danger (1994)
732,Dave (1993)
```

```
433,Heathers (1989)
651,Glory (1989)
527,Gandhi (1982)
432,Fantasia (1940)
474,Dr. Strangelove or: How I Learned to Stop Worrying and Love the Bomb (1963)
405,Mission: Impossible (1996)
550,Die Hard: With a Vengeance (1995)
588,Beauty and the Beast (1991)
435,Butch Cassidy and the Sundance Kid (1969)
```

24.6 计算相似影片

本节以星球大战（Star Wars (1977), item_id=50）为例，通过推荐模型计算与其相似的影片。推荐任务如下代码所示。

```
MemSourceBatchOp test_data = new MemSourceBatchOp(new Long[] {50L}, ITEM_COL);

new ItemCfSimilarItemsRecommender()
    .setItemCol(ITEM_COL)
    .setRecommCol(RECOMM_COL)
    .setModelData(
        new AkSourceBatchOp()
            .setFilePath(DATA_DIR + ITEMCF_MODEL_FILE)
    )
    .transform(test_data)
    .print();
```

先使用 MemSourceBatchOp 组件，构造一个只有一列"item_id"的批式数据表，包含一条数据，即所要预测的影片 ID；使用根据计算物品相似度的组件，设置用户列名、推荐结果列名，设置输出 Top10 的影片；还需要使用 setModelData 方法，从 AK 文件数据源获取模型。

输出结果如表 24-3 所示，推荐结果中包括影片 ID 和相应的推荐分值。

表 24-3 推荐结果

item_id	recomm
50	{"item_id":"[181,174,172,1,127,121,210,100,98,222]","similarities":"[0.8844757466059665,0.7648851255036908,0.7498192415368896,0.7345720560109783,0.6973318428052556,0.6928373036216864,0.6893433161183706,0.6865325410591914,0.6764284324255825,0.6739748837432759]"}

下面我们换成使用 LocalPredictor 的方式计算相似的影片，并使用 Lookup 组件将影片 ID

转化为影片名称,具体的电影名能给我们更直观的感觉,以便判断计算出来的影片是否很相似。具体代码如下,其使用方式与 24.5 节类似,这里就不详细介绍了。

```
LocalPredictor recomm_predictor = new ItemCfSimilarItemsRecommender()
    .setItemCol(ITEM_COL)
    .setRecommCol(RECOMM_COL)
    .setK(10)
    .setModelData(
        new AkSourceBatchOp()
            .setFilePath(DATA_DIR + ITEMCF_MODEL_FILE)
    )
    .collectLocalPredictor("item_id long");

LocalPredictor kv_predictor = new Lookup()
    .setSelectedCols(ITEM_COL)
    .setOutputCols("item_name")
    .setModelData(getSourceItems())
    .setMapKeyCols("item_id")
    .setMapValueCols("title")
    .collectLocalPredictor("item_id long");

String recommResultStr = (String) recomm_predictor.map(Row.of(50L)).getField(1);

List <Long> recomm_ids = JsonConverter
    .fromJson(
        JsonConverter
            . <Map <String, String>>fromJson(
                recommResultStr,
                Map.class
            )
            .get("item_id"),
        new TypeReference <List <Long>>() {}.getType()
    );

for (Long id : recomm_ids) {
    System.out.println(kv_predictor.map(Row.of(id)));
}
```

计算结果如下。

```
181,Return of the Jedi (1983)
174,Raiders of the Lost Ark (1981)
172,Empire Strikes Back, The (1980)
1,Toy Story (1995)
127,Godfather, The (1972)
121,Independence Day (ID4) (1996)
210,Indiana Jones and the Last Crusade (1989)
100,Fargo (1996)
98,Silence of the Lambs, The (1991)
222,Star Trek: First Contact (1996)
```

从图 24-12 中可以看到，推荐结果中排在第 1 位的是"星战系列"的《绝地反击》（Return of the Jedi），排在第 3 位的是"星战系列"的《帝国反击战》（Empire Strikes Back, The）。

Star Wars　　　　　　　　Empire Strikes Back, The　　　　　Return of the Jedi
(1977)　　　　　　　　　　　　(1980)　　　　　　　　　　　　(1983)

图 24-12

排在第 2 位的是《夺宝奇兵》（Raiders of the Lost Ark），其编剧乔治·卢卡斯是《星球大战》（Star Wars）的编剧和导演，主演哈里森·福特是《星球大战》的主演。

夺宝奇兵（Raiders of the Lost Ark）

导演：史蒂文·斯皮尔伯格
编剧：劳伦斯·卡斯丹/**乔治·卢卡斯**/菲利普·考夫曼
主演：**哈里森·福特**/凯伦·阿兰/保罗·弗里曼/约翰·瑞斯-戴维斯

24.7　根据影片推荐用户

在本节及 24.8 节中，我们将使用基于用户的协同过滤算法，其模型训练与前面基于物品的协同过滤算法类似。所需的训练数据格式是一样的，使用模型训练组件 UserCfTrainBatchOp，设置用户列名、影片列名和评分值，随后将模型保存为 AK 格式文件，便于在后面的推荐过程中直接导入模型进行使用。具体代码如下。

```
getSourceRatings()
    .link(
        new UserCfTrainBatchOp()
            .setUserCol(USER_COL)
            .setItemCol(ITEM_COL)
            .setRateCol(RATING_COL)
    )
    .link(
        new AkSinkBatchOp()
            .setFilePath(DATA_DIR + USERCF_MODEL_FILE)
    );
```

推荐任务如下代码所示。

```
MemSourceBatchOp test_data = new MemSourceBatchOp(new Long[] {50L}, ITEM_COL);
new UserCfUsersPerItemRecommender()
    .setItemCol(ITEM_COL)
    .setRecommCol(RECOMM_COL)
    .setModelData(
        new AkSourceBatchOp()
            .setFilePath(DATA_DIR + USERCF_MODEL_FILE)
    )
    .transform(test_data)
    .print();
```

先使用 MemSourceBatchOp 组件，构造一个只有一列"user_id"的批式数据表，包含一条数据，即所要预测的影片 ID；使用根据物品推荐用户的组件 UserCfUsersPerItemRecommender，设置用户列名、推荐结果列名，还需要使用 setModelData 方法从 AK 文件数据源获取模型。

输出结果如表 24-4 所示，推荐结果中包括用户 ID 和相应的推荐分值。

表 24-4 推荐结果

item_id	recomm
50	{"item_id":"[276,429,222,864,194,650,896,303,749,301]","score":"[0.2911498078185388,0.28980307729545673,0.2896787402948555,0.2867326943225667,0.28158684138333434,0.2815710577453545,0.2803413050267188,0.27966251684722754,0.27960936969364464,0.27907226770541305]"}

我们通过一个简单的实验来检验预测的情况，看看评分数据集中是否已有推荐用户对目标影片（ID=15）的评分。使用 filter 方法，用 Flink SQL 语法描述过滤条件即可，具体代码如下。

```
getSourceRatings()
    .filter("user_id IN (276,429,222,864,194,650,896,303,749,301) AND item_id=50")
    .print();
```

运行结果如表 24-5 所示，10 个推荐的用户都对该电影给出过评分，8 人给的是 5 分，给 4 分和 3 分的各 1 人。

表 24-5 10 个推荐的原始评分

user_id	item_id	rating	ts
429	50	5.0000	882384553
301	50	5.0000	882074647
864	50	5.0000	877214085
650	50	5.0000	891372232
276	50	5.0000	880913800
194	50	3.0000	879521396
303	50	5.0000	879466866
749	50	5.0000	878846978
222	50	4.0000	877563194
896	50	5.0000	887159211

如果我们希望将电影推荐给那些没看过该电影的新用户，那么需要在原有的推荐预测流程中配置一个新的参数 ExcludeKnown=true，该参数的默认值为 false。相关代码如下。

```
new UserCfUsersPerItemRecommender()
    .setItemCol(ITEM_COL)
    .setRecommCol(RECOMM_COL)
    .setExcludeKnown(true)
    .setModelData(
        new AkSourceBatchOp()
            .setFilePath(DATA_DIR + USERCF_MODEL_FILE)
    )
    .transform(test_data)
    .print();
```

运行结果如表 24-6 所示。

表 24-6 推荐结果

item_id	recomm
50	{"item_id":"[16,788,932,442,207,90,627,543,532,911]","score":"[0.23501681598860094,0.2342959922966516,0.2285404610116206,0.22245913526644714,0.22095385722564706,0.2198506485812082,0.21495692781445608,0.2145426338815538,0.20757353674629853,0.1931964040290373]"}

24.8 计算相似用户

本节以 user_id =1 的用户为例,通过推荐模型计算与其相似的用户。

推荐任务如下代码所示。

```
MemSourceBatchOp test_data = new MemSourceBatchOp(new Long[] {1L}, USER_COL);

new UserCfSimilarUsersRecommender()
    .setUserCol(USER_COL)
    .setRecommCol(RECOMM_COL)
    .setModelData(
        new AkSourceBatchOp()
            .setFilePath(DATA_DIR + USERCF_MODEL_FILE)
    )
    .transform(test_data)
    .print();
```

先使用 MemSourceBatchOp 组件,构造一个只有一列"user_id"的批式数据表,包含一条数据,即所要预测的用户 ID;使用计算相似用户的推荐组件,设置用户列名、推荐结果列名;还需要使用 setModelData 方法,从 AK 文件数据源获取基于用户的协同过滤模型。

输出结果如表 24-7 所示,推荐结果中包括推荐的相似用户 ID 和相应的推荐分值。

表 24-7 推荐结果

user_id	recomm
1	{"item_id":"[916,864,268,92,435,457,738,429,303,276]","similarities":"[0.5690657315279888,0.5475482621940828,0.5420770475201064,0.5405335611842348,0.5386645318853762,0.5384759750393853,0.5270310735011106,0.5259499267180840,0.525717734084985,0.5245225229720628]"}

进一步,我们可以在用户信息数据表中过滤这些推荐的用户,查看他们的信息,具体代码如下。

```
getSourceUsers()
    .filter("user_id IN (1, 916,864,268,92,435,457,738,429,303,276)")
    .print();
```

运行结果如表 24-8 所示,可以看到,1 号用户的性别为男性(Male),年龄 24 岁,职业为 technician,被推荐的相似用户绝大部分也为男性,年龄在 19~35 岁,职业方面主要是学生和工程技术人士。

表 24-8 过滤的用户信息

user_id	age	gender	occupation	zip_code
864	27	M	programmer	63021
916	27	M	engineer	N2L5N
429	27	M	student	29205
435	24	M	engineer	60007
457	33	F	salesman	30011
738	35	M	technician	95403
268	24	M	engineer	19422
276	21	M	student	95064
303	19	M	student	14853
1	24	M	technician	85711
92	32	M	entertainment	80525